DNA: Life's Instruction Manual

Salina

Table of Contents

Chapter 1:

The ribosome, initiation of protein synthesis, and internal ribosome entry sites

1.1 The ribosome is an evolutionarily conserved machine for protein synthesis

One the most fundamental truths about life is captured by the central dogma[1]. This "residue-by-residue" transfer of information from one molecule to another is at the core of all life on earth. Information usually gets transferred from DNA to mRNA to protein. While some types of DNA[2] and RNA[3] are catalytic, these two molecules serve important roles as linear forms of information storage. The vast majority of cellular processes are carried out by proteins.

Proteins are the molecular workhorses of the cell. They carry out the majority of cellular functions as enzymes, sensors, activators, regulators, transporters, cellular defenders, storage units, or structural supports. While their functional range is wide and their variety considerable, they are all built using only twenty amino acid building blocks. Proteins are linear strings of amino acids stitched together using the linear information stored in the DNA and they need to attain specific three dimensional shapes to perform their function. While the linearity of the code of life is preserved in a protein, it is its three-dimensionality which imparts each protein with its function. If a protein keeps its linear sequence of amino acids but loses its shape, it loses its function[4]. A function carried out by a protein is thus a direct and emergent property of its shape. In the cell, proteins are made by the large macromolecular machine called the ribosome.

The ribosome is a multi-megadalton sized ribonucleoprotein complex that catalyzes the synthesis of peptide bonds between amino acids as specified by an mRNA[5]. Ribosomes from all lifeforms are made up of two subunits **(Fig. 1.1 A)**. The large subunit (termed 50S in bacteria and 60S in eukaryotes) and the small subunit (termed 30S in bacteria and 40S in eukaryotes) of the ribosome come together to make the ribosome proper (termed 70S in bacteria and 80S in eukaryotes).

1.2　The ribosome is composed of two subunits that have distinct functions

The large subunit of the ribosome is made up of two or three RNA molecules adorned with numerous proteins **(Fig. 1.1 A, left)**. The largest of the large subunit RNA molecules (rRNAs) is the 28S RNA in eukaryotes or the 23S in bacteria. The 28S rRNA is highly structured. It folds into functionally critical and evolutionarily conserved core components of the ribosome such as the peptidyl-transferase center and the GTPase activation center (GAC). Another one or two shorter rRNAs (5S rRNA in bacteria and eukaryotes and 5.8S rRNA in eukaryotes) can be found on the large ribosomal subunit. In eukaryotes, about 50 large subunit specific proteins (~30 in bacteria) **(Fig. 1.1 A, left, blue)** bind to the 28S rRNA **(Fig. 1.1 A, left, cyan and gray)** at various locations and stabilize the RNA structure. Two prominent features of the large subunit are the P-stalk (L7/L12 stalk in bacteria) and the L1-stalk. Jutting out of the large subunit on either side **(Fig. 1.1 A, left)**, these ribonucleoprotein stalks are responsible for coordinating the ribosome's interactions with different translation factors and for escorting tRNAs in and out of the ribosome. The large subunit also houses the peptide exit tunnel from which the growing nascent peptide chain exits.

The small subunit of the ribosome is also made up of RNA and proteins **(Fig. 1.1 A, right)**. The small ribosomal subunit rRNA is called the 18S rRNA in eukaryotes **(Fig. 1.1 A, right, light yellow and gray)** and 16S in bacteria and it is decorated by about 32 small ribosomal subunit specific proteins in eukaryotes (~22 in bacteria) **(Fig. 1.1 A, right, yellow)**. The small ribosomal subunit's shape can be distinguished into a head and a body **(Fig. 1.1 A, right)**. The body is further distinguishable into left and right feet, a shoulder, and a platform. The head has a feature that resembles a beak and it is highly mobile and can go through distinct sets of motions. The mRNA threads into the 40S through the entrance tunnel between the beak and the shoulder, lodges into a cleft between the head and body, and then exits through the mRNA exit tunnel just above the platform.

The mRNA is decoded by the transfer RNA (tRNA) molecules which carry and deliver the amino acids to the ribosome. As the tRNAs pass through the ribosome they interact with three distinct but evolutionarily conserved binding sites. These three sites are named aminoacyl-site (A-site), peptidyl-site (P-site), and exit-site (E-site)[6] **(Fig. 1.1 A)**. tRNAs get loaded onto the A-site as dictated by the mRNA, the growing peptide chain is maintained on the tRNA at the P-site, while the deacylated tRNA exits the ribosome through the E-site. The A-, P-, and E-sites of the large subunit interact with the acceptor arms of the transiting tRNAs whereas the A-, P-, and E-sites of the small subunit interact with anticodon stem loop (ASL) ends of the tRNAs.

The small subunit A-site houses the decoding center of the ribosome **(Fig. 1.1 A, right)**. The decoding center consists of a conserved set of bases that monitor the geometry of accurate codon-anticodon base pairing between the mRNA codon and the tRNA[7]. Three nucleotides from the small subunit rRNA interact with the minor groove of the first and second codon-anticodon base pairs to check for accurate Watson-Crick base pairing geometry. The first two base pairs are strictly monitored. This strict monitoring is responsible for the fast and accurate decoding of the mRNA. The third codon-anticodon base pair is monitored less strictly and results in what is called a "wobble" base pair[8]. Once the mRNA is decoded by the proper tRNA, peptide bond formation occurs at the peptidyl-transferase center (PTC).

The PTC is the catalytic heart of the ribosome—it catalyzes the addition of amino acids to a growing peptide chain. The PTC does this by coordinating the nucleophilic attack of the aminoacyl ester on the P-site peptidyl-tRNA by the alpha-amino group on the A-site aminoacyl-tRNA. It has been calculated that the PTC enhances peptide bond formation rates by 10^7-10^9

fold[9]. While there are proteins present around the PTC, peptidyl-transfer is carried out solely by RNA[10], making the PTC and the ribosome itself a ribozyme[11,12]. As the nascent peptide chain grows at the P-site, it exits thorough a tunnel in the large subunit called the peptide exit tunnel.

1.3   The ribosome is highly dynamic machine

Specific RNA segments from the subunits called intersubunit bridges[13] interact to keep the large and small subunits of the ribosome together. The assembled ribosome is a hub of dynamic activities. Translation factors, tRNA, and mRNA interact with the ribosome and induce numerous dynamics states at any given time. The ribosome itself is undergoing numerous intrinsic movements. These dynamics are most prominent on the small subunit which undergoes two distinct set of movements.

The first of these movements is a large scale rotation of the entire small subunit with respect to the large subunit. When viewed from the solvent side of the small subunit, this movement is visible as a counterclockwise rotation of the entire small subunit **(Fig. 1.1 B)**. This rotation, or ratcheting, of the small subunit is an inherent Brownian machine like motion of the ribosome[14]. The ribosome intrinsically fluctuates between the non-rotated "classical" state and the "rotated" or "ratcheted" state[15,16] **(Fig. 1.1 B)**. Rotation of the small subunit is pivotal in moving the tRNAs and the associated mRNA codons from the A-site to the P-site and from the P-site to the E-site via a process called translocation[17]. Coupled movement of the mRNA codons and respective tRNAs needs to happen one codon at a time lest the mRNA slip and the ribosome accept an out-of-frame tRNA in the A-site. Frame maintenance is a critical function of the ribosome during protein synthesis and it happens during translocation. Translocation is catalyzed by a translocase protein factor and is coupled to the rotation of the small subunit and another particular movement of small subunit head called head swiveling, discussed below[17]. During translocation the tRNAs also occupy distinct states called classical and hybrid states, both discussed later in the chapter.

The second type of movement carried out by the small subunit is called subunit rolling[18]. Specific to eukaryotic ribosomes, subunit rolling describes the movement of the small subunit around a central helix as if it is rocking back and forth on the large subunit. The intersubunit bridges that keep the ribosome together have to constantly break and mend during both small subunit rotation and small subunit rolling.

The different parts of the small subunit also undergo a separate set of motions, most prominently the head[19]. The head of the small subunit is relatively more flexible compared to the body of the small subunit since it is connect to the body by just a single helix[20]**(Fig. 1.1 A, right, orange)**. The flexible head exhibits two different types of motions: swivel and tilt[19–21]. Head swivel is the rotation of the head around a central axis in a direction almost perpendicular to the axis of small subunit rotation[20]. Head tilt is the tilting of the head towards the intersubunit space[22].

The large subunit is mostly a stationary monolith except for the highly mobile P- or L7/L12-stalk[23] and the L1-stalk[24]. The L1-stalk moves in and out from the E-site to interact with and to assist the tRNA at numerous stages of egress from the ribosome[24,24,25]. Its movements are intimately linked with the movements of the small subunit and the states of the tRNAs on the ribosome.

A separate class of dynamics on the ribosome are its transient interactions with its factors **(Fig. 1.1 C, bottom)**. Both the small and large subunit interact with numerous factors separately and as an assembled ribosome[6,26]. A wide range of initiation factors (discussed later) interact with the small subunit at regions that are spread throughout its surface. Some factors interact with both subunits while assembling[27] or disassembling[28] the ribosome. Others yet bind as accessory factors to facilitate or regulate translation[29]. Distinct parts of the tRNAs interact with distinct sites on the ribosome during decoding and peptide synthesis[6]. The ribosome also interacts with and activates the elongation factors that deliver[30] or push[31] the tRNAs through the intersubunit space.

The dynamics of the ribosome are intricate[32] and the network of ribosome associated factors vast. Not despite these dynamics but because of them, the ribosome is a fast and accurate protein making machine[33]. An orchestral set of events are simultaneously happening on and within the ribosome when a protein is being synthetized. While there are some differences between different ribosomes from different kingdoms on how such a process is executed, the fundamental steps in protein synthesis are highly conserved.

1.4 Protein synthesis consists of four distinct and evolutionarily conserved phases

The steps that the ribosome takes to make a protein can be divided into four phases: initiation, elongation, termination, and recycling[6](**Fig. 1.1 C**).

Initiation, discussed in detail later, is the process by which a ribosome finds the correct start codon on the mRNA, loads the initiator aminoacyl-tRNA onto the start codon at the P-site, joins the two subunits, and becomes ready to elongate[26] (**Fig. 1.1 C and 1.2**). It is the more divergent phase of protein synthesis between the different kingdoms of life. Initiation results in an elongation competent ribosome that is ready to accept an aminoacyl tRNA at its A-site.

Elongation is the process by which the ribosome sequentially adds amino acids to the growing peptide chain[6]. It is highly conserved between all forms of life and it involves cyclical steps of aminoacyl-tRNA delivery, decoding, peptidyl transfer, and translocation[6] (**Fig. 1.1 D**). Elongation starts when an aminoacyl-tRNA is delivered to the vacant A-site of an elongation competent ribosome. This is carried out by the GTP-bound form of the GTPase eEF1A in eukaryotes (EF-Tu in bacteria). Different types of eEF1A-GTP bound aminoacyl-tRNAs are sampling the codon at the A-site but the ribosome only accepts the correct tRNA based on the Watson-Crick base pairing of the codon-anticodon interaction at the decoding center[34,35].

The tRNAs when first delivered to the ribosome bind in a state known as the A/T state, indicating that the mRNA-tRNA codon-anticodon interaction is located at the A-site of the small subunit while the amino acid acceptor end of the tRNA is located within the elongation (Tu) factor[36]. When the correct aminoacyl tRNA is detected by the decoding center, a series of conformational changes leads to the hydrolysis of the GTP bound to eEF1A. A segment of 28S rRNA at the GTPase center called the sarcin-ricin loop (SRL) is essential for this step[6]. An adenine base in this loop is responsible for coordinating a histidine residue in the GTPase domain of eEF1A so that a water molecule can hydrolyze the GTP[37]. After the GTP is hydrolyzed into GDP, significant domain rearrangements occur in eEF1A and it leaves the ribosome. The aminoacyl-tRNA then accommodates itself into the A-site of the ribosome **(Fig. 1.1 D, top)** and peptidyl transfer from the P-site peptidyl tRNA to the A-site aminoacyl tRNA occurs at the PTC of the ribosome[6] **(Fig. 1.1 D, right)**. This results in a ribosome with the elongating peptide attached to the A-site tRNA and the ribosome sampling between the non-rotated classical state **(Fig. 1.1 D, right)** and the rotated state **(Fig. 1.1 D, bottom)**. Both classical and rotated states have distinct arrangements of their tRNAs[17,38].

In the classical non-rotated state, the tRNAs in the A-site and the P-site assume what is known as the A/A tRNA and P/P tRNA states[6] **(Fig. 1.1 D, right)**. An A/A tRNA indicates that the mRNA-tRNA codon-anticodon interaction is located solely on the A-site of the small subunit and that the amino acid acceptor end of the tRNA is also located solely within the A-site of the large subunit. Similar logic applies for the P/P tRNA.

In the rotated state, the tRNAs assume what is known as the A/P and P/E hybrid states[17,39,40] (**Fig. 1.1 D, bottom**). A hybrid A/P state indicates that the mRNA-tRNA codon-anticodon interaction is located at the A-site of the small subunit but the amino acid acceptor ends of the tRNA is occupying the P-site of the large subunit. Hence, A/P tRNA. Similarly, the hybrid P/E tRNA indicates that the mRNA-tRNA codon-anticodon interaction is located at the P-site of the small subunit but the amino acid acceptor end of the tRNA is occupying the E-site of the large subunit.

The rotated state ribosome with its tRNAs in the hybrid state is the substrate for the translocase eEF2 (EF-G in bacteria)[6,21]. Binding of eEF2 to the ribosome its GTP-bound form stabilize the ribosome in a rotated configuration and induces an additional degree of rotation. At the same time, contacts between the domain IV of eEF2, specially around a modified histidine with diphthamide[41], stabilize the mRNA-tRNA codon-anticodon interaction throughout the process preventing loss of the correct reading frame[42]. Concurrent to the additional rotation of the small subunit, a pronounced small subunit head swivel assists the movement of the mRNA-tRNA forwards. During the process the contacts that had maintained the mRNA-tRNA interactions in the canonical state are altered[6,21].

GTP hydrolysis on eEF2 and back rotation of the ribosome is coupled with the back-swivel movement of the small subunit head[6,21]. These motions move the base paired mRNA-tRNAs forwards. The A-site tRNA with the growing peptide chain is thus moved to the P-site and the deacylated tRNA in the P-site is moved to E-site. Intermediate states that the tRNAs sample during this process are known and were defined as chimeric hybrid states[43]. One of these chimeric hybrid tRNA states, was observed when eEF2 blocks the A-site of the ribosome. The mRNA-tRNA codon-anticodon interaction of the translocating peptidyl-tRNA is then seen between the A- and P-sites of the small subunit and is indicated as ap/P tRNA. Similarly, for the deacylated-tRNA with its mRNA-tRNA codon-anticodon between the P- and the E-sites, the chimeric hybrid tRNA is indicated as pe/E tRNA.

Once the peptidyl-tRNA is in the P-site (P/P tRNA) and the deacylated-tRNA in the E-site (E/E tRNA), the ribosome attains a non-rotated state with an empty A-site, ready to add another amino acid to the peptide chain (**Fig. 1.1 D, left**). Thus the tRNAs, as discussed above, assume different conformations before, during, and after elongation as they bind to different sites of the ribosome. When an elongating ribosome encounters any of the three stop codons, elongation is complete and the ribosome enters the termination phase[28].

Termination involves recognition of the stop codon and release of the full length peptide (**Fig. 1.1 C**). The three stop codons (UGA, UAA, or UAG) are recognized at the A-site of the ribosome by the release factors[28]. A release factor such as the eukaryotic release factor eRF1 recognizes a stop codon and then cleaves and releases the peptide from the P-site of the ribosome. A new protein is thus formed.

After the peptide is released via termination, recycling follows. Recycling involves the removal of any ribosome-bound factors from the ribosome and the subsequent splitting of the subunits to prepare them for another round of protein synthesis[28,44] (**Fig. 1.1 C**). In eukaryotes recycling factors such as ABCE1 bind to an eRF1 containing post-termination complex and split the ribosome in an ATP-dependent manner[45,46]. The split subunits can now be recycled by the cell for another round of translation.

Out of the four phases, elongation is evolutionarily conserved in the basic principle, mechanisms, and associated factors between bacteria and eukaryotes. Initiation, on the other hand, has expanded, diversified, and become relatively complicated in eukaryotes[26].

1.5   Initiation of protein synthesis is evolutionarily quite divergent

Ribosomes start making proteins from an mRNA via a process called translation initiation (**Fig. 1.2**). While the details of this process differ significantly between bacteria and eukaryotes[26], both converge to the same result: a ribosome that is ready to decode the second codon of the mRNA. Such a ribosome is considered "elongation competent." In an elongation competent ribosome, the initiator tRNA is base paired with the start codon of the mRNA at the P-site of the ribosome and the A-site is vacant but ready to accept and decode the tRNA for the second codon (**Fig. 1.2 M**). While bacteria use a few factors and a relatively simple mRNA-rRNA base pairing-based mechanism[47] to initiate protein synthesis, eukaryotes have evolved a far more intricate and vastly divergent system[26,48–50].

In bacteria, a stretch of evolutionarily conserved sequence upstream of the start codon called the Shine-Dalgarno sequence base pairs with the 3'-end of the 16S rRNA where the anti-Shine-Dalgarno sequence is located[51,52]. This "molecular yardstick" interaction places the start codon near the P-site of the bacterial 30S small ribosomal subunit. Three different initiation factors then ensure the proper delivery of the initiator tRNA to the start codon at the P-site[47]. When the bacterial 50S large ribosomal subunit joins next, the three initiation factors leave. This 70S ribosome with a P-site initiator tRNA and an empty A-site becomes ready to accept and decode the tRNA for the second codon. Initiation in bacteria is completed when this elongation-ready 70S ribosome is formed.

In eukaryotes, translation initiation is vastly different[26,48–50]. Broadly, two different complexes—one with one set of initiation factors assembled on the 5′-capped mRNA (**Fig. 1.2 D**) and another with a different set of initiation factors assembled on an empty eukaryotic 40S small ribosomal subunit (**Fig. 1.2 E**)—need to be first formed and then subsequently brought together (**Fig. 1.2 F**) to start initiation. When these two complexes come together, the small ribosomal subunit scans the mRNA in a 5′ to 3′ direction looking for the proper start codon to place the initiator tRNA (**Fig. 1.2 G**). When the start codon is found within a proper sequence context (**Fig. 1.2 I**), a set of initiation factors leave (**Fig. 1.2 J**) while others bind to catalyze the proper joining of the eukaryotic 60S large ribosomal subunit (**Fig. 1.2 K**). After another round of factors leave the newly assembled 80S ribosome (**Fig. 1.2 L**), initiation is completed as the ribosome is elongation-ready (**Fig. 1.2 M**). The set of initiation factors (**Fig. 1.2 B, C, J, and N**) involved in this process range from small single peptide proteins to large multi-domain ones to even larger multi-subunit complexes that are almost as big as the small ribosomal subunit.

1.6   Eukaryote specific features and factors involved in the initiation pathway

One of the characteristic features of eukaryotic mRNAs is the 5′-cap[53,54] (**Fig. 1.2 P**). The 5′-cap is a peculiar eukaryotic specific RNA modification that is found at the 5′-terminus of a eukaryotic mRNA. It is a 7-methylguanosine residue that is covalently linked to the 5′-terminal nucleotide of the mRNA in a 5′ to 5′ manner[54]. The cap is added co-transcriptionally on the mRNA in the nucleus by a series of enzymes found in the capping complex[55]. Once the mRNA is processed and exported to the cytoplasm, a heterotrimeric complex of initiation factors called the eIF4F complex (**Fig. 1.2 D**) assembles on the 5′-cap to start initiation[26] (**Fig. 1.2 F**).

The eIF4F complex is made up of the cap binding protein eIF4E, the DEAD-box RNA helicase protein eIF4A, and the large scaffolding protein eIF4G[26] (**Fig. 1.2 A**). eIF4G, true to its function as a scaffolding protein, provides binding site to the mRNA, other components of the eIF4F complex, and numerous other initiation factors[26,48,49] **(Fig. 1.2 F)**. The cap-binding protein eIF4E binds to the 5'-cap with assistance from eIF4G **(Fig. 1.2 F)**. The DEAD-box RNA helicase eI4FA, is recruited to the mRNA and activated by the eIF4A-binding domain of eIF4G. eIF4G also interacts with the poly-A binding protein (PABP) **(Fig. 1.2 F, red)** in such a way that it brings the 5'- and 3'-end of the mRNA together to form a closed-loop mRNA protein complex (mRNP) that is thought to enhance translation[26]. In this mRNP, once the helicase eIF4A starts melting secondary structures, a stretch of the 5'-UTR of the mRNA become available for the small subunit of the ribosome to bind.

The 40S small subunit of the ribosome is converted to a 43S complex **(Fig. 1.2 E)** as preparation for binding to the above "activated" mRNP by the initiation factors eIF1, eIF1A, eIF2, the initiator aminoacyl-tRNA, eIF3, and eIF5[26] **(Fig. 1.2 B, C)**. eIF1 binds near the P-site of the 40S while eIF1A binds near the A-site and protrudes its tails into the P-site[6]. The third factor, eIF2, is composed of three subunits α, β, and γ. Together these three proteins recognize and bind to unique structural features present in the initiator methionine-tRNA and form the eIF2-GTP-aa-tRNA ternary complex (TC). The ternary complex is responsible for delivery of the initiator aminoacyl-tRNA to the P-site start codon. eIF5 interacts directly with eIF2 with both its N- and C-terminal domains.

eIF3 is the biggest translation initiation factor **(Fig. 1.2 C, purple eIF3)**. At about 800 KDa in size, it is almost as large as the 40S itself. eIF3 has twelve distinct subunits in most complex eukaryotes and binds to the solvent side of the 40S next to the platform. The common core of eIF3 is located at this site but it has numerous tendrils that bind themselves throughout the small subunit. Through distal subunits or extended tails of its core subunits, eIF3 interacts with faraway sites such as the mRNA entry channel at the beak of the 40S and features at the inter-subunit side of the 40S. With its extensive network of interactions and binding partners, eIF3 plays a key role in preparing the 40S for loading the mRNP and scanning.

eIF1, eIF1A, eIF5, and eIF3 can bind to an empty 40S directly and they additively stimulate the binding of the eIF2-GTP-aa-tRNA ternary complex to the 40S. When the ternary complex binds to the 40S in the presence of these factors, the mRNP-ready 43S-sized particle called a pre-initiation complex (PIC) is formed **(Fig. 1.2 E)**. The PIC opens the mRNA channel on the 40S (a conformation called PIC-open) and readies it for the next step in initiation which is mRNA recruitment.

How the mRNA gets recruited to the PIC **(Fig. 1.2 F)** has been a difficult question to answer. There is evidence showing that interaction between the cap-bound eIF4E, the scaffold protein eIF4G, and eIF3 is important to load the mRNA onto the PIC but we don't yet have a full picture of how this happens. There is indication that the eIF4F complex binds to the leading edge of the PIC and then threads the mRNA through the mRNA channel. This scenario present the ternary complex at the P-site with the opportunity to scan and inspect the mRNA from its very beginning. In another scenario, the eIF4F complex could bind to the PIC close to the mRNA exit mediated by subunits of eIF3. This would mean that the mRNA gets slotted into the mRNA channel from the inter-subunit side rather than threaded through the mRNA entry site. Thus in this second scenario, the ternary complex doesn't get the opportunity to scan and inspect the codons at the very beginning of the mRNA. Regardless of how the mRNA gets recruited, the scanning-competent 43S-PIC is now formed **(Fig. 1.2 H, I)**. The 43S-PIC now needs to find the proper start codon on the message via a mechanism called "scanning" **(Fig. 1.2 G)**.

It has been long hypothesized that after mRNA recruitment the PIC scans nucleotides in the 5'-UTR of an mRNA one-by-one in a 5' to 3' direction to find the correct AUG start codon. Usually, "scanning" leads to the branding of the very first AUG encountered as the legitimate start codon. This AUG is usually in a specific sequence context called the Kozak sequence[56]. When eIF2 engages the initiator tRNA with this AUG at the P-site, GTP gets hydrolyzed, eIF1 gets released, and the PIC undergoes conformational changes from an open state (PIC-open) to a closed state (PIC-closed)[26] **(Fig. 1.2 I)**. The conformational changes and the departure of eIF1, eIF2-GDP, and eIF5 **(Fig. 1.2 J)** make the PIC scanning-incompetent and the mRNA gets locked into place **(Fig. 1.2 I)**.

Now that there is an initiator tRNA at the P-site of the 40S base paired with the proper start codon **(Fig. 1.2 I)**, the PIC is ready to engage with the 60S large subunit of the ribosome and form an elongation-competent ribosome. Subunit joining is catalyzed by the checkpoint GTPase eIF5B[27] **(Fig. 1.2 J, K)**. eIF5B is a multi-domain protein that integrates signals from throughout the ribosome[57]. Using its elongated structure it verifies that the P-site tRNA is aminoacylated and properly positioned, that the 40S is in a proper configuration, and that the 60S is engaged. Only after verifying these features, it hydrolyzes the GTP and, along with eIF1A, exits the ribosome[58] **(Fig. 1.2 L)**. An elongation-ready 80S ribosome is left behind **(Fig. 1.2 M)** and it is ready to receive the aminoacyl-tRNA that codes for the second codon. Initiation is complete.

1.7   Competing models of eukaryotic initiation of translation

Initiation of translation in eukaryotes via the cap-dependent mechanism is not as uniform as outline above. Many aspects of the pathway have variations and there are competing models for how and when specific steps happen and what many of the factors do[48,50,59].

For example, there is evidence for a model where a multi-factor complex (MFC) composed of eIF1A, eIF2, eIF3, and eIF5 assembles prior to binding to an empty 40S and eIF1 to form the PIC[48]. A different model for mRNA recruitment proposes that an empty PIC first interacts with eIF4F complex and only then recruits the mRNA to the 40S[48]. It is thought that this holo-PIC unwinds and threads the mRNA into the mRNA channel.

There isn't much consensus or experimental data about how the PIC scans the mRNA either. One model proposes that scanning happens as a Brownian ratchet mechanism where unwinding of the mRNA by eIF4A is coupled with single-stranded mRNA capture by eIF4B which loops the mRNA outside the mRNA exit channel[60]. In another model for scanning, eIF4A and eIF4B are thought to first bind close to the mRNA entry channel and then mRNA spools and loops towards the inter-subunit side of the PIC[48]. The theory is that the eIF4B bound mRNA serves as the pawl in the Brownian ratchet while the unfolding and consequent refolding of the mRNA functions to prevent the complex from sliding backwards.

1.8   Non-canonical but still cap-dependent translation initiation

Given the intricacies of interactions and the multitude of factors involved in cap-dependent translation initiation, several variations of non-canonical but still cap-dependent forms of translation initiation exist[50] **(Fig. 1.3)**. Some of these mechanisms are leaky scanning, translation re-initiation, eIF4E-independent cap recognition, and ribosome shunting[59,61,62].

It is well established that the first AUG codon encountered during scanning is usually where translation gets initiated[26]. This codon however needs to be in what is considered a "strong" Kozak sequence context[63–65]. If the codon is within a "weak" or "intermediate" Kozak sequence context[64,66,67], the PIC scans past it until it finds another AUG downstream within a strong Kozak sequence context. A combination of Kozak sequence context strength and the number of different AUG codons upstream[68] of or within an open reading frame (ORF) presents opportunities for variation in translation initiation via mechanisms such as leaky scanning and translation re-initiation.

Leaky scanning is probably the most canonical of the non-canonical cap-dependent mechanisms[62] **(Fig. 1.3 A)**. It occurs on transcripts that have several start codons. All the events leading up to the formation of the scanning-competent PIC proceeds in a canonical fashion. Leaky scanning occurs when instead of selecting the first AUG as the start codon, the PIC scans past it and chooses a more downstream start codon[69] which is within a stronger Kozak sequence context. Leaky scanning occurs on the transcripts for eIF1 and eIF5 which themselves take part in regulating the start codon choice[59]. Leaky scanning can lead to interesting translational regulation schemes **(Fig. 1.3 A, top)**, production of protein isoforms **(Fig. 1.3 A, middle)**, or expansion of the coding capacity of the same mRNA **(Fig. 1.3 A, bottom)**.

Re-initiation, like leaky scanning, involves two or more start codons[59] **(Fig. 1.3 B)**. However, unlike in leaky scanning, in re-initiation one of the upstream start codon goes through a full round of translation. This means that after assembling at the 5′-end of the message, the PIC scans to find the first AUG, assembles an elongation-competent 80S ribosome, elongates, and eventually terminates **(Fig. 1.3 B, uORF)**. After termination, instead of undergoing recycling, only the 60S leaves the mRNA while the 40S is still somehow bound to it. A new PIC is then assembled and it scans to and reinitiates on a downstream ORF **(Fig. 1.3 B, main ORF)**. The regulation of the yeast stress response gene GCN4 happens via re-initiation[70].

The GCN4 gene is a master regulator transcription factor gene that responds to amino acid starvation[70,71]. GCN4 is responsible for regulating almost one tenth of the yeast genome and thus needs to be stringently regulated[72]. Yeasts do so via re-initiation[71]. The transcript of GCN4 contains four upstream ORFs (uORFs)[71,72]. The first uORF (uORF1) serves as a positive regulator while uORF4 serves as the negative regulator for re-initiation. During non-stress conditions, after the PIC is formed at the 5′-end, it scans to either the first or second start codon. There, the PIC undergoes a full round of translation and terminates. Then, the 60S is recycled while the 40S stays on the message and "traverses" to the downstream ORFs uORF3 or uORF4. The ribosomes then reinitiate on these ORFs and go through a full round of translation to be completely recycled. All these events happen upstream of the main ORF for GCN4. During non-stress conditions, the PIC never gets to the main ORF AUG to translate the main GCN4 ORF. The main GCN4 ORF only gets translated during stress conditions.

During stress conditions such as amino acid starvation, the aminoacyl tRNA delivering factor eIF2 gets inactivated via phosphorylation of its alpha subunit, limiting the availability of active eIF2[73,74]. The PIC is formed on the GCN4 transcript and it scans to and translates either uORF1 or uORF2 as usual. The 40S gets retained via re-initiation and then it traverses to uORF3 or uORF4. Here, because there is a limited pool of functional eIF2 available because of cellular stress, the PIC bypasses these two ORFs and re-initiates on the main GCN4 ORF. The main ORF of GCN4 thus gets translated and the cell is able to mount a response to the amino acid starvation stress.

Similar schemes of re-initiation have been observed in the ATF4 gene in mammals[75]. Non-canonical initiation factors like eIF2D[76] and the DENR-MCT-1 heterodimer[77] have been shown to replace eIF2 during PIC formation for re-initiation. It has been hypothesized that eIF3 stays bound to the 40S during and after the translation of the uORFs[78,79]. It is thought that eIF3 nucleates the reassembly of the PIC on the main ORF and chaperons the non-canonical initiation factors during re-initiation events.

1.9 Features on the RNA transcript can trigger non-canonical initiation events

Some non-canonical initiation events can be triggered by features on the mRNA. An eIF4E-independent has been reported for the human oncogene c-jun[80]. The 5'-UTR of c-jun inhibits recruitment of the cap-binding protein eIF4E by using a stem loop. But, as an alternative mechanism, the capped 5'-end of the c-jun transcript directly binds to eIF3d subunit of eIF3 to initiate[80].

More recently, it has been shown that the RNA modification m^6A at the 5'-UTR of genes plays crucial role in initiation[81,82] **(Fig. 1.3 C)**. Transcripts without m6A and without cap can also be translated in cell-free extracts (e.g. RRL) and such transcripts are able to assemble a PIC without recruiting the eIF4F complex[83]. m6A modified 5'-UTRs can also bind eIF3 directly to form a PIC[83].

Some viruses employ yet another set of RNA transcript based alternative modes of initiation. These RNA-based strategies fall into three main categories: ribosome shunts, cap-independent translation enhancers (CITEs), and internal ribosome entry site (IRESs).

The cauliflower mosaic virus[84] and the Sendai virus[85] use an RNA structure called a ribosome shunt **(Fig. 1.3 D)** to directly move a PIC from its assembly site to the start codon, essentially skipping all or most of scanning[59]. In doing so, the shunted PIC can skip over large distances, complex folded regions, or upstream start codons. Sometimes, however, the ribosome translates a short uORF but when it reaches the shunt, it splits and re-initiates at another start codon downstream. Ribosome shunting has been proposed as a mechanism for the translation for some cellular transcripts[86].

The cap-independent translation enhancers (CITEs) **(Fig. 1.3 E)** found in some plant viruses carry out a 5'-cap independent but a 5'-UTR dependent initiation of their uncapped messages[59]. CITEs usually have T-, Y-, or L-shaped dimensions and are located at the 3′-UTR of viral mRNAs[87,88]. They use their pseudoknotted structures to recruit eIF4F. Some cellular RNAs have also been shown to contain CITEs[89].

An interesting case of RNA feature based exploitation of translation initiation are viral internal ribosome entry sites (IRESs) which subvert the ground rules of translation initiation and its regulation **(Fig. 1.4)**.

1.10   IRESes are the Trojan horses of translation initiation

Translation regulation plays a crucial role in the maintenance of cellular homeostasis and cellular functions. Regulatory checks and balances on the translation machinery and associated parts are used by the cell so that the resources dedicated to protein synthesis—which uses a large fraction of all available cellular resources—do not go to waste. The cell also needs to make sure opportunistic agents aren't siphoning off precious cellular resources for nefarious means. Viruses, however, have evolved strategies to do exactly that.

Viruses have relatively small genomes and as such they have to maximize the use of this scarce resource. As a means to this end, they have evolved numerous ways to hijack and steal the host cell protein production machinery to make viral proteins. Some positive strand RNA viruses have evolved intricate RNA based mechanisms to hijack and steal ribosomes from the host cell. These RNA sequences found in the 5′-UTR or in the intergenic region of such viruses are the Trojan horses of the cellular world **(Fig. 1.4 A)**. These viruses use their RNA structure, sometimes accompanied by viral and/or host factors, to manipulate, hijack, and steal the host cell ribosomes for viral protein production. These viral RNA structures are called internal ribosome entry sites (IRESs).

1.11   Discovery of "a conspicuously unusual eukaryotic mRNA"

IRESs were first reported in two different viruses from the *Picornaviridae* family[90,91].
Picornaviridae genomes are single-stranded positive-sense RNA genomes. The viral genome is a
single ORF that gets translated as one long poly-peptide which gets cleaved by viral proteases
into individual proteins. Very early-on, researchers working on the poliovirus, the prototypical
picornavirus, noticed its mRNA genome lacks a 5′-cap[90]. The 5′ terminus of the genome was
instead covalently linked to a small viral peptide. Despite lacking the 5′-cap structure, the
genomes of these viruses seemed to be translated very rapidly and efficiently in an infected host
cell. Efficient translation of viral genes was coupled with almost-complete shut-off of translation
in the infected cell[90]. These disparate observations lead researchers to postulate that the 5′ region
of the viral genome was in some way responsible for an alternate 5′-cap independent mechanism
which also suppressed translation of the host cell mRNAs.

The 5′ region of the poliovirus is long, structured, and untranslated (**Fig. 1.5**). This 5′-
untranslated region (5′-UTR) was, however, littered with unused AUG codons[90]. These
observations lead the researchers to reason that instead of a 5′-cap dependent recruitment of the
ribosome, some sort of internal ribosome recruitment mechanism was at play.

Further research into the poliovirus 5′-UTR led to the observation that during viral
infection eIF4G was proteolytically cleaved such that it no longer interacted with the cap binding
protein eIF4E[92]. These observation and subsequent data from bicistronic reporters showed that
the poliovirus 5′-UTR indeed contained an internal ribosome entry site. The existence of 5′-cap
independent IRESs was further confirmed by a concurrent study that showed the
encephalomyocarditis virus (EMCV) genome also contained an IRES element in its 5′-UTR[91]

(Fig. 1.6). Soon, more IRESs were discovered in other positive sense single-stranded RNA genome viruses.

The human pathogen hepatitis C virus (HCV) was shown to contain an IRES in its 5'-UTR region[93]. This IRES was much shorter than the picornavirus IRESs and seemed to only require a subset of the initiation factors **(Fig. 1.7)**. An even shorter IRES was later discovered in the intergenic region of an insect infecting picorna-like virus[94] that initiated protein synthesis from a non-AUG codon[95,96] **(Fig. 1.8)**.

While all of these IRESs are able to initiate and drive translation in a 5'-cap independent manner, they are diverse in length, flexibility, secondary structure features, and factor requirements. Taking these factors into consideration, IRESs were classified into four different types[97] **(Fig. 1.4 B)**.

1.12   Four different families of IRESs

IRESs were categorized into four different groups according two main criteria: canonical initiation factor requirement and the degree of RNA structure[97]**(Fig. 1.4 B)**. While some IRESs have low structural complexity (Type I) others are highly structured with tightly folded modules (Type IV). Some IRESs require most or at least some of the canonical initiation factors (Type I or II) while others require a few (Type III) or none (Type IV). From this classification system, it becomes evident that factor requirement and degree of RNA structure are inversely related—the more compactly folded the IRES, the lesser the factor requirement **(Fig. 1.4, arrows on the left)**.

1.12.1 Type I IRES: the poliovirus IRES family

Type I IRESs are IRESs that are similar to the poliovirus IRES[97] **(Fig. 1.5)**. These IRESs are long and have relatively unstructured topology. They require an almost complete set of initiation factors and undergo scanning to locate and initiate translation from an AUG codon[98]. They are 5′-cap independent as they eschew the cap-binding protein eIF4E and use a cleaved version of the scaffolding protein eIF4G to implement a corrupted version of scanning[98,99]. One key feature of type I IRESs is that they complement the cellular initiation factors with viral or cellular IRES trans-activating factors (ITAFs)[100,101] such as an isoform of the poly(rC)-binding protein 2 (PCBP2) or the poly-pyrimidine tract binding protein (PTB) **(Fig. 1.4 A, 1.5)**.

Structural probing of the poliovirus IRES revealed that it folds into multiple domains with distinct motifs[102]. The minimal IRES from poliovirus was shown to be about 650 nucleotides long and span from domains II to V in the 5′-UTR region[103] **(Fig. 1.5, dashed box)**. The various domains contains many conserved RNA structures, motifs, and ITAF binding sites **(Fig. 1.5)**. One of the RNA domain was shown to be important for neurovirulence caused by the poliovirus[104] **(Fig. 1.5, green shading)**.

The current model for how the type I IRESs work is as follows[105]. A type I IRES is a highly dynamic entity which interacts with ITAFs at specific regions which help it fold into an "activated" state. The IRES then recruits the viral protease cleaved eIF4G and eIF4A using domain V and assembles a PIC-like structure on the uAUG embedded in domain VI **(Fig. 1.5, AUG$_1$)**. Then it uses the canonical factors eIF1 and eIF1A to scan along the mRNA to locate and load the initiator tRNA onto the main ORF AUG downstream **(Fig. 1.5, AUG$_2$)**. Thus, the poliovirus IRES achieves initiation in a 5′-cap independent manner with an almost canonical PIC.

While there are NMR structures[105] of segments of the type I IRESs are available, due to their large size, inherent flexibility, and multiple factor and ITAF requirements, they haven't been structurally characterized at high resolution in their entirety.

1.12.2 Type II IRESs: the EMCV IRES family

Type II IRESs such as the EMCV-IRES **(Fig. 1.4 B, 1.6)** are similar to type I IRESs[97]. They are long and relatively unstructured like the type I IRESs and use a similar set of initiation factors and ITAFs. They differ from type I IRES in how they place the start codon into the P-site of the ribosome. While type I IRESs scan the mRNA to locate the AUG but type II IRESs place the AUG codon in the P-site of the PIC without any scanning[97,105]. Type II IRESs were first discovered in EMCV[91] around the same time as the type I IRESs were discovered in poliovirus. Two well studied IRESs from this class are the EMCV-IRES and the FMDV-IRES[106].

Sequence-wise, type II IRESs are quite different from type I IRESs but structural probing and NMR studies of type II IRES segments have shown that the two share a similar RNA domain structures. Some of those domains share conserved folds. Like the type I IRESs, EMCV-IRES recruits a viral protease cleaved eIF4G **(Fig. 1.6, green)** along with eIF4A and eIF4B (to domain V), but not the cap-binding protein eIF4E, to form a PIC which assembles directly on the AUG with no scanning.

Like the type I IRESs, because of their large size, flexibility, and multiple factor requirements, we lack structural information on type II IRESs. While there are some NMR data and X-ray structures of segments from these IRESs[107], a complete picture is missing here too.

1.12.3 Type III IRES: the HCV-like family

Type III IRESs on the other hand are shorter, more compact, and more structured than either type I or type II[97] **(Fig. 1.4 B, 1.7)**. They have a reduced set of initiation factor requirement and they do not use any ITAFs **(Fig. 1.4 B)**. Type III IRESs can function with a minimal set of just two initiation factors—eIF2 and eIF3[97].

Type III IRESs like the HCV-IRES, found in the 5′-UTR of the human pathogen hepatitis C virus (HCV)[93], fold into four domains of which domains II and IV consist of the minimal IRES[108] **(Fig. 1.7 A, dashed box)**. Type III IRESs can recruit the 40S small subunit of the ribosome on their own without any need for the eIF4 and eIF1 factors. They use their high binding affinity to bind to the 40S and use only eIF3 and eIF2 to load the AUG into the P-site of the ribosome[109,110]. Because of its high binding affinity of type III IRESs to the 40S subunit and the 80S ribosomes numerous structural studies have elucidated the mechanism with which the HCV-IRES and HCV-like IRESs interact with 40S and eIF3 to hijack and steal ribosomes[111–115].

The most common mechanism of action for type III IRESs integrates information from structures of the HCV-IRES bound to 40S[112], the HCV-like CSFV-IRES bound to the 40S and eIF3[115], and the HCV-IRES on the 80S ribosome[111]—all obtained via cryo-EM experiments. A mechanism for their working is as follows[105]: type III IRESs first bind to the 40S IRES on their own without any factors. In this binary complex the IRES binds to both rRNA and proteins using a large area on the solvent side of the small subunit right behind the platform. Since this is the usual binding site of eIF3, it was hypothesized that eIF3 cannot bind to the 40S when HCV is bound to it. Indeed, structure of the CSFV-IRES bound to the 40S in presence of eIF3 showed that eIF3 is displaced from its usual 40S binding site[115]. Instead of binding directly to the 40S eIF3 is bound to domain III of the IRES **(Fig. 1.7 B)**. Domain II of the IRES extends from the back of the 40S and reaches into the E-site of the ribosome to base pair with residues in the E-site. The head of the 40S is constrained and the AUG codon gets situated into the P-site[112]. Then eIF2, without help from eIF1 or eIF1A, but facilitated by eIF5, loads the initiator tRNA onto the P-site forming a 48S-like complex. eIF5B stimulates subunit joining[116] and the ribosome becomes elongation-competent and ready to receive the aminoacyl-tRNA coded by the second codon. While this is a generally accepted mode of initiation, the HCV-IRES has been shown to initiate via alternative pathways that use eIF3 and eIF5B only[117], or use eIF2A[118] or eIF2D[76,119,120] as replacements for eIF2.

The HCV-IRES and the CSFV-IRES are well studied as a model type III IRESs. There are, however, IRESs from the 5′-UTR region of dicistroviruses that are predicted to be type III IRESs but share very little sequence or structural similarity with either. Recent analysis of the cricket paralysis virus 5′UTR-IRES (CrPV-5′-UTR-IRES) classified it as a type III IRES and predicted a 2D structure based on selective 2′-hydroxyl acylation analyzed by primer extension assays (SHAPE) and dimethyl sulphate (DMS) probing assays[121]. What this newly characterized type III IRES looks like on the ribosome, its initiation factor requirements, its initiation factor recruitment strategy, its scanning mode, all are, however, completely unknown.

1.12.4 Type IV IRES: the CrPV-like IRESs

Of the four different types of IRESs, type IV IRESs follow the most minimalistic approach to hijacking ribosomes[97] **(Fig. 1.8, 1.9, 1.10)**. The previous three classes of IRESs are at least 350 nucleotides long and require all or a subset of initiation factors to function. Type IV IRESs, on the other hand, are at most 250 nucleotides long and they require absolutely none of the initiation factors to initiate protein synthesis[122]. Rather than assemble an initiation complex on an AUG start codon like the other three classes, the type IV IRESs jumpstart translation from a non-AUG codon[95,96] **(Fig. 1.8 A, GCU)**. And unlike the other three, which go through a virally perturbed version of the canonical initiation pathway, type IV IRESs exploit the elongation step in translation to hijack and steal host cell ribosomes[123–126] **(Fig. 1.12)**.

Type IV IRESs were discovered from genome sequencing studies of several viruses that infect insects[127] **(Fig. 1.13 B)**. The single-stranded positive-sense RNA genomes of these viruses are dicistronic **(Fig. 1.13 A)**. The first ORF (ORF1) of their genome codes for non-structural proteins while the second (ORF2) codes for the structural ones[125,128,129]. The two ORFs are driven by two different IRESs. ORF1, at the 5'-end of the genome, is preceded by a 5'-UTR region that contains an IRES similar to type III IRESs[121]. ORF2, which initiates from a non-AUG codon, is preceded by a type IV IRES[122].

Toe-printing experiments, in vitro reconstitution assays, and bicistronic reporter assays were instrumental in validating the factor requirements of the type IV IRESs[124,125,128–130]. These biochemical studies showed that the type IV IRESs do not use any initiation factors. All that a type IV IRES needs to initiate protein synthesis are the two ribosomal subunits and the elongation factors eEF2 and eEF1A[131–134].

Sequence and SHAPE based analysis of the full length type IV IRESs indicated that type IV IRESs share a conserved structure despite sequence divergence[122]. This was further corroborated by NMR and X-ray crystallography studies[105] of different domains of the IRES. The structural studies showed that the IRES folds into a compact shape that consists of three domains—each of which contains a pseudoknot[122,123,135–137] **(Fig. 1.8 A)**. Domain 1 contains pseudoknot II (PKII) and the L1.1 region **(Fig. 1.8 A, blue box)**. Domain 2 contains pseudoknot III (PKIII) and the SLIV and SLV stemloops **(Fig. 1.8 A, red box)**. Domain 3 contains pseudoknot I (PKI) and the variable loop region (VLR) **(Fig. 1.8 A, green boxes)**. The IRES uses domains 1 and 2 to bind to the ribosome[123,132,136] **(Fig. 1.8 B, gray shading)**. Domain 1 interacts with the L1-stalk of the 60S subunit while domain 2 uses the two stemloops to bind to the head region of the 40S small subunit[138,139,134]. Domain 3 contains PKI includes the codon-anticodon mimicking part of the IRES and it is important for the initiation of protein synthesis and frame maintenance[140,141] **(Fig. 1.8 B, dashed box)**. The type IV IRESs all share the above set of core features. There is, however, still some differences and unique features found in some

members.

The Dicistroviruses which harbor these IRESs have been divided into two genera: the

Cripaviruses and the Aparaviruses[127]. Following this classification, the type IV IRESs are also

divided into two classes[142]**(Fig. 1.8 A, bottom green boxes)**. Class I consists of the prototypical

CrPV-IGR-IRES and other related IRESs **(Fig. 1.8 A, bottom right)**. Class II consists of a

closely clustering set of IRESs, some of which belong to bee infecting viruses such as the Israeli

Acute Paralysis Virus (IAPV) **(Fig. 1.8 A, bottom left)**. The class II IRESs have an additional

stem loop (SL III) in their domain 3 and an elongated L1.1 while class I lacks these

features[122,142].

There has been some biochemical work on the class II IRESs[141,143–146] but most of what

we know about how type IV IRESs work has been derived using CrPV-IRES as a model.

Biochemical and X-ray crystallography[136,147] have been used to structurally inform the

mechanism of this IRES. Because of the inherent flexibility and size of this IRES, these methods

could only be used in truncated pieces of the IRES which also truncates their mechanistic

understanding. There is a structure of a type IV IRES on a bacterial ribosome but it is trapped in

a non-functional state[148]. To fully understand the mechanism of type IV IRESs structures of a

full length IRES on the ribosome are essential. Cryo-EM based structural studies have been

instrumental in visualizing such structures and arriving at a mechanistic understanding of type IV

IRESs.

1.13   Low and medium resolution structures of the type IV IRESes on the ribosome

Since the mid-2000s, the CrPV-IRES has been used as a model system for structural

determination of IRESs on the ribosome **(Fig. 1.9)**. The first structural study of the CrPV-IRES showed how it binds to the 40S subunit and to the 80S ribosomes[149]. Using a full length IRES and 40S and 60S subunits from HeLa cells, Spahn and colleagues assembled the binary complex and imaged it using cryo-EM to give a global view the IRES on the 40S subunit **(Fig. 1.9 A)** and on the 80S ribosome. The CrPV-IRES was observed bound to the 40S at the cleft between the head and the body and was seen to maintain that binding in the 80S ribosome too. From the density maps, comparisons with 80S ribosomes without any factors, and fitting densities and models from both prokaryotic and eukaryotic ribosomes, the study showed that the IRES occupies the inter-subunit space usually occupied by the tRNAs. Though due to the limited resolution of these structures, many questions remained unanswered. It was, however, clear that this was a different binding mechanism compared to one followed by the type III HCV-IRES[150].

Further structural exploration of the CrPV-IRES by Schüler et al. gave us a more detailed view of the IRES on the ribosome[139] **(Fig. 1.9 B)**. These set of structures distinctly visualized the compact structure of the IRES and showed how the three pseudoknots interacted with the different parts of the ribosome. This study placed Domain 3/PKI in-between the A- and P-sites of the ribosome. From their low resolution of the density, the authors predicted that domain 3 has to move out of the A-site to accommodate the Ala-tRNA coded by the first codon. They, however, proposed that inherent dynamics of the IRES afforded it the ability to undergo conformational re-arrangement to move to the P-site without any external intervention.

While these two studies and subsequent fragment NMR and crystal structures of the CrPV-IRES and related IRESs[143,147,151] shed more light on the mechanism many questions remained unanswered. Unambiguous data about the position of PKI and residue level information about contacts between the IRES and the ribosome were still missing. The mechanisms of how the IRES moved through the ribosome and out of it was unknown. Whether the movement of the IRES was similar to or different from the movement of tRNAs and factors during translocation elongation was also unknown.

1.14 The resolution revolution in cryo-EM lead to the first set of high resolution structures of the type IV IRESes

In early 2010s, fueled by advances in detector technology, new data processing algorithms and software, cryo-EM (see chapter 2) became a powerful tool to dissect biological mechanisms that were untenable via X-ray crystallography or NMR[152]. High resolution structural determination of inherently dynamic macromolecular complexes such as the ribosome became feasible[153]. The resulting high resolution structures of type IV IRESes bound to the ribosomes helped us understand their mechanism in unprecedented detail **(Fig. 1.9 C, D)**.

The first set of high resolution structures of the CrPV-IRES bound to an 80S ribosome were solved at resolutions beyond 4 Å[134]. This helped to unambiguously position and build models for all three domains of the IRES **(Fig. 1.9 E)**. The models clarified the contacts the different domains made with different parts of the ribosome **(Fig. 1.9 F)**. These structures corroborated the interactions of the IRES domain 2 with a cluster of proteins on the head of the 40S, including the eukaryote specific protein eS25[154]. Domain 1 along with L1.1 and L1.2 was observed stabilizing the L1-stalk in an extremely outward conformation. It was pushed further out than its furthest known position with an E-site tRNA during canonical translation. A surprising result from these structures, however, was the position of domain 3/PKI and the implications thereof.

The CrPV-IRES PKI contains the codon-anticodon mimicking stem loop **(Fig. 1.8 A)** and was thought to occupy the P-site[125,126,139,149,155]. This would mean that the IRES-80S binary complex would resemble an elongation-competent ribosome, ready to accept the incoming tRNA in the A-site. However, from these new high resolution structures it was clear that PKI was unambiguously in the A-site. The PKI was inserted into the decoding center at the A-site of the 40S. The decoding bases that check for the accuracy of the mRNA-tRNA codon-anticodon interaction were engaged with the PKI indicating that this pseudoknot was inducing a normal decoding event. Consequently, this placed the first coding codon not at the A-site of the ribosome but just outside it. This meant that the IRES had to undergo one translocation event before it could accept the first aminoacyl-tRNA.

It was also interesting that the PKI was anchored to the decoding center in both the canonical and ratcheted state of the ribosome[134]. The 80S-IRES binary complex was thus mimicking a pre-translocated state. This implied that the IRES needed to be translocated from the A-site to the P-site to accept the first aminoacyl-tRNA. Indeed, previous biochemical experiments with another related type IV IRES from the *Plautia stali* intestine virus (PSIV) had shown that translocation of the IRES from the A-site to the P-site by eEF2 was essential for loading of the first aminoacyl-tRNA[133].

Type IV IRESs thus use their PKI to engage with the decoding center in the A-site, maintain that interaction in both the canonical and rotated states of the ribosome, mimic a pre-translocated state of the ribosome, and need to undergo one round of eEF2 mediated translocation to accept the first aminoacyl tRNA delivered by eEF1A. All these observations outline how the type IV IRESs use only a minimal set of factors (two: eEF2 and eEF1A) and mimic a translation elongation intermediate rather than an initiation intermediate to hijack and steal host cell ribosomes for viral protein production.

The placement of the PKI at the decoding center, the range of contacts made by the IRES with ribosomal proteins and the L1 stalk, and the IRES promoting different rotational states of the ribosome was further corroborated in another cryo-EM reconstruction of the type IV IRES from the Taura syndrome virus (TSV) bound to the 80S ribosome[156].

Biochemical studies coupled with these high resolution structures and follow-up structures of the IRES/PKI in the P-site of the ribosome resolved by cryo-EM have further clarified how the type IV IRES transits through the ribosome.

1.15   Type IV IRES in the P-site of the ribosome

A post translocated IRES-80S binary complex with the PKI at the P-site proved difficult to study owing to frequent back-translocation exhibited by the CrPV-IRES[134,157]. Previous toeprinting experiments had utilized a strategy where they mutated the first codon to a stop codon to reduce back-translocation events[131]. This meant that a ligand such as a defective translation termination factor that stalls at the A-site could be utilized to trap the IRES in the P-site in a post translocated state[157].

Employing this strategy, Muhs and colleagues were able to trap the CrPV-IRES in the P-site of the ribosome and characterize it using cryo-EM[157]. From their sub-nanometer resolution density maps the authors were able to deduce that the IRES moved from the A-site to the P-site as a rigid body. The maps also showed that as the IRES moved it lost its contacts with the cluster of proteins on the head of the 40S. The PKI was seen to still mimic a codon-anticodon interaction in the P-site. The L1-stalk was pushed further outward here than with a pre-translocated IRES.

Higher resolution cryo-EM structures of the CrPV IRES and the related TSV-IRES on the 80S that traced the path of the IRESs from the A-site to the P-site soon became available[42,158] **(Fig. 1.10)**. These structures captured the IRES in numerous intermediate conformations indicating how the type IV IRESs translocated in a dynamic fashion.

A cryo-EM study trapped the CrPV-IRES with its PKI in an chimeric hybrid ap/P tRNA-like state using a GDPCP bound eEF2 with the ribosome in a rotated state[42] **(Fig. 1.10 A, B, C)**. The codon-anticodon interaction in the PKI is conserved in this structure. Domain IV of the eEF2 was seen to interact with PKI in a manner similar to how it interacts with the tRNA-mRNA codon-anticodon during canonical translocation. As the PKI has disengaged from the A-site, the decoding bases have flipped back potentially because of the histidine modification found at the tip of domain IV of eEF2[159]. The L1-stalk gets further pushed even further outward compared to its position in the initial A-site bound IRES.

The path that an IRES takes from such an intermediate state until it places its PKI solely into the P-site was traced in detail using ensemble cryo-EM[158]. Using maximum likelihood approaches to classify the data, the authors were able to capture many different intermediates in the path taken by the translocating IRES. They showed that the related type IV TSV-IRES moved from the A-site to the P-site like an inchworm **(Fig. 1.10 D)**.

Using five different intermediates of the eEF2 directed translocation step, Abeyrathne and colleagues[158] showed that the IRES rearranges itself from extended to bent to extended shape to reach the P-site. This inchworm-like movement is coupled with large scale movements of the 40S such as back rotation and head swivel **(Fig. 1.10 D, bottom text)**. The IRES moves forward as the 40S is back rotating from its ratcheted state to a more canonical-like POST state. Four out of the five states are back-rotating with different head swivel positons. During those movements the PKI of the IRES samples different intermediate positions between the A- and the P-site of the ribosome. The fifth state captures the IRES in a fully translocated position with the PKI in the P-site of the ribosome.

When the PKI is finally placed in the P-site of the ribosome, the head is least swiveled and the 40S is in an almost canonical (non-rotated) state[158]. The IRES has undergone dynamic conformational changes within its highly structured shape and has now disengaged its contacts with the cluster of proteins on the 40S head. Both SLIV and SLV do not contact any ribosomal proteins or rRNA and are exposed to the solvent **(Fig. 1.10 D, V)**. The L1-stalk, however, is still maintaining its interactions with the L1.1 and L1.2 regions. It gets most displaced in the final P-site occupied state compared to the either the pre-translocated IRES or the intermediate states induced by eEF2.

1.16   The CrPV-IRES in the E-site of the ribosome

Experiments designed to trap the CrPV-IRES in a double translocated position i.e. with the PKI in the E-site, revealed two different states[160]. The first was an intermediate state **(Fig. 1.11, top left)** where the PKI of the IRES is in between the P- and E-sites and the aminoacyl-tRNA in between the A- and P-sites. Most of the IRES in this intermediate is outside the E-site, exposed to the solvent. The SLIV and SLV stemloops are still detached from the head region of the 40S. There is distinct head swivel similar to one seen during late stages of the TSV-IRES translocation from the A- to the P-site.

In the second state, the IRES is completely translocated into the E-site **(Fig. 1.11, top right)**. Here, the ribosome is in a non-rotated state and there is no head-tilt or head-swivel. Two sets of interactions in the IRES are drastically remodeled when it enters the E-site. The sugar pucker and A-minor interactions of PKII and PKIII are destroyed leading to a widening of the IRES. As a result the IRES reattaches itself to the 40S head via an interaction with eS25 but via a different binding modality. The codon-anticodon interaction at the ASL mimicking part of the PKI is also broken. This interaction is religiously maintained by the IRES in free solution, on the

40S, on the 80S at the A-site, P-site and even during intermediate transition steps. But at the E-site the PKI undergoes a drastic and surprising rearrangement.

At the E-site the PKI disassembles and flip up into the 60S to resemble the anticodon stem loop of an E-site tRNA. This was a previously unseen and unexpected state occupied by PKI. This meant that PKI mimics a codon-anticodon stem loop of the tRNA-mRNA at the A- and P-sites but switches to mimic the amino-acid acceptor stem at the E-site. An unprecedented dual mimicry of the anticodon stem loop and the amino-acceptor end of the tRNA by a viral pseudoknot. Though an unpredicted state, comparison of the IRES at the E-site with an HCV-IRES bound to the ribosome showed striking similarities between the PKI and HCV-IRES domain II[111,112,116] **(Fig. 1.11, bottom left and right)**. Two different types of IRESs have seemingly arrived at very similar evolutionary solutions when it comes to their interactions with the ribosome at the E-site.

1.17 The model for IRES transit through the ribosome

From a trove of biochemical, single molecule, and structural data, we can now trace the path a type IV IRES takes as it transits through the ribosome in great detail **(Fig. 1.12)**.

The type IV IRES is pre-formed into its distinct shape when it is in solution[122,161] **(Fig. C1)**. We can then imagine that as soon as the virus injects its genome into the cell, the preformed IRES assembles itself into an 80S-IRES binary complex. It can do so via two pathways. In one pathway, it can bind to the 40S first and then recruit the 60S[162] **(Fig. 1.12 A)**. In another yet unexplained pathway, as seen in recent single molecule experiments, the IRES can directly hijack the 80S ribosome[162,163] **(Fig. 1.12 B)**.

In the 80S-IRES binary complex, the PKI is inserted into the decoding center at the A-site of the 80S[134,156]. In this pre-translocated complex, the 40S is ratcheting between its canonical and rotated states and the IRES tracks that movement as a rigid body **(Fig. 1.12 C)**. Then, eEF2 binds to the binary complex and pushes the IRES **(Fig. 1.12 D)** from the A-site to the P-site[42] **(Fig. 1.12 E)**. The IRES moves like an inchworm during this transition and samples numerous intermediate conformations[158] **(Fig. 1.10 D)**. At the P-site, in what is termed the single translocated state, the IRES domain 3 loses most of its contacts with the 40S[42,158].

With the PKI in the P-site the IRES can now accept the first aminoacyl-tRNA. After the tRNA is loaded by eEF1A **(Fig. 1.12 F)**, the complex undergoes another round of eEF2 catalyzed translocation **(Fig. 1.12 G, H)**. This translocation of the PKI from the P-site to the E-site results in the double-translocated state of the IRES[160] **(Fig. 1.12 I)**. During this transition, the IRES samples at least one intermediate state[160]. At the E-site the IRES undergoes major structural changes **(Fig. 1.12 I)**. The PKI disassembles and flips into the E-site as a tRNA amino-acceptor end mimic **(Fig. 1.12 J to K)**. The ribosome is ready to accept the second tRNA and form the very first peptide bond.

We can see that the IRES actively models and remodels itself to sample numerous intermediate states and distinct conformations as it transits through the A-, P-, and E-site of the ribosome. This pathway highlights the delicate balance between structural rigidity and dynamics of the IRES and how it couples the two its advantage. Type IV IRESs have one more trick up their sleeve that they sometimes use to gain an extra advantage against their hosts.

1.18   Type IV IRESs can undergo a +1 frameshifting event to produce ORFx

Some of the type IV IRESs have been shown to undergo a +1 frameshift event at their first codon[164,165] **(Fig. 1.13 C)**. First identified via bioinformatics analyses of viruses with type IV IRESs, this +1 frameshifting event codes for a genuine protein product termed ORFx[143] **(Fig. 1.13 C)**. Extensive mutational studies performed on the +1 frameshifting Israeli Acute Paralysis

Virus IRES (IAPV-IRES) revealed that a wobble base pairing at the start site causes this frameshift event[144,145]. Further studies revealed that features distal from the start site could be perturbed to couple and decouple the 0 and +1 frames[146]. While extensive biochemical data characterizing the IAPV-IRES is available, structural data of this frameshifting IRES is scarce.

Hybrid small angle X-ray scattering (SAXS) and NMR analysis of the PKI of the IAPV-IRES showed that it assumes the shape of a tRNA[143] **(Fig. 1.13 D)**. The model also indicated that SLIII **(Fig. 1.13 C, yellow arrow)**, unique to the class II of type IV IRESs **(Fig. 1.8 A, bottom left)**, folded like a tRNA acceptor stem. When this model was docked into a ribosome with the structure of the CrPV-IRES as a reference, the tRNA like shape of the PKI occupied the space occupied by the acceptor stem of a tRNA at the A-site of the ribosome **(Fig. 1.13 E)**. The PKI was further postulated to be at a dynamic equilibrium of switching between conformations that drive either the 0 or the +1 frame.

Another interesting feature of the IAPV-IRES is a 5′ stem loop (SLVI) **(Fig. 1.13 C, green arrow)**. The SLVI has been shown to be increase both 0 and +1 frame translation[146]. While the exact role of the SLVI is unknown it was predicted to mimic an E-site tRNA and thus help position the tRNA-like SLIII in the A-site of the ribosome[144,146]. Structural data to support this prediction however is unavailable. This leads to one of the goals of this body of work.

1.19   Known Unknowns and Unknown Unknowns of IRES biology

As discussed earlier in the chapter, mechanistic insights of type IV IRESs have been almost exclusively derived from the structures of the prototypical CrPV-IRES. There is also some structural data about the Taura syndrome virus IRES (TSV-IRES). The Taura syndrome virus occupies a singular branch in the family tree of dicistroviruses and it is also the only known Dicistrovirus that infects a marine invertebrate[142]. It diverges from other insect infecting viruses in either the Cripavirus or Aparavirus genus[127] **(Fig. 1.13 B)**. Indeed, depending upon the region of the genome used for phylogenetic analysis, it can be placed in either genus[142]. It has been proposed that TSV is the first member of a new Dicistrovirus genus[142].

Metagenomic analysis of viruses from a wide range of habitats has revealed a tremendous diversity of viruses[166–168]. Of these, Dicistroviruses have emerged as one of the most prominent and diverse[169]. Using structural methods to visualize and further explore ribosome hijacking methods employed by a diverse set of IRESs will supplement current bioinformatics and biochemical data. Such an endeavor would also help contextualize the effects of different biochemically characterized mutations. Models obtained from structural studies can either confirm or refute hypotheses about what different parts of an IRES look like or what critical bases of the IRES are responsible for. Further, since the type IV IRESs are broadly related, insights gleaned from the structural characterization of a previously understudied[143–145] or new IRESs[170] can be extrapolated for others. Structural characterization can also provide us critical insights about canonical cellular pathways involved in type IV IRES mechanisms. The type IV class II intergenic IRES from IAPV and the 5′-UTR-IRES from CrPV, the main subjects of study in this body of work, provide us with opportunities to do just that.

A rich trove of biochemical data, discussed earlier, shows that IAPV-IRES has a unique +1 frameshifting ability. When this frameshifting event is triggered and what are the structural elements responsible for it are unknown. Cryo-EM structures from the TSV-IRES and CrPV-IRES trace the path of an IRES from the A-site to the E-site[42,134,156,158,160]. How the distantly related IAPV-IRES transits through the ribosome is unknown. Whether this IRES behaves similarly to the others at each ribosome site is unknown. How does the IAPV-IRES bind to the 40S or the 80S is unknown. How elements such as the SLVI and SLIII—unique to the IAPV-

IRES and other related IRESs—work and how they contribute to ribosome hijacking is unknown. The role of another structural element of an IRES, the variable loop region (VLR), is unclear since it hasn't been visualized at high resolutions[134]. In what three dimensional context of the IAPV-IRES do the mutations that decouple the +1 frameshifting exist is unknown. These known unknowns about the IAPV-IRES can only be answered with structural study of the IRES captured in different states of binding and transitions. The IAPV-IRES also has the potential to provide us information about unknown unknowns.

One key piece of insight completely missing from the current understanding of the type IV IRESs is about how they exit from the ribosome. When they first bind the ribosome, they assume an extended conformation that spans all three tRNA binding sites[134]. Each successive translocation step winds the IRES up like a spring[158]. The IRESs reach the E-site and can undergo remarkable structural rearrangements as seen in the case of CrPV-IRES[160]. What happens next is a true mystery. How does the IRES deal with another translocation step? Given the bulkiness of the IRES and the extensive contacts it is making with the L1-stalk of the large subunit and the head of the small subunit, the egress of the IRES might not be as simple as it is for an E-site tRNA which—for all intents and purposes—just falls off the ribosome. How does the ribosome clear such a bulky and highly structured RNA from its inter-subunit space? In the case of IAPV-IRES, the bulkiness of the RNA is further exaggerated by the presence of two extra stem loops i.e. SLIII and SLVI and a longer L1.1 region[122,144]. What rearrangements, if any, such an IRES needs to undergo before, during, or after it exits the E-site? All this is completely unknown. A single molecule study showed that elongation happens at non-canonical rates for tRNAs added after the IRES has presumably left the E-site of the ribosome[171]. This indicates a non-canonical egress of the IRES form the ribosome. What is causing this slower rate of elongation and how or if the IRES at the E-site acts as a hindrance for incoming tRNAs is unknown. Another unanswered question is what happens to the IRES once it exits the E-site. Does the IRES detach completely from the ribosome to float freely in the solution or does it stay bound to the ribosome for around for the next round of translation? We do not have answers to any of these questions and they can only be answered with 3D models of the IRES in different stages of egress. This brings us to the first goal of this body of work. **The first goal of this body of work is to complete our understanding of the mechanism of type IV IRESs by answering the questions raised above.** This has been attempted using cryo-electron microscopy to trace the trajectory of the IAPV-IRES as it binds to and transits through the ribosome. The results of the first goal of this body of work are presented in chapter 3.

       The second goal of this body of work is to structurally characterize a novel 5′-UTR-IRES. Found in the 5′-UTR region of the dicistronic CrPV genome, the 5′-UTR-IRES is provisionally characterized as a type III IRES—a class that includes the HCV-IRES from the monocistronic genome of HCV[97]. While 5′-UTR-IRESs have been identified in all dicistroviruses, none of them have been structurally characterized. The structural information about type III IRESs comes from either HCV- or CSFV-IRES[111,112,115,116]. A recent study established the minimal CrPV 5′-UTR-IRES, its factor requirements, and a secondary structure model[121] **(Fig. 1.14)**. A key structural peculiarity of the CrPV 5′-UTR-IRES identified from the secondary structure is the presence of a pseudoknot which was implicated in proper folding of the IRES and in ribosome recruitment. The secondary structure of this IRES, however, was determined from the IRES in solution and not bound to either the small subunit or the ribosome. Whether the structure of the IRES on the ribosome is the same as it is in solution is unknown. The IRES was shown to directly recruit eIF3, similar to other type III IRESs such as HCV-

IRES[112] and the CSFV-IRES[115]. Whether this is a conserved interaction between the three and within the type III IRESs remains to be seen. Another feature of the CrPV 5′-UTR-IRES is the presence of a start-stop upstream AUG codon (uAUG)[121].

uAUGs have been enigmatic as ribosome profiling experiments have shown that they do recruit ribosomes but peptide products of such recruitments have remained elusive[68,69,172]. Many uAUGs have been well characterized as essential regulators of stress related genes in yeasts or humans[49]. How the CrPV 5′-UTR-IRES recruits a 40S, where it places its start codon, how it places the initiator Met-tRNA$_i^{Met}$ on that codon, whether the initiator tRNA is placed at the genuine start codon, and whether the upstream start-stop uAUG plays any role for this IRES, all are unknown. To this end, this body of work uses biochemical techniques and cryo-EM to characterize the novel 5′-UTR-IRES and how it binds to the 40S of the ribosome. The results of this second goal is presented in chapter 4.

As cryo-EM has been a critical component of this body of work a generalized and conceptual introduction to the method is presented in the next chapter.

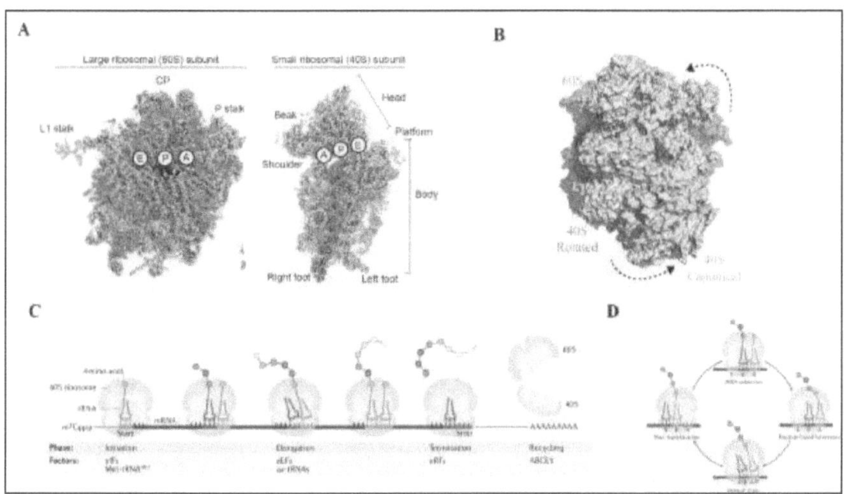

Figure 1.1 | The ribosome and the steps of protein synthesis

A: The large 60S subunit (cyan) and the small 40S subunit (yellow) of the eukaryotic ribosome as seen from their inter-subunit side. Prominent features of each are marked. Also marked are the A-, P-, and E-sites on each subunit. CP= central protuberance
B: The counterclockwise rotated (gray) and canonical (yellow) positions of the small subunit of the ribosome during ratcheting, as seen from the solvent side of the ribosome (modified from Fernandez et al[134]).
C: The four steps of translation shown as a schematic (modified from Schuller and Green[44]). Table below the schematic indicates the names of each step and the factors involve during each. eIFs = eukaryotic Initiation Factors, eEFs = eukaryotic Elongation factors, eRFs = eukaryotic Release Factors/Translation termination factors, ACBE1 = ATP-binding cassette sub-family E member 1 (Reprinted with permission from Springer Nature from Schuller and Green[44]).
D: An overview of the elongation cycle of protein synthesis shows how elongation proceeds in a cyclical manner. Of note are the hybrid states of the tRNA at the bottom. Not shown are the translation elongation factors eEF1A and eEF2 (Reprinted with permission from Springer Nature from from Schuller and Green[44]).

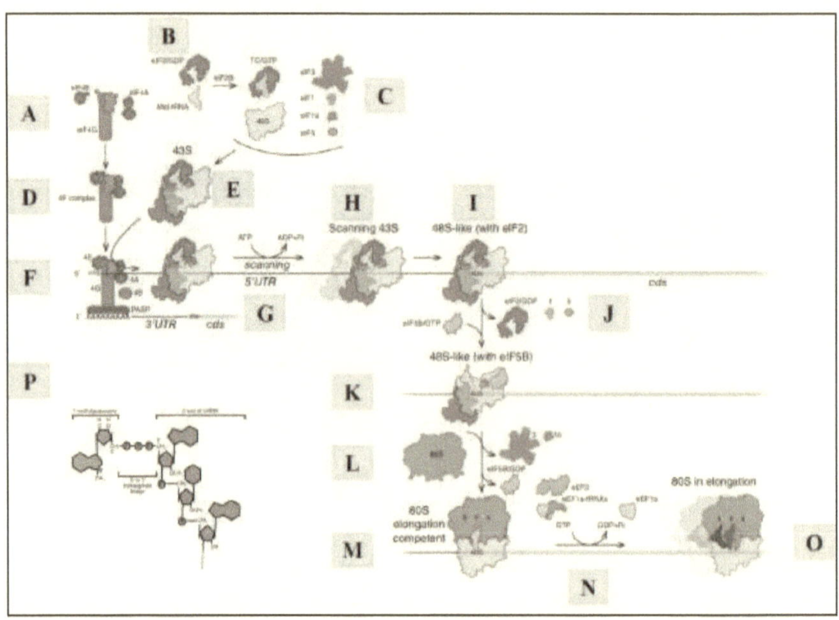

Figure 1.2 | Overview of translation initiation in eukaryotes (Reprinted with permission from John Wiley and Sons from Mailliot et al.[105])
A-O: Please refer to the text in sections 1.5, 1.6, and 1.7.
P: Structure of the 5'-cap. The 7-methylguanosine (red) is attached to the 5' terminus of the mRNA by a 5'- to 5'- triphosphate linkage. The first few nucleotides (blue) of the mRNA are usually 2'-methylated as shown.

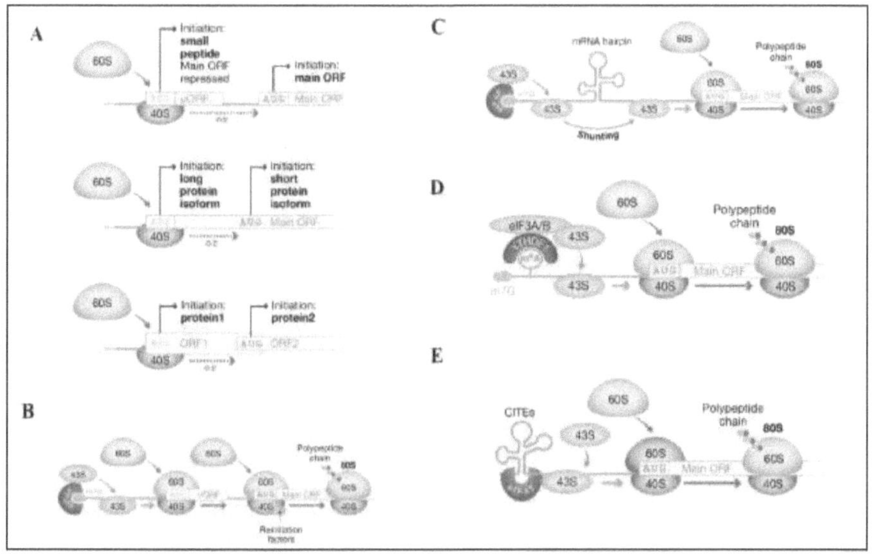

Figure 1.3 | Some alternative mechanisms of translation initiation in eukaryotes (Reprinted with permission from John Wiley and Sons from Sriram et al.[59])
A: Different variations of leaky scanning found in different organisms. 5′-cap dependent.
B: An overview of the re-initiation mechanism. Dependent upon 5′-cap.
C: m6A-dependent alternative initiation mechanism. 5′-cap dependent but RNA feature based.
D: Overview of the ribosome shunting mechanism found in viruses. Dependent on 5′-cap.
E: CITE-based initiation overview. Still cap dependent but RNA structure based.

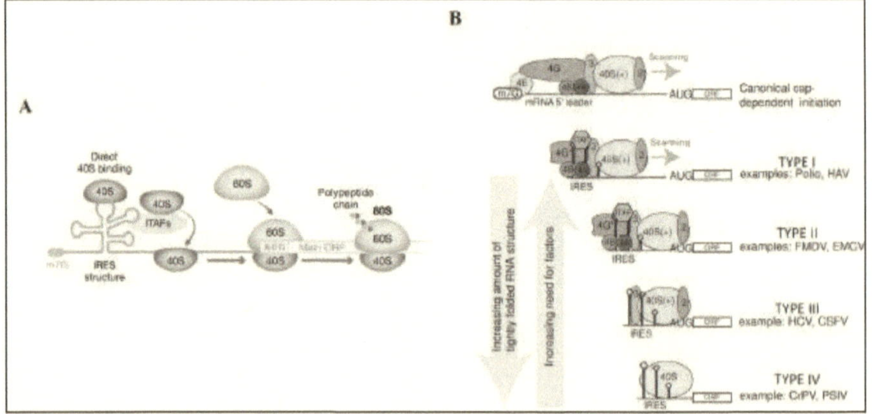

Figure 1.4 | The four types of IRES are classified according to their factor requirements and RNA structure

A: Overview of IRES-based alternative initiation of protein synthesis. IRESs can be found either at the 5'-end of a viral genome or the intergenic region (Reprinted with permission from John Wiley and Sons from Sriram et al.[59]).

B: The four types of IRESs classified according to their structure and factor requirements (Reprinted with permission from Elsevier from Filbin and Keift[97]).

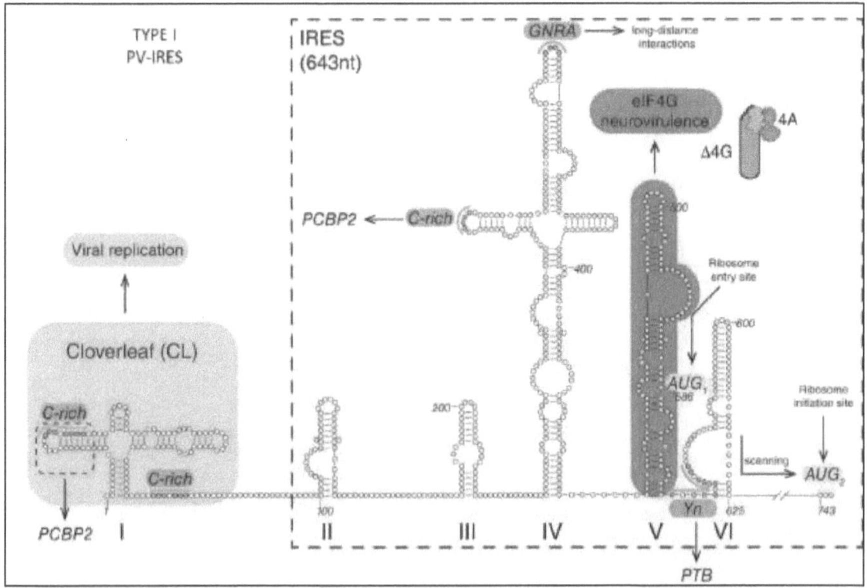

Figure 1.5 | Secondary structure schematic of the type I poliovirus IRES (Reprinted with permission from John Wiley and Sons from Mailliot and Martin[105])
Shown are prominent RNA features, the 5'-end cloverleaf structure, the GNRA loop used for long range interactions, the pyrimidine tract (Y_n) and the C-rich regions where ITAFs bind. Also shown (in green) the neurovirulence site where the viral protease cleaved eIF4G binds. The minimal PV-IRES is boxed within the dashed lines. Critical nucleotides shown in orange.

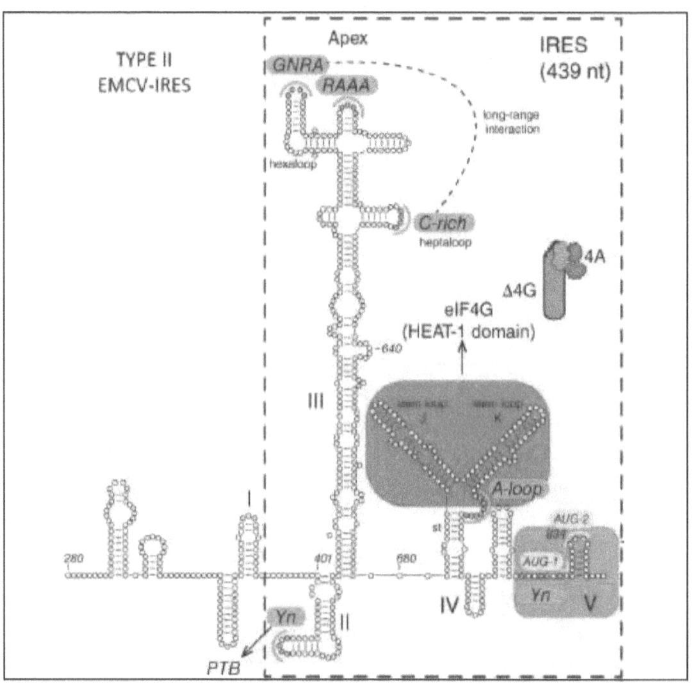

Figure 1.6 | Secondary structure schematic of the type II Encephalomyocarditis virus IRES (Reprinted with permission from John Wiley and Sons from Mailliot and Martin[105])
Shown are prominent RNA features, the GNRA and the C-rich loops which make long range interactions, the pyrimidine tract (Y_n) where ITAFs bind, and the cleaved eIF4G binding site (green). The minimal EMCV-IRES is boxed within the dashed lines. Critical nucleotides shown in orange.

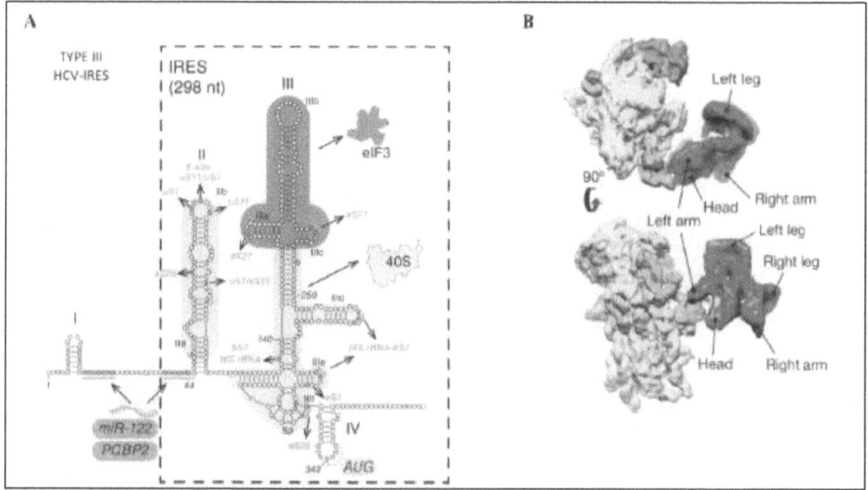

Figure 1.7 | The type III Hepatitis C virus IRES and the Classical Swine Fever virus IRES
A: Schematic of the secondary structure of the HCV-IRES showing prominent RNA features, various interaction sites for various ribosomal proteins and rRNA expansion segments, a microRNA target site, and the eIF3 binding site (purple). The minimal HCV-IRES is boxed within the dashed lines. Critical nucleotides shown in orange (Reprinted with permission from John Wiley and Sons from Mailliot and Martin[105]).
B: The HCV-like type III CSFV-IRES (blue) bound to the 40S (yellow) and eIF3 (red) in a displaced position with an accessory protein DHX29 shown (in green) (Reprinted with permission from Springer Nature from Hashem et al.[115]).

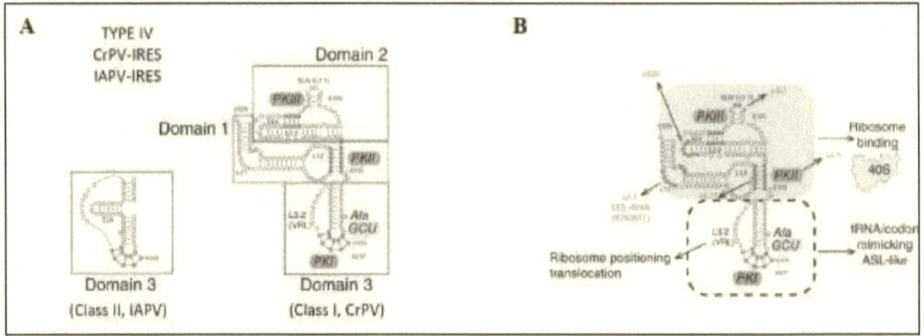

Figure 1.8 | The type IV Cricket Paralysis virus IRES and the Israeli Acute Paralysis Virus IRES (Reprinted with permission from John Wiley and Sons from Mailliot and Martin[105])
A: A schematic of the overall organization of the Type IV IRESs. PKI = pseudoknot I, PKII = pseudoknot II, PKIII = pseudoknot III, SLIII = stem loop III, SLIV = stem loop IV, SLV = stem loop V, VRL = variable loop region.
The two different subclasses are differentiated by a difference in the Domain 3 as shown.
B: A schematic of the different interactions of this minimal IRES with different regions of the ribosome.

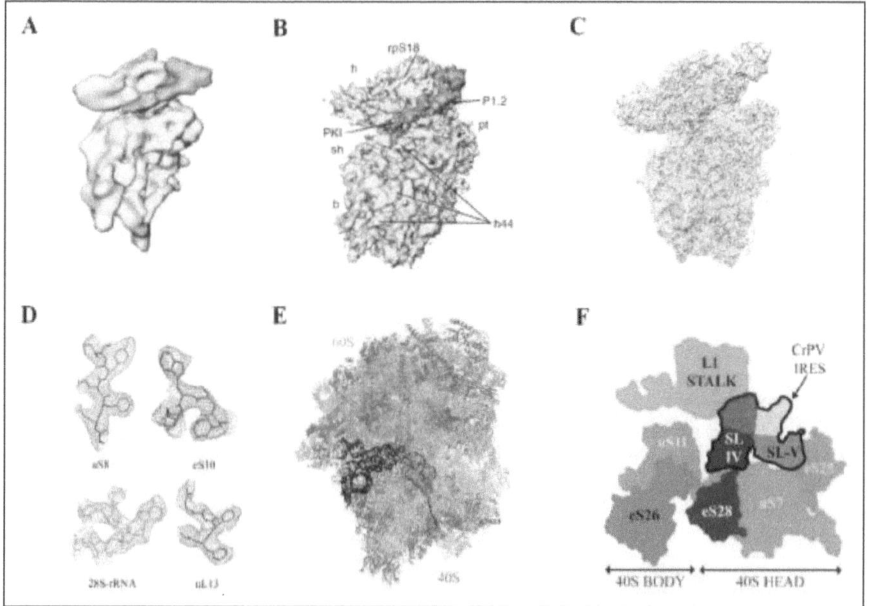

Figure 1.9 | Studies of the type IV CrPV-IRES using cryo-EM

A: Early cryo-EM studies were able to resolve the CrPV-IRES (purple) on the 40S (yellow) at very low resolution showing how it binds to the cleft between the head and the body of the small subunit of the ribosome (Reprinted with permission from Elsevier from Spahn et al.[149]).

B: Improvement in resolution of the CrPV-IRES (purple) on the 40S (yellow) helped locate some of the features of the IRES. The PKI of the IRES was thought to bind to the P-site of the 40S (Reprinted with permission from Springer Nature from Schuler et al. [139]).

C: High resolution map of the CrPV-IRES (purple) the 40S (yellow) unambiguously positioned the PKI of the IRES on the A-site of the ribosome. The CrPV-IRES is represented here as a ribbon diagram model built from the cryo-EM density (panels C-F modified from Fernandez et al. [134]).

D: Details of the high resolution map in C.

E: Model built from the map in C.

F: Interactions between the IRES and the ribosome as observed in the high resolution study.

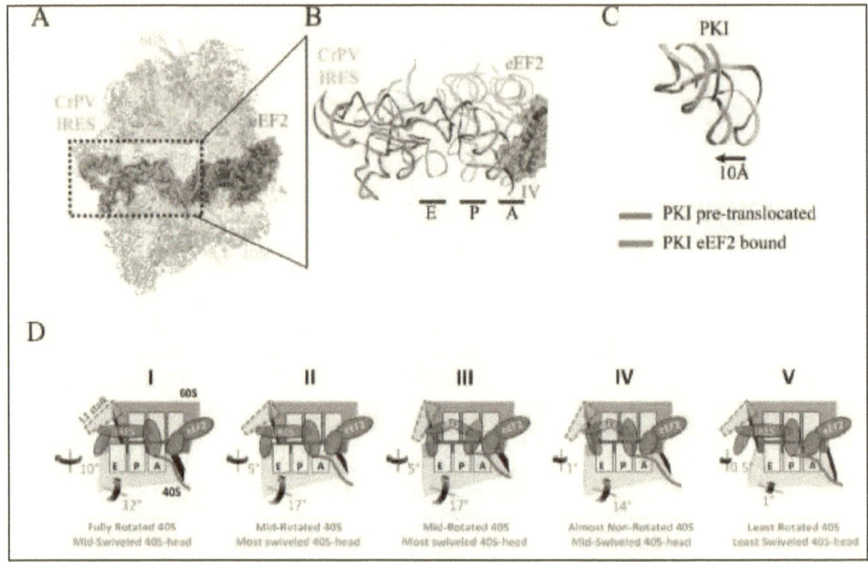

Figure 1.10 | Cryo-EM studies of the CrPV-IRES on the P-site of the ribosome
A-C: The CrPV-IRES (green) in an intermediate state of being translocated from the A-site of the ribosome to the P-site by eEF2 (red). In this intermediate state, the PKI of the IRES is displaced by 10 A (modified from Murray et al.[42]).
D: Another type IV IRES, the TSV-IRES translocates like an inchworm from the A-site of the ribosome to the P-site (modified from Abeyrathne et al.[158]).

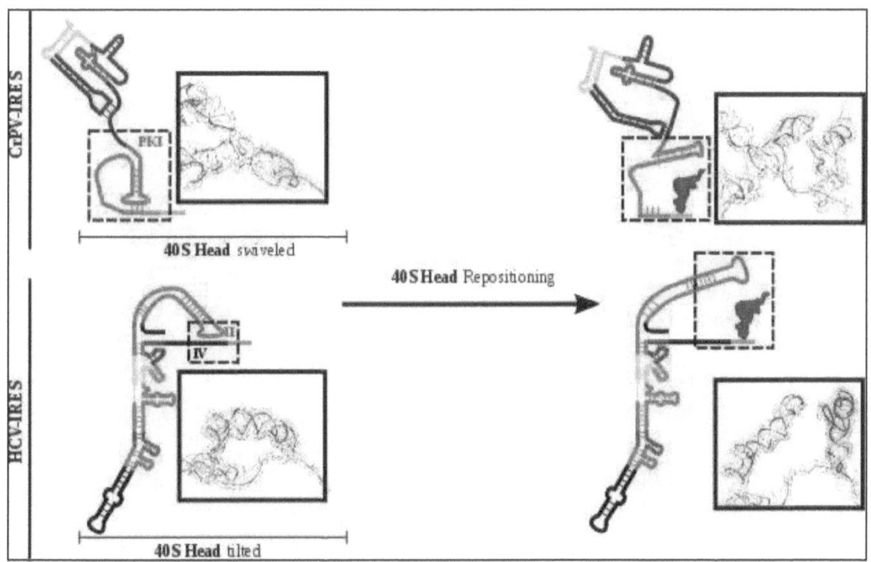

Figure 1.11 | The CrPV-IRES on the E-site of the ribosome adopts a conformation similar to that of the HCV-IRES (from Pisareva et al.[160])

At the E-site of the ribosome drastic structural re-arrangement of the codon:anticodon mimicking PKI domain **(top, left)** of the CrPV-IRES results in a conformation where it mimics an aminoacyl acceptor end of a tRNA **(top, right)**. This conformation of the CrPV-IRES is similar to that assumed by domain II of HCV-IRES on the 80S ribosome. **(bottom, left and right)**.

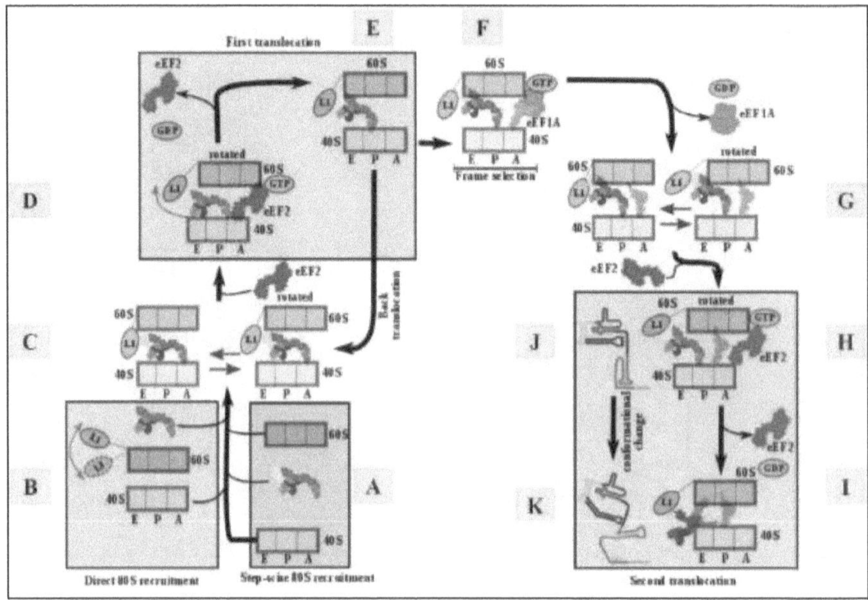

Figure 1.12 | Current model for the transit of the type IV IRESs through the ribosome
A-J: Please refer to the text in section 1.17 (modified from Pisareva et al.[160]).

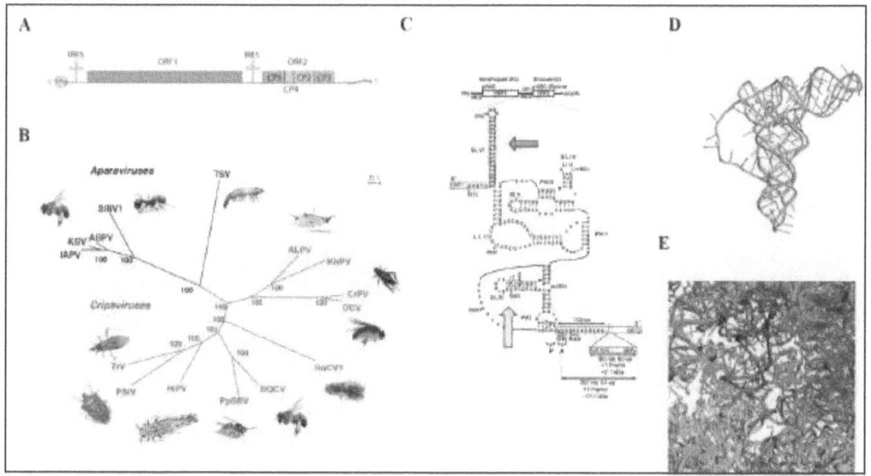

Figure 1.13 | Diversity within the type IV IRESs from Dicistroviruses
A: Genome organization of the Dicistroviruses showing the 5′-UTR-IRES and the intergenic
IRES (modified Viralzone: https://viralzone.expasy.org/resources/Dicistroviridae%5Fgenome.jpg).
B: Phylogeny of Dicistroviruses and their respective host animals (Reprinted with permission
from Elsevier from Hertz and Thompson[142]).
C: 2D structure diagram of the type IV IAPV-IRES with class II features, stem loop III (yellow
arrow) and stem loop VI (green arrow) pointed out (modified from Ren et al.[144]).
D: A model of the PKI of IAPV-IRES derived from a hybrid-NMR approach (orange) resembles
a tRNA (green) (modified from Au et al.[143]).
E: Predicted binding modality of the IAPV-IRES SLIII (red) within the A-site of the ribosome
(green and blue) (modified from Au et al.[143]).

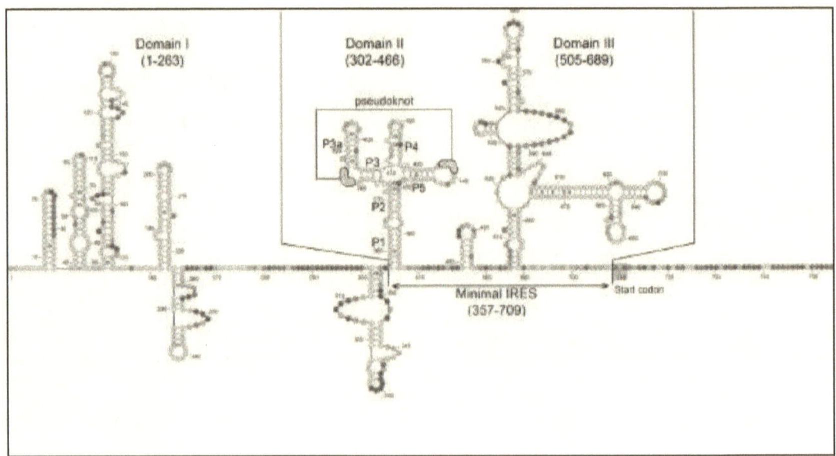

Figure 1.14 | The 2D structure of the 5′-UTR-IRE from the Cricket Paralysis virus derived using biochemical methods. (Reprinted with permission from Oxford University Press from Gross et al.[121])

1.21 References

1.   Crick, F. Central Dogma of Molecular Biology. *Nature* **227**, 561–563 (1970).

2.   Breaker, R. R. DNA enzymes. *Nature Biotechnology* **15**, 427–431 (1997).

3.   Doherty, E. A. & Doudna, J. A. Ribozyme Structures and Mechanisms. *Annual Review of Biophysics and Biomolecular Structure* **30**, 457–475 (2001).

4.   Anfinsen, C. B., Haber, E., Sela, M. & White, F. H. THE KINETICS OF FORMATION OF NATIVE RIBONUCLEASE DURING OXIDATION OF THE REDUCED POLYPEPTIDE CHAIN. *Proc Natl Acad Sci U S A* **47**, 1309–1314 (1961).

5.   Frank, J. & Gonzalez, R. L. Structure and Dynamics of a Processive Brownian Motor: The Translating Ribosome. *Annual Review of Biochemistry* **79**, 381–412 (2010).

6.   Voorhees, R. M. & Ramakrishnan, V. Structural Basis of the Translational Elongation Cycle*. *Annual Review of Biochemistry* **82**, 203–236 (2013).

7.   Ogle, J. M. *et al.* Recognition of Cognate Transfer RNA by the 30S Ribosomal Subunit. *Science* **292**, 897–902 (2001).

8.   Crick, F. H. Codon--anticodon pairing: the wobble hypothesis. *Journal of Molecular Biology* **19**, 548–555 (1966).

9.   Sievers, A., Beringer, M., Rodnina, M. V. & Wolfenden, R. The ribosome as an entropy trap. *PNAS* **101**, 7897–7901 (2004).

10.   Noller, H. F., Hoffarth, V. & Zimniak, L. Unusual resistance of peptidyl transferase to protein extraction procedures. *Science* **256**, 1416–1419 (1992).

11.   Ban, N., Nissen, P., Hansen, J., Moore, P. B. & Steitz, T. A. The Complete Atomic Structure of the Large Ribosomal Subunit at 2.4 Å Resolution. *Science* **289**, 905–920 (2000).

12.   Nissen, P., Hansen, J., Ban, N., Moore, P. B. & Steitz, T. A. The Structural Basis of Ribosome Activity in Peptide Bond Synthesis. *Science* **289**, 920–930 (2000).

13.   Liu, Q. & Fredrick, K. Intersubunit Bridges of the Bacterial Ribosome. *Journal of Molecular Biology* **428**, 2146–2164 (2016).

14.   Frank, J. & Agrawal, R. K. A ratchet-like inter-subunit reorganization of the ribosome during translocation. *Nature* **406**, 318–322 (2000).

15.   Agirrezabala, X. & Frank, J. Elongation in translation as a dynamic interaction among the ribosome, tRNA, and elongation factors EF-G and EF-Tu. *Quarterly Reviews of Biophysics* **42**, 159 (2009).

16.   Julián, P. *et al.* Structure of ratcheted ribosomes with tRNAs in hybrid states. *PNAS* **105**, 16924–16927 (2008).

17.   Noller, H. F., Lancaster, L., Zhou, J. & Mohan, S. The ribosome moves: RNA mechanics and translocation. *Nature Structural & Molecular Biology* **24**, 1021–1027 (2017).

18.	Regulation of the Mammalian Elongation Cycle by Subunit Rolling: A Eukaryotic-Specific Ribosome Rearrangement. *Cell* **158**, 121–131 (2014).

19.	Guo, Z. & Noller, H. F. Rotation of the head of the 30S ribosomal subunit during mRNA translocation. *PNAS* **109**, 20391–20394 (2012).

20.	Mohan, S., Donohue, J. P. & Noller, H. F. Molecular mechanics of 30S subunit head rotation. *Proc Natl Acad Sci U S A* **111**, 13325–13330 (2014).

21.	Ratje, A. H. *et al.* Head swivel on the ribosome facilitates translocation by means of intra-subunit tRNA hybrid sites. *Nature* **468**, 713–716 (2010).

22.	Nguyen, K. & Whitford, P. C. Steric interactions lead to collective tilting motion in the ribosome during mRNA–tRNA translocation. *Nature Communications* **7**, 10586 (2016).

23.	The L7/L12 stalk, a conserved feature of the prokaryotic ribosome, is attached to the large subunit through its N terminus - ScienceDirect. https://www.sciencedirect.com/science/article/pii/0022283681903272.

24.	Fei, J., Kosuri, P., MacDougall, D. D. & Gonzalez, R. L. Coupling of Ribosomal L1 Stalk and tRNA Dynamics during Translation Elongation. *Molecular Cell* **30**, 348–359 (2008).

25.	Trabuco, L. G. *et al.* The Role of L1 Stalk–tRNA Interaction in the Ribosome Elongation Cycle. *Journal of Molecular Biology* **402**, 741–760 (2010).

26.	Hinnebusch, A. G. The Scanning Mechanism of Eukaryotic Translation Initiation. *Annu. Rev. Biochem.* **83**, 779–812 (2014).

27.	Pestova, T. V. *et al.* The joining of ribosomal subunits in eukaryotes requires eIF5B. *Nature* **403**, 332–335 (2000).

28.	Hellen, C. U. T. Translation Termination and Ribosome Recycling in Eukaryotes. *Cold Spring Harb Perspect Biol* a032656 (2018) doi:10.1101/cshperspect.a032656.

29.	Fredrick, K. & Ibba, M. The ABCs of the ribosome. *Nature Structural & Molecular Biology* **21**, 115–116 (2014).

30.	Krab, I. M. & Parmeggiani, A. Mechanisms of EF-Tu, a pioneer GTPase. in (ed. Biology, B.-P. in N. A. R. and M.) vol. 71 513–551 (Academic Press, 2002).

31.	Yamamoto, H. *et al.* EF-G and EF4: translocation and back-translocation on the bacterial ribosome. *Nature Reviews Microbiology* **12**, 89–100 (2014).

32.	Blanchard, S. C. Single-molecule observations of ribosome function. *Current Opinion in Structural Biology* **19**, 103–109 (2009).

33.	Wohlgemuth, I., Pohl, C., Mittelstaet, J., Konevega, A. L. & Rodnina, M. V. Evolutionary optimization of speed and accuracy of decoding on the ribosome. *Philos Trans R Soc Lond B Biol Sci* **366**, 2979–2986 (2011).

34.	Loveland, A. B., Demo, G., Grigorieff, N. & Korostelev, A. A. Ensemble cryo-EM elucidates the mechanism of translation fidelity. *Nature* **546**, 113–117 (2017).

35.   Agirrezabala, X. *et al.* Structural insights into cognate versus near-cognate discrimination during decoding. *The EMBO Journal* **30**, 1497–1507 (2011).

36.   Stark, H. *et al.* Visualization of elongation factor Tu on the Escherichia coli ribosome. *Nature* **389**, 403–406 (1997).

37.   Shi, X., Khade, P. K., Sanbonmatsu, K. Y. & Joseph, S. Functional Role of the Sarcin–Ricin Loop of the 23S rRNA in the Elongation Cycle of Protein Synthesis. *Journal of Molecular Biology* **419**, 125–138 (2012).

38.   Zhou, J., Lancaster, L., Donohue, J. P. & Noller, H. F. How the ribosome hands the A-site tRNA to the P site during EF-G-catalyzed translocation. *Science* **345**, 1188–1191 (2014).

39.   Moazed, D. & Noller, H. F. Intermediate states in the movement of transfer RNA in the ribosome. **342**, 7 (1989).

40.   Frank, J. Intermediate states during mRNA–tRNA translocation. *Current Opinion in Structural Biology* **22**, 778–785 (2012).

41.   Schaffrath, R., Abdel-Fattah, W., Klassen, R. & Stark, M. J. R. The diphthamide modification pathway from *S accharomyces cerevisiae* – revisited. *Molecular Microbiology* **94**, 1213–1226 (2014).

42.   Murray, J. *et al.* Structural characterization of ribosome recruitment and translocation by type IV IRES. *ELIFE* **5**, (2016).

43.   Zhou, J., Lancaster, L., Donohue, J. P. & Noller, H. F. Spontaneous ribosomal translocation of mRNA and tRNAs into a chimeric hybrid state. *PNAS* **116**, 7813–7818 (2019).

44.   Schuller, A. P. & Green, R. Roadblocks and resolutions in eukaryotic translation. *Nature Reviews Molecular Cell Biology* **19**, 526–541 (2018).

45.   Pisarev, A. V. *et al.* The Role of ABCE1 in Eukaryotic Posttermination Ribosomal Recycling. *Molecular Cell* **37**, 196–210 (2010).

46.   Schuller, A. P. & Green, R. The ABC(E1)s of Ribosome Recycling and Reinitiation. *Molecular Cell* **66**, 578–580 (2017).

47.   Laursen, B. S., Sørensen, H. P., Mortensen, K. K. & Sperling-Petersen, H. U. Initiation of Protein Synthesis in Bacteria. *Microbiol. Mol. Biol. Rev.* **69**, 101–123 (2005).

48.   Aitken, C. E. & Lorsch, J. R. A mechanistic overview of translation initiation in eukaryotes. *Nature Structural & Molecular Biology* **19**, 568–576 (2012).

49.   Jackson, R. J., Hellen, C. U. T. & Pestova, T. V. The mechanism of eukaryotic translation initiation and principles of its regulation. *Nat Rev Mol Cell Biol* **11**, 113–127 (2010).

50.   Jackson, R. J. Alternative mechanisms of initiating translation of mammalian mRNAs. *Biochemical Society Transactions* **33**, 11 (2005).

51.   Shine, J. & Dalgarno, L. The 3′-Terminal Sequence of Escherichia coli 16S Ribosomal RNA: Complementarity to Nonsense Triplets and Ribosome Binding Sites. *PNAS* **71**, 1342–1346 (1974).

52.   Steitz, J. A. & Jakes, K. How ribosomes select initiator regions in mRNA: base pair formation between the 3′ terminus of 16S rRNA and the mRNA during initiation of protein synthesis in Escherichia coli. *PNAS* **72**, 4734–4738 (1975).

53.   Shatkin, A. J. Capping of eucaryotic mRNAs. *Cell* **9**, 645–653 (1976).

54.   Banerjee, A. K. 5′-terminal cap structure in eucaryotic messenger ribonucleic acids. *Microbiology and Molecular Biology Reviews* **44**, 175–205 (1980).

55.   Ramanathan, A., Robb, G. B. & Chan, S.-H. mRNA capping: biological functions and applications. *Nucleic Acids Res* **44**, 7511–7526 (2016).

56.   Kozak, M. Compilation and analysis of sequences upstream from the translational start site in eukaryotic mRNAs. *Nucleic Acids Res* **12**, 857–872 (1984).

57.   Huang, B. Y. & Fernández, I. S. Long-range interdomain communications in eIF5B regulate GTP hydrolysis and translation initiation. *PNAS* **117**, 1429–1437 (2020).

58.   Wang, J. *et al.* Structural basis for the transition from translation initiation to elongation by an 80S-eIF5B complex. *Nature Communications* **11**, 5003 (2020).

59.   Sriram, A., Bohlen, J. & Teleman, A. A. Translation acrobatics: how cancer cells exploit alternate modes of translational initiation. *EMBO Rep* **19**, (2018).

60.   Spirin, A. S. How Does a Scanning Ribosomal Particle Move along the 5′-Untranslated Region of Eukaryotic mRNA? Brownian Ratchet Model. *Biochemistry* **48**, 10688–10692 (2009).

61.   Kozak, M. Pushing the limits of the scanning mechanism for initiation of translation. *Gene* **299**, 1–34 (2002).

62.   Hinnebusch, A. G. Molecular Mechanism of Scanning and Start Codon Selection in Eukaryotes. *Microbiology and Molecular Biology Reviews* **75**, 434–467 (2011).

63.   Kozak, M. Possible role of flanking nucleotides in recognition of the AUG initiator codon by eukaryotic ribosomes. *Nucleic Acids Res* **9**, 5233–5252 (1981).

64.   Kozak, M. Point mutations define a sequence flanking the AUG initiator codon that modulates translation by eukaryotic ribosomes. *Cell* **44**, 283–292 (1986).

65.   Li, J., Liang, Q., Song, W. & Marchisio, M. A. Nucleotides upstream of the Kozak sequence strongly influence gene expression in the yeast S. cerevisiae. *Journal of Biological Engineering* **11**, 25 (2017).

66.   Dunston, J. A. *et al.* The human LMX1B gene: transcription unit, promoter, and pathogenic mutations. *Genomics* **84**, 565–576 (2004).

67.   Kochetov, A. V. *et al.* uORFs, reinitiation and alternative translation start sites in human mRNAs. *FEBS Letters* **582**, 1293–1297 (2008).

68.   Calvo, S. E., Pagliarini, D. J. & Mootha, V. K. Upstream open reading frames cause widespread reduction of protein expression and are polymorphic among humans. *Proceedings of the National Academy of Sciences* **106**, 7507–7512 (2009).

69.   Wang, X.-Q. & Rothnagel, J. A. 5′-Untranslated regions with multiple upstream AUG codons can support low-level translation via leaky scanning and reinitiation. *Nucleic Acids Res* **32**, 1382–1391 (2004).

70.   Hinnebusch, A. G. Gene-specific translational control of the yeast GCN4 gene by phosphorylation of eukaryotic initiation factor 2. *Molecular Microbiology* **10**, 215–223 (1993).

71.   Hinnebusch, A. G., Jackson, B. M. & Mueller, P. P. Evidence for regulation of reinitiation in translational control of GCN4 mRNA. *Proceedings of the National Academy of Sciences of the United States of America* **85**, 7279 (1988).

72.   Hinnebusch, A. G. & Natarajan, K. Gcn4p, a Master Regulator of Gene Expression, Is Controlled at Multiple Levels by Diverse Signals of Starvation and Stress. *Eukaryotic Cell* **1**, 22–32 (2002).

73.   Proud, C. G. eIF2 and the control of cell physiology. *Seminars in Cell & Developmental Biology* **16**, 3–12 (2005).

74.   Pakos-Zebrucka, K. *et al.* The integrated stress response. *EMBO reports* **17**, 1374–1395 (2016).

75.   Sonenberg, N. & Hinnebusch, A. G. Regulation of Translation Initiation in Eukaryotes: Mechanisms and Biological Targets. *Cell* **136**, 731–745 (2009).

76.   Dmitriev, S. E. *et al.* GTP-independent tRNA Delivery to the Ribosomal P-site by a Novel Eukaryotic Translation Factor. *J. Biol. Chem.* **285**, 26779–26787 (2010).

77.   Schleich, S. *et al.* DENR–MCT-1 promotes translation re-initiation downstream of uORFs to control tissue growth. *Nature* **512**, 208–212 (2014).

78.   Pisarev, A. V., Hellen, C. U. T. & Pestova, T. V. Recycling of eukaryotic post-termination ribosomal complexes. *Cell* **131**, 286 (2007).

79.   Kim, T.-H., Kim, B.-H., Yahalom, A., Chamovitz, D. A. & Arnim, A. G. von. Translational Regulation via 5′ mRNA Leader Sequences Revealed by Mutational Analysis of the Arabidopsis Translation Initiation Factor Subunit eIF3h. *The Plant Cell* **16**, 3341–3356 (2004).

80.   Lee, A. S. Y., Kranzusch, P. J., Doudna, J. A. & Cate, J. H. D. eIF3d is an mRNA cap-binding protein that is required for specialized translation initiation. *Nature* **536**, 96–99 (2016).

81.   Niu, Y. *et al.* N6-methyl-adenosine (m6A) in RNA: An Old Modification with A Novel Epigenetic Function. *Genomics, Proteomics & Bioinformatics* **11**, 8–17 (2013).

82.   Wang, X. *et al.* N6-methyladenosine Modulates Messenger RNA Translation Efficiency. *Cell* **161**, 1388–1399 (2015).

83.   Meyer, K. D. *et al.* 5′ UTR m6A Promotes Cap-Independent Translation. *Cell* **163**, 999–1010 (2015).

84.   Pooggin, M. M., Fütterer, J., Skryabin, K. G. & Hohn, T. Ribosome shunt is essential for infectivity of cauliflower mosaic virus. *PNAS* **98**, 886–891 (2001).

85.   Latorre, P., Kolakofsky, D. & Curran, J. Sendai Virus Y Proteins Are Initiated by a Ribosomal Shunt. *Molecular and Cellular Biology* **18**, 5021 (1998).

86.   Sherrill, K. W. & Lloyd, R. E. Translation of cIAP2 mRNA Is Mediated Exclusively by a Stress-Modulated Ribosome Shunt. *Molecular and Cellular Biology* **28**, 2011–2022 (2008).

87.   Truniger, V., Miras, M. & Aranda, M. A. Structural and Functional Diversity of Plant Virus 3′-Cap-Independent Translation Enhancers (3′-CITEs). *Front. Plant Sci.* **8**, (2017).

88.   Simon, A. E. & Miller, W. A. 3′ Cap-Independent Translation Enhancers of Plant Viruses. *Annu. Rev. Microbiol.* **67**, 21–42 (2013).

89.   Andreev, D. E., Dmitriev, S. E., Terenin, I. M. & Shatsky, I. N. Cap-independent translation initiation of Apaf-1 mRNA based on a scanning mechanism is determined by some features of the secondary structure of its 5′ untranslated region. *Biochemistry Moscow* **78**, 157–165 (2013).

90.   Pelletier, J. & Sonenberg, N. Internal initiation of translation of eukaryotic mRNA directed by a sequence derived from poliovirus RNA. *Nature* **334**, 320–325 (1988).

91.   Jang, S. K. *et al.* A segment of the 5′ nontranslated region of encephalomyocarditis virus RNA directs internal entry of ribosomes during in vitro translation. *Journal of Virology* **62**, 2636–2643 (1988).

92.   Belsham, G. J. & Sonenberg, N. RNA-protein interactions in regulation of picornavirus RNA translation. *Microbiol. Mol. Biol. Rev.* **60**, 499–511 (1996).

93.   Tsukiyama-Kohara, K., Iizuka, N., Kohara, M. & Nomoto, A. Internal ribosome entry site within hepatitis C virus RNA. *Journal of Virology* **66**, 1476–1483 (1992).

94.   Reinganum, C., O'Loughlin, G. T. & Hogan, T. W. A nonoccluded virus of the field crickets Teleogryllus oceanicus and T. commodus (Orthoptera: Gryllidae). *Journal of Invertebrate Pathology* **16**, 214–220 (1970).

95.   Sasaki, J. & Nakashima, N. Translation Initiation at the CUU Codon Is Mediated by the Internal Ribosome Entry Site of an Insect Picorna-Like Virus In Vitro. *Journal of Virology* **73**, 1219–1226 (1999).

96.   Sasaki, J. & Nakashima, N. Methionine-independent initiation of translation in the capsid protein of an insect RNA virus. *Proceedings of the National Academy of Sciences* **97**, 1512–1515 (2000).

97.   Filbin, M. E. & Kieft, J. S. Toward a structural understanding of IRES RNA function. *Current Opinion in Structural Biology* **19**, 267–276 (2009).

98.   Sweeney, T. R., Abaeva, I. S., Pestova, T. V. & Hellen, C. U. T. The mechanism of translation initiation on Type 1 picornavirus IRESs. *The EMBO Journal* **33**, 76–92 (2014).

99.   Breyne, S. de, Yu, Y., Unbehaun, A., Pestova, T. V. & Hellen, C. U. T. Direct functional interaction of initiation factor eIF4G with type 1 internal ribosomal entry sites. *PNAS* **106**, 9197–9202 (2009).

100.   King, H. A., Cobbold, L. C. & Willis, A. E. The role of IRES trans-acting factors in regulating translation initiation. *Biochem Soc Trans* **38**, 1581–1586 (2010).

101.   Komar, A. A. & Hatzoglou, M. Cellular IRES-mediated translation. *Cell Cycle* **10**, 229–240 (2011).

102.   Skinner, M. A. *et al.* New model for the secondary structure of the 5′ non-coding RNA of poliovirus is supported by biochemical and genetic data that also show that RNA secondary structure is important in neurovirulence. *Journal of Molecular Biology* **207**, 379–392 (1989).

103.   Burrill, C. P. *et al.* Global RNA Structure Analysis of Poliovirus Identifies a Conserved RNA Structure Involved in Viral Replication and Infectivity. *Journal of Virology* **87**, 11670–11683 (2013).

104.   Guest, S., Pilipenko, E., Sharma, K., Chumakov, K. & Roos, R. P. Molecular Mechanisms of Attenuation of the Sabin Strain of Poliovirus Type 3. *Journal of Virology* **78**, 11097–11107 (2004).

105.   Mailliot, J. & Martin, F. Viral internal ribosomal entry sites: four classes for one goal: Viral internal ribosomal entry sites. *WIREs RNA* **9**, e1458 (2018).

106.   BELSHAM, G. & BRANGWYN, J. A REGION OF THE 5′ NONCODING REGION OF FOOT-AND-MOUTH-DISEASE VIRUS-RNA DIRECTS EFFICIENT INTERNAL INITIATION OF PROTEIN-SYNTHESIS WITHIN CELLS - INVOLVEMENT WITH THE ROLE OF L-PROTEASE IN TRANSLATIONAL CONTROL. *Journal of Virology* **64**, 5389–5395 (1990).

107.   Lozano, G., Francisco-Velilla, R. & Martinez-Salas, E. Deconstructing internal ribosome entry site elements: an update of structural motifs and functional divergences. *Open Biol.* **8**, 180155 (2018).

108.   BROWN, E., ZHANG, H., PING, L. & LEMON, S. SECONDARY STRUCTURE OF THE 5′ NONTRANSLATED REGIONS OF HEPATITIS-C VIRUS AND PESTIVIRUS GENOMIC RNAS. *Nucleic Acids Research* **20**, 5041–5045 (1992).

109.   Kieft, J. S., Zhou, K., Jubin, R. & Doudna, J. A. Mechanism of ribosome recruitment by hepatitis C IRES RNA. *RNA* **7**, 194–206 (2001).

110.   Pestova, T. V., Shatsky, I. N., Fletcher, S. P., Jackson, R. T. & Hellen, C. U. T. A prokaryotic-like mode of cytoplasmic eukaryotic ribosome binding to the initiation codon during internal translation initiation of hepatitis C and classical swine fever virus RNAs. *Genes & Development* **12**, 67–83 (1998).

111.   Yamamoto, H. *et al.* Molecular architecture of the ribosome-bound Hepatitis C Virus internal ribosomal entry site RNA. *EMBO J* **34**, 3042–3058 (2015).

112.   Quade, N., Boehringer, D., Leibundgut, M., van den Heuvel, J. & Ban, N. Cryo-EM structure of Hepatitis C virus IRES bound to the human ribosome at 3.9-Å resolution. *Nat Commun* **6**, 7646 (2015).

113.   Yokoyama, T. *et al.* HCV IRES Captures an Actively Translating 80S Ribosome. *Molecular Cell* **74**, 1205-1214.e8 (2019).

114.   Boehringer, D., Thermann, R., Ostareck-Lederer, A., Lewis, J. D. & Stark, H. Structure of the hepatitis C virus IRES bound to the human 80S ribosome: Remodeling of the HCVIRES. *Structure* **13**, 1695–1706 (2005).

115.   Hashem, Y. *et al.* Hepatitis-C-virus-like internal ribosome entry sites displace eIF3 to gain access to the 40S subunit. *Nature* **503**, 539–543 (2013).

116.   Yamamoto, H. *et al.* Structure of the mammalian 80S initiation complex with initiation factor 5B on HCV-IRES RNA. *Nat Struct Mol Biol* **21**, 721–727 (2014).

117.   Terenin, I. M., Dmitriev, S. E., Andreev, D. E. & Shatsky, I. N. Eukaryotic translation initiation machinery can operate in a bacterial-like mode without eIF2. *Nat Struct Mol Biol* **15**, 836–841 (2008).

118.   Kim, J. H., Park, S. M., Park, J. H., Keum, S. J. & Jang, S. K. eIF2A mediates translation of hepatitis C viral mRNA under stress conditions. *Embo Journal* **30**, 2454–2464 (2011).

119.   Pestova, T. V., de Breyne, S., Pisarev, A. V., Abaeva, I. S. & Hellen, C. U. T. eIF2-dependent and eIF2-independent modes of initiation on the CSFV IRES: a common role of domain II. *EMBO J* **27**, 1060–1072 (2008).

120.   Jaafar, Z. A., Oguro, A., Nakamura, Y. & Kieft, J. S. Translation initiation by the hepatitis C virus IRES requires eIF1A and ribosomal complex remodeling. *Elife* **5**, e21198 (2016).

121.   Gross, L. *et al.* The IRES5′UTR of the dicistrovirus cricket paralysis virus is a type III IRES containing an essential pseudoknot structure. *Nucleic Acids Research* **45**, 8993–9004 (2017).

122.   Kieft, J. S. Comparing the three-dimensional structures of Dicistroviridae IGR IRES RNAs with other viral RNA structures. *Virus Research* **139**, 148–156 (2009).

123.   Jan, E. & Sarnow, P. Factorless Ribosome Assembly on the Internal Ribosome Entry Site of Cricket Paralysis Virus. *Journal of Molecular Biology* **324**, 889–902 (2002).

124.   Jan, E. *et al.* Initiator Met-tRNA-independent translation mediated by an internal ribosome entry site element in cricket paralysis virus-like insect viruses. *COLD SPRING HARBOR SYMPOSIA ON QUANTITATIVE BIOLOGY* **66**, 285–292 (2001).

125.   Wilson, J. E., Pestova, T. V., Hellen, C. U. T. & Sarnow, P. Initiation of Protein Synthesis from the A Site of the Ribosome. *Cell* **102**, 511–520 (2000).

126.   Pestova, T. V. & Hellen, C. U. T. Translation elongation after assembly ribosomes on the Cricket paralysis virus internal ribosomal entry site without initiation factors or initiator tRNA. *Genes & Development* **17**, 181–186 (2003).

127.   Bonning, B. C. & Miller, W. A. Dicistroviruses. *Annual Review of Entomology* **55**, 129–150 (2010).

128.   Wilson, J. E., Powell, M. J., Hoover, S. E. & Sarnow, P. Naturally Occurring Dicistronic Cricket Paralysis Virus RNA Is Regulated by Two Internal Ribosome Entry Sites. *Molecular and Cellular Biology* **20**, 4990–4999 (2000).

129.   Pisarev, A., Shirokikh, N. & Hellen, C. Translation initiation by factor-independent binding of eukaryotic ribosomes to internal ribosomal entry sites. *COMPTES RENDUS BIOLOGIES* **328**, 589–605 (2005).

130.   Deniz, N., Lenarcic, E. M., Landry, D. M. & Thompson, S. R. Translation initiation factors are not required for Dicistroviridae IRES function in vivo. *RNA* **15**, 932–946 (2009).

131.   Jan, E., Kinzy, T. G. & Sarnow, P. Divergent tRNA-like element supports initiation, elongation, and termination of protein biosynthesis. *Proceedings of the National Academy of Sciences* **100**, 15410–15415 (2003).

132.   Nishiyama, T. Structural elements in the internal ribosome entry site of Plautia stali intestine virus responsible for binding with ribosomes. *Nucleic Acids Research* **31**, 2434–2442 (2003).

133.   Yamamoto, H., Nakashima, N., Ikeda, Y. & Uchiumi, T. Binding Mode of the First Aminoacyl-tRNA in Translation Initiation Mediated by *Plautia stali* Intestine Virus Internal Ribosome Entry Site. *J. Biol. Chem.* **282**, 7770–7776 (2007).

134.   Fernandez, I. S., Bai, X.-C., Murshudov, G., Scheres, S. H. W. & Ramakrishnan, V. Initiation of Translation by Cricket Paralysis Virus IRES Requires Its Translocation in the Ribosome. *CELL* **157**, 823–831 (2014).

135.   Kanamori, Y. & Nakashima, N. A tertiary structure model of the internal ribosome entry site (IRES) for methionine-independent initiation of translation. *RNA* **7**, 266–274 (2001).

136.   Costantino, D. A preformed compact ribosome-binding domain in the cricket paralysis-like virus IRES RNAs. *RNA* **11**, 332–343 (2005).

137.   Pfingsten, J. S., Costantino, D. A. & Kieft, J. S. Conservation and diversity among the three-dimensional folds of the Dicistroviridae intergenic region IRESes. *Journal of Molecular Biology* **370**, 856–869 (2007).

138.   Spahn, C. M. *et al.* Domain movements of elongation factor eEF2 and the eukaryotic 80S ribosome facilitate tRNA translocation. *EMBO J* **23**, 1008–1019 (2004).

139.   Schüler, M. *et al.* Structure of the ribosome-bound cricket paralysis virus IRES RNA. *Nat Struct Mol Biol* **13**, 1092–1096 (2006).

140.   Jang, C. J., Lo, M. C. Y. & Jan, E. Conserved Element of the Dicistrovirus IGR IRES that Mimics an E-site tRNA/Ribosome Interaction Mediates Multiple Functions. *Journal of Molecular Biology* **387**, 42–58 (2009).

141.   Au, H. H. T. & Jan, E. Insights into Factorless Translational Initiation by the tRNA-Like Pseudoknot Domain of a Viral IRES. *Plos One* **7**, e51477 (2012).

142.   Hertz, M. I. & Thompson, S. R. Mechanism of translation initiation by Dicistroviridae IGR IRESs. *Virology* **411**, 355–361 (2011).

143.   Au, H. H. *et al.* Global shape mimicry of tRNA within a viral internal ribosome entry site mediates translational reading frame selection. *Proc Natl Acad Sci USA* **112**, E6446–E6455 (2015).

144.   Ren, Q. *et al.* Alternative reading frame selection mediated by a tRNA-like domain of an internal ribosome entry site. *Proceedings of the National Academy of Sciences* **109**, E630–E639 (2012).

145.   Ren, Q., Au, H. H. T., Wang, Q. S., Lee, S. & Jan, E. Structural determinants of an internal ribosome entry site that direct translational reading frame selection. *Nucleic Acids Res* **42**, 9366–9382 (2014).

146.   Au, H. H. T., Elspass, V. M. & Jan, E. Functional Insights into the Adjacent Stem-Loop in Honey Bee Dicistroviruses That Promotes Internal Ribosome Entry Site-Mediated Translation and Viral Infection. *JOURNAL OF VIROLOGY* **92**, (2018).

147.   Kieft, J. S., Chase, E., Costantino, D. A. & Golden, B. L. Identification and characterization of anion binding sites in RNA. *Rna* **16**, 1118–1123 (2010).

148.   Colussi, T. M. *et al.* Initiation of translation in bacteria by a structured eukaryotic IRES RNA. *Nature* **519**, 110–113 (2015).

149.   Spahn, C. M. T. *et al.* Cryo-EM Visualization of a Viral Internal Ribosome Entry Site Bound to Human Ribosomes: The IRES Functions as an RNA-Based Translation Factor. *Cell* **118**, 465–475 (2004).

150.   Spahn, C. M. T. Hepatitis C Virus IRES RNA-Induced Changes in the Conformation of the 40S Ribosomal Subunit. *Science* **291**, 1959–1962 (2001).

151.   Costantino, D. A., Pfingsten, J. S., Rambo, R. P. & Kieft, J. S. tRNA–mRNA mimicry drives translation initiation from a viral IRES. *Nat Struct Mol Biol* **15**, 57–64 (2008).

152.   Kühlbrandt, W. The Resolution Revolution. *Science* **343**, 1443–1444 (2014).

153.   Bai, X., Fernandez, I. S., McMullan, G. & Scheres, S. H. Ribosome structures to near-atomic resolution from thirty thousand cryo-EM particles. *eLife* **2**, e00461 (2013).

154.   Landry, D. M., Hertz, M. I. & Thompson, S. R. RPS25 is essential for translation initiation by the Dicistroviridae and hepatitis C viral IRESs. *Genes & Development* **23**, 2753–2764 (2009).

155.   Zhu, J. *et al.* Crystal structures of complexes containing domains from two viral internal ribosome entry site (IRES) RNAs bound to the 70S ribosome. *Proc Natl Acad Sci USA* **108**, 1839–1844 (2011).

156.   Koh, C. S., Brilot, A. F., Grigorieff, N. & Korostelev, A. A. Taura syndrome virus IRES initiates translation by binding its tRNA-mRNA-like structural element in the ribosomal decoding center. *Proceedings of the National Academy of Sciences* **111**, 9139–9144 (2014).

157.   Muhs, M. *et al.* Cryo-EM of Ribosomal 80S Complexes with Termination Factors Reveals the Translocated Cricket Paralysis Virus IRES. *Molecular Cell* **57**, 422–432 (2015).

158.   Abeyrathne, P. D., Koh, C. S., Grant, T., Grigorieff, N. & Korostelev, A. A. Ensemble cryo-EM uncovers inchworm-like translocation of a viral IRES through the ribosome. *ELIFE* **5**, (2016).

159.   Argüelles, S., Camandola, S., Cutler, R. G., Ayala, A. & Mattson, M. P. Elongation factor 2 diphthamide is critical for translation of two IRES-dependent protein targets, XIAP and FGF2, under oxidative stress conditions. *Free Radical Biology and Medicine* **67**, 131–138 (2014).

160.   Pisareva, V. P., Pisarev, A. V. & Fernández, I. S. Dual tRNA mimicry in the Cricket Paralysis Virus IRES uncovers an unexpected similarity with the Hepatitis C Virus IRES. *eLife* **7**, e34062 (2018).

161.   Pfingsten, J. S., Costantino, D. A. & Kieft, J. S. Structural Basis for Ribosome Recruitment and Manipulation by a Viral IRES RNA. *Science* **314**, 1450–1454 (2006).

162.   Petrov, A., Grosely, R., Chen, J., O'Leary, S. E. & Puglisi, J. D. Multiple Parallel Pathways of Translation Initiation on the CrPV IRES. *Molecular Cell* **62**, 92–103 (2016).

163.   Johnson, A. G., Grosely, R., Petrov, A. N. & Puglisi, J. D. Dynamics of IRES-mediated translation. *Phil. Trans. R. Soc. B* **372**, 20160177 (2017).

164.   Firth, A. E., Wang, Q. S., Jan, E. & Atkins, J. F. Bioinformatic evidence for a stem-loop structure 5 '-adjacent to the IGR-IRES and for an overlapping gene in the bee paralysis dicistroviruses. *Virology Journal* **6**, 193 (2009).

165.   Sabath, N., Price, N. & Graur, D. A potentially novel overlapping gene in the genomes of Israeli acute paralysis virus and its relatives. *Virol J* **6**, 144 (2009).

166.   Dolja, V. V. & Koonin, E. V. Metagenomics reshapes the concepts of RNA virus evolution by revealing extensive horizontal virus transfer. *Virus Research* **244**, 36–52 (2018).

167.   Zhang, Y.-Z., Chen, Y.-M., Wang, W., Qin, X.-C. & Holmes, E. C. Expanding the RNA Virosphere by Unbiased Metagenomics. *Annu. Rev. Virol.* **6**, 119–139 (2019).

168.   Greninger, A. L. A decade of RNA virus metagenomics is (not) enough. *Virus Research* **244**, 218–229 (2018).

169.   Shi, M. *et al.* Redefining the invertebrate RNA virosphere. *Nature* **540**, 539–543 (2016).

170.   Ng, T. F. F. *et al.* High Variety of Known and New RNA and DNA Viruses of Diverse Origins in Untreated Sewage. *Journal of Virology* **86**, 12161–12175 (2012).

171.   Zhang, H., Ng, M. Y., Chen, Y. & Cooperman, B. S. Kinetics of initiating polypeptide elongation in an IRES-dependent system. *eLife* **5**, e13429 (2016).

172.   Brar, G. A. & Weissman, J. S. Ribosome profiling reveals the what, when, where and how of protein synthesis. *Nat. Rev. Mol. Cell Biol.* **16**, 651–664 (2015).

Chapter 2:

A conceptual introduction to cryogenic electron microscopy

2.1   Cryo-EM is an increasingly popular method of choice for structural biology

In the last decade, cryogenic electron microscopy (cryo-EM) has emerged as a powerful tool to determine the structure of biological specimens[1]. As a structural biology technique, cryo-EM is able to overcome limitations inherent to X-ray crystallography or nuclear magnetic resonance spectroscopy (NMR)[2]. Structure determination by X-ray crystallography is impossible if the specimen of interest does not crystallize or crystallizes poorly[3]. NMR can overcome some of these limitations since structure of dynamic specimens can be determined using this method[4]. There is, however, a limit to how large an NMR specimen can be. Both methods require large quantities of the sample and they usually also require engineering of the specimen and optimization of buffer conditions to suit the technique—interventions that create a non-native sample or artificial sample buffer conditions. Given the limitations of the two traditional structural biology techniques, cryo-EM offers us an almost sample agnostic approach to determine structures of a wide range of biological specimens in their native states in an almost native sample buffer.

Cryo-EM includes three distinct approaches to structure determination: single-particle cryo-EM, cryo-electron tomography (cryo-ET), and electron crystallography[5]. All three techniques are transmission electron microscopy based but each is directed towards a specific type of sample. Electron crystallography is a method that is complementary to X-ray crystallography[6]. Using electron diffraction in a transmission electron microscope (TEM), structures of samples that form small[7] or 2D crystals[8] that aren't amenable to X-ray crystallography can be determined in electron crystallography. Cryo-ET[9], on the other hand, is largely used as a complement to and extension of light and fluorescence microscopy. Using entire cells or organisms as specimens, the structures of macromolecular complexes, organelles, or cellular interactions can be determined via cryo-ET[10]. The third of cryo-EM approaches, single-particle cryo-EM[5], deals with structure determination of biological particles of interest embedded in a thin layer of glass-like ice. The basic principle behind single-particle cryo-EM is that given a large number of two dimensional projections of a three-dimensional macromolecule from all possible angles, it is possible to use those images to computationally arrive at a three dimensional model.

In a single-particle cryo-EM workflow, thousands of images that correspond to two dimensional projections of a three dimensional particle that have been flash frozen in a thin layer of vitreous ice are first taken using a TEM[5,11,12]. The flash freezing of the particle traps it in numerous states of tumble thus providing all possible views or at least the minimum required number of views. Then, the position and orientation of each particle with respect to all other particles is computationally determined. And finally, if there are enough views to cover the entire particle and if the positions and orientations of the particles can be accurately assigned, the three dimensional particle is computationally reconstructed. Such particles can range from small proteins[13] or RNA[14] about 50 kDa in size to mega-Dalton macromolecular complexes[15] to viruses[16] to cellular megastructures tens of nanometers in diameter[17] **(Fig. 2.1)**. Such particles have been resolved at near atomic[11] (<4 Å) to atomic resolutions[18] (~1.2 Å) using single particle cryo-EM.

Single particle cryo-EM (referred simply as cryo-EM hereafter) exists at the confluence of sample preparation, data collection using a TEM, and data processing using dedicated algorithms and software. A general overview of structure determination using cryo-EM starting from sample preparation to data collection on a TEM to computational data processing to generate a cryo-EM density map is summarized below.

2.2   A pure and homogeneous sample is a critical starting point

A typical workflow (**Fig. 2.2 A**) for cryo-EM starts with purification of native or
recombinant samples[19]. Unlike sample preparation for NMR or X-ray crystallography, samples
for cryo-EM are usually generated without isotopes or extensive modifications intended to
stabilize sample. The buffers used are kept simple as possible. In most cases the final sample
buffers contain a pH buffer that maintains a range close to native or cellular pH, a monovalent or
divalent salt, or both as in the case for samples with nucleic acids, and a reducing agent. The
purity and integrity of the samples are checked with SDS-PAGE or UREA-PAGE, and gel
filtration chromatography. If available, techniques like multiple angle light scattering (MALS)
can be used to further verify the homogeneity of the sample. Since inherently low contrast
images are the starting point for cryo-EM, ensuring that the sample is as pure and as
homogeneous as possible is of utmost importance. A heterogeneous sample hinders downstream
steps by making data collection and processing unnecessarily complicated. In some cases, it can
even prevent structural determination altogether so great care must be taken to eliminate or
significantly reduce sample heterogeneity.

Sample heterogeneity exists in two forms: compositional heterogeneity or conformational
heterogeneity. Compositionally heterogeneous samples contain one or more subcomponent in
sub-stoichiometric proportion resulting in particles of different sizes and shapes. Unstable or
degrading components also cause compositional heterogeneity. Conformationally heterogeneous
samples, on the other hand, are stable and stoichiometric but have flexible parts that undergo a
range of motions which are hard to discretely separate.

Compositional heterogeneity can be reduced by first optimizing buffer conditions.
Testing buffers at different pH and salt concentrations or adding detergents can help stabilize the

complex of interest. Conformational heterogeneity is more difficult to address. If some domains are known to be highly flexible *a priori* they can be truncated or removed during construct design. In cases where such information isn't available and the sample is still heterogeneous, it can be chemically stabilized using cross-linking agents such as glutaraldehyde. Methods such as gradient centrifugation based GraFix[20] or agar matrix based AgarFix[21] or on-column cross-linking of the complex during gel filtration chromatography[22] are available to reduce sample heterogeneities. Chemical fixing of samples can be useful in stabilizing the complex but can also result in non-native conformations or compositions which might lead to incorrect biological insights. These methods should only be used after all other approaches to reduce heterogeneity have been exhausted.

Biochemical approaches to stabilize the sample and reduce heterogeneity should always be coupled with on-scope screening of the sample using negative stain electron microscopy[23,24] **(Fig. 2.2 B)**. Negative stain is a rapid, inexpensive, and highly informative technique to diagnose the stability, heterogeneity, and concentration of the sample[25]. For negative staining of samples, a heavy metal salt is applied onto a sample on a grid. The salt stains the background, usually a layer of thick carbon foil, but leaves the sample particles unstained as white spots. The contrast provided by such stain is high and can help users evaluate the overall size, shape, diameter, concentration, distribution, and heterogeneity of the particles.

A stable and homogeneous sample can be readily recognized by negative staining. After collecting a small set of images of negatively stained particles, the particles are first evaluated by eye. If they look of similar size and shape, they are picked, and 2D class averages are calculated from the picked particles using reference free alignments methods[26]. (These data pre-processing and processing steps are described in more detail later in this chapter.) A stable and homogeneous sample will present with a set of class averages that has uniform features. Since negative stain results in high contrast images, diagnosing samples using 2D class averages and even possibly 3D densities can be achieved using relatively low powered or inexpensive electron microscope setups.

Recognizing that any EM project is only as good as the sample is crucial for successfully executing any cryo-EM project. An iterative feedback loop of negative stain EM coupled with bench-work to optimize sample stability and heterogeneity is a prudent first step to take towards that goal.

2.3   Screening is the most important step in cryo-EM

After establishing the stability and homogeneity of the sample using biochemistry and negative stain EM, a user moves towards evaluating it using cryo-EM **(Fig. 2.2 C)**. For cryo-EM, the sample is flash frozen into a thin layer of vitreous ice on a small circular metal mesh called a grid. A grid is a thin, circular, 3 mm wide piece of metal that holds the sample. It is crisscrossed with metal bars that form a grid with numerous squares. Inside such squares are microns wide perforations which hold the sample in suspension **(Fig. 2.2 D)**. The perforations can be irregular or semi-regular or highly regular depending upon the material and the grid-making method used. A typical grid currently in use is made up of either copper or gold and has a uniformly sized and regularly spaced circular holes perforated in either a thin layer of carbon or gold.

A small amount of sample is applied to the grid and any excess blotted away before it is rapidly plunge frozen to create a thin layer of vitreous ice. Vitreous ice is an absolutely essential component of cryo-EM as crystalline forms of ice destroy biological samples. The electron beam inside the electron microscope requires high vacuum conditions. Since biological samples are unable to cope with such high vacuum, they need to be embedded in a medium that keeps them intact but is transparent enough for the electron beam to pass through. The solution for this as devised by Dubochet et al. was to embed the sample in a thin layer of vitreous ice[27]. Vitreous ice is a non-crystalline form of ice. It is achieved by rapidly plunge freezing the sample in liquid ethane (-180 °C) which provides the essential cooling rates of thousands of degrees per second. Jacques Dubochet was awarded a share of the 2017 Nobel Prize in Chemistry for his contributions in sample vitrification for cryo-EM. Sample vitrification is virtually unchanged since it was invented in the 1980s. There are however a few devices[28,29] in use that help the user control parameters such as temperature and humidity during sample application, blots the grid for a defined interval, and then rapidly plunge freezes the grid into liquid ethane. Once the sample has been vitrified, it can be imaged in the microscope.

At this initial stage, the user creates a set of grids with different vitrification and blotting parameters. The first goal of this step is to obtain parameters that result in a thin and uniform layer of vitreous ice on the grids. Usually the blotting time is critical for obtaining a thin layer of ice. It needs to be optimized according to the device used to blot the grids and also the type of grid used. Parameters that can be optimized are how long the grids get blotted for and how forceful that blotting is. Sometimes, variations in grid preparation methods result in differences in the quality and uniformity of the ice. A widely used techniques to achieve uniform ice is to deposit a thin layer of amorphous carbon tens of angstroms thick onto the grid before sample application[19]. This makes the ice layer uniform throughout the grid but results in images that have less contrast. While the blotting condition for a specific type of grid on a specific device is more or less uniform, it is very important to use grids from the same batch or the same preparation. Routine maintenance of blotting devices and regular tests with a well behaving sample are very important in ensuring reproducibility of blotting between different batches of grids and of uniformity of the ice within grids.

As the grid making parameters are being screened for, the user can concurrently screen for particle concentration, stability, distribution, and orientation. The concentration of the sample used for negative stain is a good reference point for making cryo grids. Usually, the amount of sample needed for cryo-EM grids is three to five times more than the concentration used for negative stain grids. In our laboratory, we usually make a set of grids around that concentration and screen for a concentration where the particles are packed just right **(Fig. 2.2 D)**. The goal here is to maximize number of particles per field of view without overcrowding or overlapping of particles on top of each other. It is a fine balance.

Stability of the frozen particles can be first evaluated done by eye. A good indicator of stability is whether the particles have shapes similar to ones seen via negative stain. Large areas of sample aggregation in the ice can be easily spotted in the images. Sometime particles that are stable in negative stain can disassemble in cryo conditions. If drastic, the disassembly can be easily seen by eye. If subtle, disassembly of particles can be diagnosed by picking the particles and running a reference free 2D classification. Uniform class averages indicate stable and intact particles.

As structure determination of a sample by cryo-EM relies on of images of that sample from all possible angle of views, proper distribution and random orientation of the particles in the ice are fundamental. The ideal condition is where the particles mono-disperse in random orientations throughout the circular hole of the grid **(Fig. 2.2 D)**. Numerous factors influence particle distribution and their orientation. Affinity of the particles to the air-water interface, to the grid material, to the grid support film, the hydrophobicity of the grid, the thickness of the ice, buffer conditions, all play a role[30].

The various affinities between the sample, the grid, and the support film can result in a skewed distribution of the particles. The particles can attach themselves to the edge of the hole or to the grid bars or support film. This reduces the number of particles per field of view and can also damages the particles. Sometimes, because of the meniscus like property of the thin layer of ice in the holes, the particles are evicted from the center to the edges. In extreme cases, the particles never occupy the holes. These issues are easily diagnosed by eye.

In some extreme cases, the particles are monodispersed in a uniform layer of ice but only occupy very few orientations[31] **(Fig. 2.2 E)**. This means that though the particles can be picked and processed, there isn't enough coverage of orientations throughout the complex to calculate a three dimensional model of the complex. Strong preferred orientations of large molecules can be easily seen by eye in the images but for smaller particles it might not be so obvious. While some symmetric particles can overcome slight orientation bias, preferred orientation becomes especially pronounced when dealing with asymmetric particles.

A quick way to diagnose preferred orientation is, again, to perform reference free 2D classification. If the particles only populate a few class averages with obviously oriented views, then preferred orientation is a suspect. If further diagnosis is required, one can attempt to generate a 3D model. Failure of the particles to generate an initial 3D model or to properly refine into an existing reference is characteristic of significant preferred orientation. Slight or moderate preferred orientation of the particles is visible as smearing of the 3D density in a particular direction and can be diagnosed by looking at the distribution of particles assigned to specific positions or angles within the 3D model. The preferably oriented particles usually cluster around a discrete set of positions and angles rather than being uniformly distributed.

2.4   Particle distribution and orientation issues can be resolved to an extent

There are two approaches that might resolve particle distribution and preferred orientation issues. One approach is grid based while the other is sample based. In the gird based approach the user optimizes the grid type and support film material. In the sample based approach the user optimizes the sample or sample buffer contents to achieve a mono-dispersed and randomly oriented distribution of particles.

The most common grids in use have circular holes of a defined diameter which are regularly spaced. If there are issues with particle distribution, a user can try grids with holes with either a smaller or larger diameter. Smaller holes results in more uniform but slightly thicker ice. Ice in larger holes retain more of the convex meniscus shape of the sample buffer. This results in thinner ice in the middle and thicker ice towards the edges of the hole. Unfortunately, this can sometimes crush the sample and disassemble it. Depending upon the size of the sample the user can dictate if they want thinner or thicker ice.

The fluidic environments differ in different sized holes and can contribute to properly distribute the particles[30]. In our experience, thickness of the ice plays a critical role in both the orientation of the particles and the integrity of the complex of interest. Sometimes, using different sized holes doesn't solve the problem. In such cases, the user can try grids with different types of support films such as lacey or holey carbon. While these types of grids might solve particle distribution or orientation issues, the user should keep in mind that both lacey and holey carbon grids cannot be used for fully automated data collection.

Different types of support films can also be used to overcome sample distribution and orientation issues. In our experience, a thin carbon support film laid over the grid helps the some samples to evenly distribute and reduce preferred orientation. We have used this approach in our lab to mitigate the slight preferred orientation seen in eukaryotic ribosomes[32]. Other choices for support films are graphene[33,34], graphene oxide[35], and affinity monolayers[36,37]. Such support films have been shown to drastically improve particle distribution and orientation in many cases[34]. For some samples, particular support films have rescued an otherwise intractable project.

Another approach to make particles tumble more and thus adopt more random orientations is to use detergents[38] or other additives in the sample buffer. Our lab has used various detergents such as NP-40, Tween 20, or Triton-X with varying results. Recently, the zwitterionic detergent CHAPSO[39] was shown to improve particle orientation distribution in cryo grids for numerous RNA polymerases. Care should be taken when adding detergents to cryo-EM samples. A range of different concentrations should be tested to make sure that the detergent isn't denaturing the sample. Of critical consideration when using detergents is to make sure that their concentration is well below the critical micellar concentrations, a number specific to each detergent. Sometimes, adding small concentrations of glycerol (<5%) can also help with overall sample distribution and particle orientations. It should be noted that adding carbon rich molecules such as detergents and glycerol significantly reduces image contrast.

If all above approaches still cannot resolve particle orientation issues, the user can tilt the stage during data collection[40]. Tilting is an extreme measure for overcoming preferred orientation of particles but it has been used to reconstruct particles of various sizes and origin. Stage tilting, though useful, not only amplifies many inherent issues that occur during data collection but also adds extra complications. A tilted stage is less stable than an un-tilted one and it thus amplifies sample motion and increases the blurriness of the images[40]. Because of the tilt

angle, the specimen gets exposed at their thickest edge and the defocus of the image gets

unevenly distributed across the tilt gradient[41]. All these factors degrade the quality of the image

and result in a loss of high resolution information. In some cases, when all other options have

been exhausted, the sample has to be tilted. The user should test a range of tilt angles and use the

lowest tilt angle that provides the missing views to minimize the artifacts of tilting. In our

experience, coupling one tilt angle $(+\theta)$ with the opposite tilt angle $(-\theta)$ provides some benefits.

Accurate estimation of the contrast transfer function and defocus parameters, a key data

processing step discussed later, is crucial for accurately processing a tilted data set.

2.5   Recent advances in sample preparation, grid making, and screening are promising

Despite the meteoric rise of cryo-EM, sample preparation for cryo-EM has essentially

been stuck in the 1980s. The fundamental methods used for preparing samples and grids for high

resolution has remained essentially unchanged. Recently, however, several promising

technologies have emerged.

One of the most successful technology to be introduced in sample preparation is the gold

support grid[42,43]. Instead of the widely used carbon foil support film on copper grids, these grids

use a gold support film over a gold grid **(Fig. 2.3 A, top)**. Gold-on-gold grids were shown to

drastically reduce sample motion to less than 2 Å and nearly eliminate the vertical grid

movement—a 40 to 60-fold reduction **(Fig. 2.3 A, bottom)**. This reduction in movement greatly

increases the quality of the collected images which in turn increases the accuracy of downstream

data processing steps such as angular assignments during 3D reconstruction. Indeed, the use of gold grids led to the first intermediate resolution reconstruction of the now ubiquitous test molecule apo-ferritin[42], which was intractable by cryo-EM before their introduction.

Traditional grid preparation has been based on the plunge-freeze method[27,44] **(Fig. 2.2 C)**. This method is only quasi-reproducible and is full of various hard to control parameters. It also isn't user agnostic since any slight modifications or improper use by one user will change the nature of the plunge-freeze machine for the subsequent user. Also, though grid making only requires microliters of samples, for a precious samples the blotting step wastes a majority of the sample. To overcome these issues, grid making automation via instruments such as the Vitrojet[45] and Spotiton[46] have been recently introduced **(Fig. 2.3 B, C)**. These instruments employ automation and picoliter dispensing to make grids. They promise ease of use, sample savings, and, most importantly, reproducibility. These devices are about to be commercially introduced while other low-cost lab-buildable grid-making devices[47] are on the horizon.

Another promising advance is in the field of grid screening. Low resolution screening of grids with optical or fluorescent cameras have been suggested. Automated screening has been implemented in some scopes with an autoloader[48]. Here, the user loads a set of grids to be screened and the software collects a set of low resolution images such as the atlas and the grid images. The user can then check such images from all the grids and decide which ones have desired ice distribution and ice thickness properties and proceed with data collection. By combining a robust automated data collection programs[48–50] and on-the fly data processing[51,52], a user can readily asses if high resolution data collection is possible from a certain grid or not.

These new sample preparation and screening tools and devices promise tremendous improvements in a traditionally underdeveloped part of cryo-EM. However, sometimes nothing works for a given sample and the user has move on to another project. As much as it is important to properly and systematically screen and optimize a sample for cryo-EM, it is also very important to realize when a sample is intractable and then stop. If the sample is properly optimized using the above discussed principles and considerations, then it is ready for prime time on the electron microscope.

2.6   The electron microscope is an extremely capable instrument

Recently, humans observed the making and breaking of individual bonds for the first time ever[53]. This would not have been possible without an electron microscope. Since its inception, the transmission electron microscope has opened a window into nature through which we have seen things that are impossible to see with light. Using such microscopes, we have seen biological molecules impossible to see by other techniques—the Zika virus[54], an entire axoneme[17], a pressure sensing ion channel[55], to name a few. Though a transmission electron microscope is drastically different from a light microscope, the basic set up of both is very similar[12].

Like a light microscope, an electron microscope (**Fig. 2.4 A**) consists of a source, a set of condenser lenses, different apertures, the sample stage, an objective lens, a projector/magnifying lens, and an image recording system (**Fig. 2.4 B**). Unlike a light microscope, the source is an electron source and the lenses are magnetic fields generated by coils situated throughout the microscope. Also unlike a light microscope, all of these components are maintained at high vacuum inside a column to avoid scattering of the electron beam by air.

At the top of the column is an electron source (**Fig. 2.4 B**). Traditionally a heated filament of tungsten or LaB_6 was used as a source of electrons. By heating the filaments to 2000-3000 °C, the electrons are thermo-ionically extracted and accelerated towards the sample. Filament based electron microscopes still see wide use as screening microscopes but they have been largely supplanted by microscopes with "cold" electron sources. A benefit of a cold electron source is a more stable, coherent, and parallel electron beam. One such cold source is the field emission gun (FEG)[56,57]. The field emission gun consists of a very thin and sharp tungsten crystal through which electrons are extracted by applying a relatively low amount of current. The extracted electrons are then accelerated, depending upon the microscope type or the needs of the user, to 100 to 300 kilovolts (kV) as the electron beam descends the column.

As the beam descends towards the sample, a set of condenser lenses and apertures shape it into a usable electron beam[12] **(Fig. 2.4 B)**. The condenser lens takes the emerging beam and shapes it into a parallel beam while the condenser aperture, which is a physical hole in a plate of metal, removes any electrons in the beam that have high degree of spread. Usually, two condenser lenses are used to achieve a parallel and coherent beam. Sometimes, a third[58] or even fourth[59] condenser lens—implemented in high end microscopes—maintains the parallelism and phase coherence of the beam at wider range of scope settings. Image formation is highly sensitive to the nature of incident beam and a parallel and coherent beam ensures that the incident electrons interacts with the sample as coherently and parallel as possible.

The sample is situated directly below the condenser lens system in a stage that can be maintained at either liquid nitrogen or at room temperatures **(Fig. 2.4 B)**. The stage can be moved in all three directions and be tilted in one or more axes when necessary. In side-entry microscopes access to the column is via an airlock located on the side of the column. A single grid is clamped onto a cryo-holder maintained at liquid nitrogen temperatures and loaded into the microscope via the airlock. This method by nature is a low-throughput and contaminant-prone way of loading grids into the microscope. However, because of their easier availability and relative low cost side-entry microscopes are widely used as screening microscopes. High-end microscopes include an automated grid loading system that can hold and handle multiple grids at the same time in a virtually contaminant free manner and can keep them at liquid nitrogen temperatures for multiple days or even a couple of weeks. Using such autoloader systems, numerous grids can be screened and when a suitable grid is found data can be collected for multiple days.

The stage is followed by the objective lens and the objective aperture **(Fig. 2.4 B)**. The objective lens collects the beam emerging from the sample and focuses it at the back focal plane of the microscope. The objective aperture removes any inelastically scattered electrons from the beam, reducing noise and increasing contrast in the image. Then, a set of magnifying lenses magnify the image and project it onto the image recording system. Some high end microscopes have a phase plate and an energy filter before the image recording system, both of which help increase contrast albeit via different mechanisms discussed later in this chapter.

2.7 Interactions of the electron beam with the sample cause different types of scattering

The electron beam in the electron microscope can interact with the sample in multiple ways **(Fig. 2.4 C)**. Some of the electrons pass through the sample unimpeded (unscattered), others collide with the sample atoms and lose their energy (inelastically scattered), while others still are deflected by the sample atoms without any energy loss (elastically scattered)[60]. The unscattered and the elastically scattered electrons are responsible for creating the image whereas the inelastically scattered electrons contribute to noise in the image and cause sample damage.

Inelastically scattered electrons are the main source of sample damage as they create X-rays, ionize sample atoms, damage chemical bonds, create free radicals, and generate highly scattered or backscattered electrons. This radiation damage by inelastically scattered electron is the main limitation in cryo-EM of biological samples. While techniques such as negative staining reduces sample damage, they also prohibit high resolution structure determination. To image radiation-sensitive biological samples embedded in a very thin film of ice at high resolutions a low dose imaging regimen is required.

For non-biological samples, such as those studied in materials science, a very high electron dose can be used to generate high resolution images. The high atomic weight atoms (e.g. metals such as Pt, Au etc.) present in such samples make the sample radiation-insensitive for the purpose of the experiment and these electron dense samples strongly scatter or absorb a majority of the incident electron beam. Such highly scattered or any inelastically scattered beams form the sample either never make it to the image or are physically removed. This absence of majority of the scattered electrons along with the unscattered electrons creates what is known as amplitude contrast in the image.

Biological samples, on the other hand, are very thin and made up of low atomic number elements (H, C, O, P, and N). Such atoms neither significantly absorb nor highly scatter the incident electron beam. They produce very little amplitude contrast. This relatively low amplitude contrast is further degraded by the low dose imaging requirements of biological sample which reduces the signal to noise ratio in the images further reducing overall contrast of the images. The incident electron beam, however, experiences a phase shift as it interacts with the sample. In such a situation where the amplitude contrast negligible, the phase shifting behavior of biological specimens is exploited to generate contrast in cryo-EM.

2.8   Contrast in cryo-EM has to be generated by deliberate defocusing of the image

When the incident electron beam interacts with biological samples, the sample induces mostly low angled elastic scattering. While this low degree of scattering doesn't produce enough amplitude contrast, it changes the path length of the incident beam and causes a phase shift in the emergent beam. The phase shifted electrons in the emergent beam then interfere with the unscattered electron beam to create what is known as phase contrast.

Thin biological samples scatter electrons weakly and the extent to which they do so can be estimated using the weak phase approximation. For weak phase objects, if the image is taken at exact focus then the phase contrast is only generated by the aberrations in the objective lens, specifically the spherical aberration. Spherical aberration is the inability of a lens to focus the incident beams at its periphery and the incident beams at its center to the same focal point **(Fig. 2.5 A)**. Because of spherical aberration, beams at the periphery of the lens are refracted more strongly as opposed to those that are closer to the optical axis of the lens which creates a blurred image of an object with some contrast. Such phase contrast, however, is virtually indistinguishable from the background noise. But if the nature of the scattered electron beam can somehow be phase shifted by an additional ±90° then its interference with the unscattered beam will create maximum possible phase contrast in the image. This shift in the phase of the scattered beam is obtained in cryo-EM by deliberately under-focusing (i.e. defocus) the objective lens to generate an image with enough usable contrast. This defocus generated image, however, is not a true representation of the object as it is affected by aberrations of the microscope.

The electron microscope lens system has a set of aberrations similar to the ones found in a light microscope. Along with the spherical aberration mentioned above, chromatic aberration, and astigmatism of the objective lenses are of import. These aberrations along with the applied defocus result in a blurring of the image such that a point object gets imaged as an airy disc. This blurring of the image is inherent to transmission electron microscopes and can be represented by a convolution of the object with a point spread function (PSF). In Fourier space, where most of the cryo-EM theory is grounded and data processing is performed, the point spread function is represented by the contrast transfer function (CTF).

2.9 A contrast transfer function (CTF) describes how an image is degraded by the transmission electron microscope

The CTF at a certain defocus setting describes in Fourier space how the applied defocus and the spherical aberration of the microscope contribute to the image formed on the detector. The mathematical formulation of CTF is:

$$T(k) = -\sin\left[\pi\Delta z\lambda k^2 + \frac{\pi}{2}C_s\lambda^3 k^4\right]$$

Eq. (1)

Where, $T(k)$ is the CTF, C_s is the spherical aberration of the objective lens, λ is the wavelength of the electron beam, k is the spatial frequency, and Δz is the applied defocus value.

In the above equation, we can see that because of the sine operator, the CTF is an oscillating function **(Fig. 2.5 B, C)**. This oscillation begins as a negative value and as the function progresses from low spatial frequency to high spatial frequency, it oscillates above and below the zero. For a 2D image, this oscillatory nature of the CTF can be easily observed in a 2D Fourier transform of the image called a power spectrum **(Fig. 2.5 B, right)**. The power spectrum displays the characteristic concentric and oscillating light and dark rings of the CTF. A side-on slice of the power spectrum reveals critical information about the image **(Fig. 2.5 C)**.

We see that the signal oscillates slowly at low spatial frequency and rapidly at higher spatial frequencies **(Fig. 2.5 C)**. We also see that at particular positions the signal is inverted (i.e. negative) and at others it is totally lost (i.e. zero). This means that at some spatial frequencies the signal is taking away from the contrast of image (by inverting it) while at other spatial frequencies the signal is not contributing any contrast at all. The image thus becomes a highly degraded representation of the object. This presents one of the most fundamental problems in transmission electron microscopy: how does one retrieve a truer representation of the object if the image produced by the microscope is always degraded by the CTF?

2.10   Defocus modulation and phase flipping correct for image degradations caused by the CTF

Looking at the equation for the CTF we see that it is dependent upon the two main variables: spherical aberration and defocus. Spherical aberration is the inability of any lens to focus all the rays passing through it at a single focal point. Because of the non-zero width of the lens, rays that are passing through the edges of the lens are focused at a point before the actual focus of the lens. This results in a diffuse plane of focus rather than a sharp single point of focus **(Fig. 2.5 A)**. Spherical aberration is a constant value and it can be determined for a particular lens. For electron microscopes, the spherical aberration of the objective lens is a number unique to each microscope and is calculated by the manufacturer and provided to the user. This number can be easily fed into the equation while calculating the CTF. This leaves us with the other variable in the CTF: defocus.

Defocus as described above is the deliberate under-focusing of the objective lens to generate contrast in the image and is essential in conventional transmission electron microscopy.

Defocusing changes the phase of the electrons exiting the sample and this property can be exploited to retrieve the zeros of the CTF.

If all the image from an experiment is collected at a single defocus then the CTF curves for all the images will virtually be the same. All such images will have the same CTF curve that passes through zero at exactly the same spots. Thus, all the images will be missing all the information about the object at those particular spatial frequencies. The object cannot be recreated from such dataset. Even if it is recreated it will have significant chunks of density missing. To solve this problem, rather than picking a single defocus value a range of defocus values is used during data collection.

Using a range of defocus values results in a dataset with images that have CTF curves with a range of different phases **(Fig. 2.5 C)**. As phases of the CTF curve change from one image to another, its zeros do too. Thus, the contrast information about spatial frequencies missing in one image will be present in another image taken at a different defocus. Taking an image at a high defocus, i.e. away from focus, creates a sharper looking image with higher contrast **(Fig. 2.5 C, blue curve)**. This sharpness is because the lower spatial frequencies are better represented than the higher spatial frequencies in such image. High defocus images are useful in seeing the particles by eye and can be more easily aligned during data processing. Images taken at low defocus, i.e. close to focus, have less contrast and the particles are harder to see by eye **(Fig. 2.5 C, red curve)**. Such images can be harder to align too but that doesn't mean that they aren't useful. In fact, low defocus images carry more high spatial frequency information that contributes high resolution information during reconstruction of the object. Thus, by modulating the defocus throughout an experiment, information about the object can be generated at all spatial frequencies and the zeros of the CTF curve can be virtually eliminated. This leaves us with the inverted/flipped values in the CTF curve that take away contrast from the image/contribute negative contrast.

The inverted signal in the CTF, i.e. the negative values, can be treated in a relatively straightforward way during CTF estimation. In a widely implemented method, the regions of the CTF curve with negative values are multiplied by -1 to "flip" the sign of the values to a positive one. While the sign of the CTF values changes, the amplitude of the curve in such cases stays the same and leads to a better estimate of contrast transfer.

2.11 The CTF dampens at higher spatial frequencies

The contrast transfer of a hypothetical perfect microscope is a linear 1:1 transfer from the object to the image. As we have seen in the preceding sections, the principles upon which the transmission electron microscope is built and the defocus based imaging conditions required for biological samples result in a non-linear contrast transfer described by the CTF. This unique contrast transfer regime is peculiar but still mathematically treatable and can be estimated from the data itself in a straightforward manner. An ideal CTF **(Fig. 2.5 C)** would oscillate uniformly throughout the full range of spatial frequencies providing similar intensity of information at both low and high spatial frequencies. This, however, is not the case.

In reality, the CTF dampens at high spatial frequencies. The signal is strong in the lower spatial frequencies but degrades rapidly at higher frequencies **(Fig. 2.5 D)**. This dampening of the contrast signal is caused by slight phase differences at high spatial frequencies which cause significant destructive interferences. Such phase differences can be because of the spherical aberration of the lens or because of electron beam incoherence. Electron beam incoherencies can be mathematically described as envelope functions that modulate the CTF curve.

88

The first of these is partial spatial coherence. In this case, two different electrons of equal energy strike the sample at an angle to each other i.e. they are not parallel. When the exit waves formed by such incident beams interact to form an image they have slightly different phases. In this interaction most of the low frequency signal is retained but the higher frequency signal is degraded. Another source of dampening of the CTF is partial temporal coherence of the beam. It occurs when the incident electrons are parallel but have slightly different energies because of issues such as an unstable source. The differences in energies mean that two different electrons beams will form images at two different planes and blur the image. In this case too, higher spatial frequencies are more dampened than lower frequencies. Both these issues can be modeled in the CTF by envelope functions:

$$T(k)_{eff} = T(k)\, E_{sc} E_{tc}$$

Eq. (2)

Where $T(k)_{eff}$ is the effective contrast function, $T(k)$ is the CTF that is dampened by the envelope function caused by partial spatial coherence (E_{sc}) and by partial temporal coherence (E_{sc}). The above equation gives a more accurate model of the CTF than Equation 1.

A properly maintained and accurately aligned microscope is critical for accurate CTF estimation. For data collected from such a microscope, CTF estimation is essentially effortless and highly accurate. Over the years, many hardware interventions on the microscope have been implemented to diminish or eliminate phenomena that degrade contrast transfer.

2.12   Hardware interventions in the scope improve the contrast transfer

Inelastically scattered electrons are the main source of noise and contrast degradation in a cryo-EM images. Various parts of the microscope are geared towards removing such electrons from the final image. The objective aperture is crucial in this regard. Situated immediately below the objective lens **(Fig. 2.4 B)** at the back focal plane, this aperture can be set at specific widths and any electrons that are scattered beyond the chosen width are removed from the beam. Care must be taken while choosing the width of the objective aperture. Since this aperture removes all highly scattered electrons from the beam, a small aperture width can also remove elastically scattered electrons with high angles of scatter thus removing high resolution information from the images.

An energy filter is another device implemented in most high end microscopes of today for contrast improvement[61]. Energy filters can be of two types: in-column[62] or post-column[63]. Either type is situated before the detector with the goal of removing inelastically scattered electrons. As the electrons pass through the sample, the inelastically scattered electrons have less energy than the elastically scattered or unscattered electrons. Energy filters use a slit and a slightly modified path to the detector to gate for energy of the electrons. Any electron that has energy lower than the specified range of energy at the gate/slit is removed which improves the contrast in the images.

Phase plates are also available in high end microscopes for improving contrast[64]. While conventional cryo-EM is based on defocus generated phase contrast, in phase plate based cryo-EM, contrast is generated without defocus and images are collected at focus. A phase plate is a thin film of carbon with or without holes located after the objective lens and it changes the phase exit beam. Phase plate induced phase shift fundamentally changes the nature of the CTF from a sine function that starts negative to a cosine function[64]. Phase plates also improve the amplitudes of the higher frequency components. There are many types of phase plates but the Zernike phase plate[65] and the Volta phase plate[66] are the most widely used ones. While phase plates hold a lot of promise[13], their actual usage is tricky as they require constant vigilance and intervention during data collection and a modified protocol for data processing[67].

Another high end microscope hardware intervention for improving contrast is a spherical aberration (C_s) corrector[68,69]. A C_s corrector eliminates spherical aberration from the microscope system. Addition of such correctors adds significant cost to an already expensive microscope and can create unwanted complications during data collection. The benefit of using such a corrector is also negligible if biological questions of interest can be answered without a reconstruction better than 2 Å.

A third condenser lens implemented in high end microscopes has become a very useful tool in achieving high resolution reconstructions[58]. With a third condenser lens, the microscope can generate a highly parallel beam at a wider range of scope settings than a two condenser lens system where the beam is parallel at a very narrow range of settings. This third condenser virtually eliminates partial spatial coherency of the beam and leads to images with better contrast and higher resolution information.

While the above mentioned hardware interventions have unique contributions towards improving the quality of the image, the detector plays the most vital role.

2.13   Direct electron detectors ushered in the resolution revolution in cryo-EM

The earliest electron microscopes used a fluorescent screen to view images of the sample. The electrons hit the fluorescent material and produce a low resolution and transient image. Television screens and silver halide based film were subsequently used to display and record images form an electron microscope[5,70]. Film is a laborious and time consuming image recording system but had been used for a majority of cryo-EM experiments until recently. Digital recording systems such as a charge-coupled device (CCD) camera were the next generation of detectors used in cryo-EM[71]. CCD cameras were much easier to use than film **(Fig. 2.6 A, left)**. They

provided real-time feedback about the collected images and enabled early versions of automated

data collection. They, however, weren't able to beat film in generating images that led to high

resolution reconstructions. Because of the large noise and signal degradation inherent in such

scintillator based detection systems, CCDs were mostly limited to screening purposes while film

was the method of choice for higher resolution 3D reconstructions. This was the status quo in

image recording until the development of direct electron detectors (DED)[72] about a decade ago.

DED cameras subsequently ushered in the current resolution revolution in cryo-EM[1].

DED cameras **(Fig. 2.6 A, right)** are complementary metal-oxide semiconductor

(CMOS)[73] based electron detectors that, as the name implies, can directly detect electrons

without the use of a scintillator such as ones found in CCD cameras. Compared to CCD cameras,

DED cameras have a very high signal to noise ratio and very fast readout rates[74]. A comparison

of DED cameras with CCD cameras and film show that they have a very high detective quantum

efficiency (DQE)[75] **(Fig. 2.6 B)**.

DQE of a detector is defined as:

$$DQE = \frac{\left(\frac{S}{N}\right)^2_{OUT}}{\left(\frac{S}{N}\right)^2_{IN}} \qquad \text{Eq. (3)}$$

where, $\frac{S}{N}$ is the signal-to-noise ratio. DQE thus informs us how much the input signal is degraded

by the noise or the signal conversion mechanism of the detector when producing the output.

When plotted against the spatial frequency, the DQE provides us with very valuable information

about the performance of a detector at different spatial frequencies **(Fig. 2.6 B)**. A spatial

frequency value known as the Nyquist frequency represents the edge of the DQE curve. Since two pixels are required to completely sample a wave, the wave with the shortest wavelength that can be sampled is twice the size of the pixels. This represents the Nyquist frequency. A DQE of 1 means that all the detector does not add any noise to the image. The higher the DQE of a camera at each spatial frequency, the higher the signal to noise ratio of recorded images at that spatial frequency. A DQE curve that is taller than that of either film or a CCD camera means that single electron events can be easily recorded by a DED. With their fast frame rates, even a short exposure on a DED can be subdivided into multiple frames to generate a movie of the particles[76,77] rather than one static image.

DED cameras can be operated in two different modes[72]. An integrating mode where the signal is captured as an integral of the charge deposited by an electron over the area where it hits the detector is one mode. This mode, however, is noisier than the counting mode of a DED camera. In counting mode, by using an even lower incident dose, single electron events can be detected without coincidence loss and the location of such event can be precisely mapped onto the pixels of the detector. Counting mode is slower than integrating mode but the DQE of counting mode data is higher at low and intermediate frequencies. Low and intermediate frequency information is essential to accurately determine the orientations of the particles and generate higher resolution reconstructions. In a modified counting mode collection strategy called super-resolution mode[78], even the sub-pixel location of an electron detection event can be precisely determined.

DED cameras are widely available now a days as most new microscopes are equipped with one and older microscopes are being upgraded with one. DED cameras have become capable of generating many thousands of images per day[79]. An image dataset of a well behaved sample from a DED camera has a high probability of providing a high resolution 3D reconstruction. Before attempting to generate a 3D reconstruction of a complex from such dataset a number of pre-processing steps have to be carried out **(Fig. 2.7 A)**.

2.14 Pre-processing: from images to stacks of particles

The very first step of preprocessing is to remove any unusable images. Any images with dirt, significant movement of particles, empty holes, broken ice, crystalline ice, or damaged foil are discarded. This step cuts down on cleanup in the later stages and also saves valuable computational time. While manually cleaning up the dataset is time consuming, some amount of clean up during data collection is a prudent choice. Since thousands of images per day from a fast readout DED camera are a norm these days, automated or semi-automated tools[51] are available with data collection packages to speed image curation. Many new programs and plug-in tools[80,81] make it easier to exclude undesirable images on the fly using a range of preset cut-off parameters or global views of the dataset. After image curation, the first processing step called motion correction is implemented.

2.15 Motion correction using movies from DED cameras vastly improves data quality

Particles embedded in vitrified ice move when they interact with the beam[77]. This beam induced motion and other motions caused by deformations of the ice during imaging is captured by direct electron detectors in a stack of images i.e. a movie. Since these movies are dose fractionated stack of images of the particles, the frames can be used to estimate and correct for the movement of the particles throughout the duration of the movie[82]. This step of preprocessing is motion correction **(Fig. 2.7 B)**.

Motion correction software packages perform motion correction in three main ways: whole frame alignment[83,84], patched alignments[85,86], and per-particle alignments[87]. For whole frame alignments, subsequent frames are cross correlated with the previous frames to come up with an aligned average of all the frames. Patched alignment divides the field of view into patches as specified by the user, aligns those patches, averages them, and forms a final aligned image by stitching the aligned patches. Since particle movements throughout the images aren't uniform, per-patch alignments give a more accurate estimation of the motions of the particles within an image compared to whole frame alignment. This is slightly more computationally expensive than whole frame alignment but yields better results. An even more accurate estimation of motion can be performed by doing per-particle motion correction[88]. To do this however, location of the particles is required—information that a user doesn't have during the first pass of data processing. A per-patch motion correction is good enough for the preliminary round of data processing. When the user has selected particles and maybe even obtained a preliminary 3D density, then they can perform per particle motion correction to obtain a more accurate estimate of particle motion and then use those particles to improve subsequent data processing steps and possibly improve the resolution of the reconstructions.

Two other operations are usually implemented during motion correction: discarding early frames from the movie and dose weighting[84,89]. It has been shown that the first few frames in a movie show much larger sample movements. Discarding early frames with large beam induced motions helps to better estimate and correct for motion in the later frames. As the sample is exposed, it gets progressively damaged by the effects of radiation. The later images in a movie stack thus contain snapshots of a deteriorated particle. Because the particle is damaged and the high resolution information has been lost, dose weighting of the later images is performed. Dose weighting down weighs the high frequencies components of the later frames of the movies with damaged particles. After motion correction and dose weighting, CTF correction is performed.

2.16 CTF estimation and correction

As discussed earlier, images captured by an electron microscope are blurred by the CTF of the microscope[12]. However, usable information about the object is still present within such blurred images. This information, is buried in the image when the object gets convoluted with the CTF and is masked by the noise generated by the microscope. The buried information about the object can be recovered by accurately estimating the CTF of an image and then using it to correct the blurry image. A CTF-corrected version of an image thus is a more accurate representation of the object[5].

While the CTF is dependent upon fixed quantities such as acceleration voltage, spherical aberration, and amplitude contrast, these quantities stay constant for a given microscope[90]. Defocus and astigmatism, on the other hand, vary from image to image and play a critical role in CTF estimation. CTF estimation is indeed, at the very fundamental level, the calculation of these two parameters for a given image. The parameters that are estimated for CTF are the maximum defocus, the minimum defocus, and the astigmatism of the image.

Various programs and packages are available for CTF correction. Some are integrated within data-processing packages while others are available separately but can be integrated into the preferred package of choice. CTF estimation is computationally intensive but fast GPU-accelerated CTF correction programs such as gCTF[91] are available. Recent improvements to CPU-based program such as CTFFIND4[92] have made it almost as fast as GPU-accelerated ones.

On-the-fly CTF correction during data collection has become an important tool to make critical decisions about data collection parameters, microscope alignments, and image quality[51].

Any irregularities is immediately picked up by the CTF of the collected images and this information can be used to fine tune data collection. CTF correction post data collection can also be used to exclude unwanted images. Images with hexagonal or non-vitreous ice or images of the gold or carbon support films have distinct power spectra that can be detected and removed manually or automatically[11].

The CTF also provides information about the resolution content of the images. By evaluating the power spectrum of an image, a user can easily tell whether high resolution information is being collected in the images[12]. A resolution based filtering can be implemented after CTF collection to remove images with only very low resolution information[86,87]. If the data is collected from a tilted stage, then special care must be taken while estimating the CTF[40]. Tilting the specimen creates a geometrical thickening of the samples that results in larger beam induced motions. Tilting also creates a defocus gradient parallel to the tilt axis. These complications should be kept in mind and a representative sample of the CTF estimation should be manually inspected to ensure proper fit of the CTF curve. Recently, CTF estimation programs specifically geared towards tilted specimens have been introduced[41,85].

After the CTF is accurately estimated, the image is corrected for the CTF. This newly generated CTF-corrected image contains a truer representation of the object and is used next to pick particles.

2.17 Particle Picking is as critical as screening of samples

If screening is the most important step in a cryo-EM project, a properly picked set of particles is probably the most important part of cryo-EM data processing. The picked particles need to be free of artifacts such as dirt, ice, clipped holes, and fringes. They also need to be accurately centered and non-overlapping with neighboring particles.

Particles can be picked from an image either manually or automatically. A user can go from image to image looking for particles and manually mark their position. Manual particle picking by eye can be accurate for large or previously seen or distinctly shaped particles. However, since humans are prone to recognizing familiar shapes and patterns they might miss less frequent views of the particles. Also, particles in images very close to focus are almost impossible to distinguish from the background noise by the naked eye. Centering of particles during manual picking can also be inaccurate. Manual picking is feasible for a small or preliminary data set but with the ever expanding number of images per dataset one has to use fully automated or semi-automated particle picking approaches.

For automated particle picking, users have a myriad of options[11]. All of these options employ one of two main particle picking techniques: templated and non-templated. In templated particle picking the user provides either a 2D or a 3D reference to the software[87,93]. In one implementation of templated particle picking, the particle picking software takes the user provided 2D template (or creates 2D projections from the user provided 3D reference) and then cross-correlates it with the images to find spots where the particles are located. Templated particle picking performed in the Fourier space can be very fast. It is, however, prone to reference bias and frequently fails to pick rare or infrequent views missing in the supplied reference templates. Templated particle picking is also susceptible to what is known as the Einstein from Noise problem[94]. Using a template such as a picture of Einstein can pick false positives from random noise images. Such false positive particles when processed further can lead to a model looking similar to the reference (i.e. similar to Einstein) even when there are no such particles in the image[95,96]. Reference free algorithms avoid such issues and are widely used by the cryo-EM community.

Reference free particle pickers use methods ranging from blob detection to convolutional neural networks. The "simpler" blob detection algorithms use sliding windows coupled with a Laplacian of Gaussian[97] or Difference of Gaussian[98] based approaches. Such particle pickers are quite robust and work extremely well for spherical particles. But such pickers based on traditional algorithms work poorly with non-spherical particles and are prone to produce significant number of false positives. They are not suited to particle picking from images that contain a very high contrast area such as a small bit of gold foil, an edge of the grid hole, or a fleck of dust. They also tend to over-pick when some of the particles are aggregated in the image. The most recent generation particle pickers are based on convolutional neural nets[99] and they try to overcome such issues with traditional pickers.

2.18   New generation of particle pickers are driven by convolution neural nets

Convolutional neural net driven object detection algorithms are used across various disciplines[100]. They have also seen widespread use in our day to day lives as part of social media networks, machine learning algorithms, and even self-driving car technology[101,102]. Particle pickers based on such object detection algorithms have recently become available to the cryo-EM community. crYOLO[103], Topaz[104], and Warp[86] are a few of such particle pickers released in the past two years. While they differ in the minutiae, they have similar working principles.

The neural net based pickers are first trained on a large set of curated data. The curated data contains images, validated particle locations on those images, and can also have useless images without any particles or images of ice and other artifacts. By systematically "looking at" the training data the neural net algorithm discovers salient features of a cryo-EM particle and learns to distinguish it from random noise, imaging artifacts, or sample artifacts. By looking at more and more training data the neural net generates a model for a wide range of particles. When the neural net encounters a previously unseen image with previously unseen particles it uses its model to pick particles.

Neural nets have been shown to be faster than traditional particle picking approaches, they require minimal intervention (in most cases), avoid picking contaminants such as ice or dirt, pick infrequent views more reliably, pick particles almost to completion, obviate 2D classification, and improve resolution of the 3D reconstruction[104]. A drawback to neural net pickers is some datasets will require training of the neural net which is extremely time consuming and can be technically challenging for an average user. Another glaring drawback is that a neural net is essentially a black box and any insight it learns about how to distinguish and pick particles is almost impossible to extract to repurpose for other uses.

Whatever the method used to pick particles the end goal is to have a set of well picked particles that has a high signal to noise ratio, has low background noise, and doesn't have any dirt, ice, grid foil or other contaminations masquerading as genuine particles. After a satisfactory set of particles are picked they are extracted with a large enough box size and normalized (i.e. set the mean value of the pixels to zero and the variance to 1). The contrast of the particles is inverted as per convention (density is white, background is black) and the entire set of particles is recorded in a database file referred to as a stack. This stack of particles contains all relevant information such as the image collection parameters, the location and position of each particle in its source image, the calculated CTF values, the defocus values, etc[105]. Creation of this stack of particles marks the end of pre-processing stage of data processing in cryo-EM. Next, the classification and alignment steps of data processing is performed **(Fig. 2.7 A)**.

2.19   Reference-free 2D classification is both a data processing step and a data cleaning tool

After a particle stack is generated, it is put through a reference-free 2D classification step **(Fig. 2.7 C)**. In 2D classification, the particles are sorted into a number of user defined classes with the goal of (a) maximizing differences between different classes while at the same time (b) minimizing differences between particles within the same class. At the end of 2D classification, all the particles from any given class are averaged to create a high signal to noise ratio image called a class average **(Fig. 2.7 A)**.

The class averages obtained from 2D classification are pivotal in cleaning and assessing not just the picked particles but also the entire dataset **(Fig. 2.7 C, left)**. Damaged particles, fringes, ice, dirt, and other artifacts in the particle stack usually get sorted into separate 2D classes and thus can be easily discarded. The overall shape and integrity of the particles can be easily assessed from the class averages. The accuracy of picking and centering of particles is also evident in the class averages. Particles that are picked off-center become clearly visible in the class averages. Any mistakes made in selecting the diameter of the particle during particle picking or in selecting the box size for particle extraction are easily identified.

The class averages also provide a good indication of particle orientation. If just a few views dominate most of the class averages then the dataset has preferred particle orientation. If the sub-components of the molecule of interest are large and distinct enough then 2D classification can be used to separate distinct stoichiometric states of the particle even before having to perform a 3D classification step. The presence or absence of such large sub-components can also be easily determined. 2D classification has also been used to determine the stoichiometry of a sub-components in an experiment.

Putting a set of particles through a few iterations of 2D classification and removing any non-particles or bad particles can result in a clean and homogeneous set of particles. Any easily distinguishable subclasses of particles can also be differentiated during this step **(Fig. 2.7 C, right)**. Such a set will provide a more accurate initial models or 3D reconstructions compared to a set with artifacts. Removing artifacts from the particle set also significantly speeds up subsequent processing steps as the total number of particles that need to be computed upon is reduced and time lost trying and failing to align artifacts with true particles is recouped.

While 2D classification is pivotal step in data processing, it is also one of the most computationally intensive. With high-throughput image collection becoming more widespread and standard, a typical experiment can contain upwards of a million particles. In such cases, the computational load can be reduced by using two- or four-time binned particles[89]. Binned particles can speed up the 2D classification step by hours or even days. However, care must be taken while picking the correct box size and the binning parameter so as to not clip the particle or to unnecessarily leave any usable information outside the extraction box limits. Another strategy to reduce computational load is to start the classification with smaller—and thus faster—subsets and gradually add in more and more of the particles until all the particles are included[106,107]. Binning and subsetting of particles can be combined to further speed up the process. After a satisfactory 2D classification is performed, particles in desired classes are selected and fed into 3D reconstruction algorithms.

2.20   3D reconstruction is underpinned by the central slice theorem

The next step in data processing is 3D reconstruction of the particle **(Fig. 2.7 A)**. It is here that users encounter the problem that lies at the heart of cryo-EM: for a given a set of randomly oriented 2D projections how does one determine where they fit into the 3D volume of the object? To correctly slot a 2D projection into the 3D volume of an object we need to determine a set of parameters—three Euler angles, two in-plane positions and defocus—of that particular projection. 3D reconstruction algorithms use the central slice theorem[5] to perform this task **(Fig. 2.8 A)**.

The central slice theorem states that the Fourier transform of a 2D projection of a 3D object represents a central slice through the Fourier transform of that 3D object[5]. This means that given the Fourier transform of any two particles there exists at least one shared set of amplitudes and phases, i.e. the central slice, which can be used to calculate the relative orientations and positions of the two particles within the Fourier transform of the 3D object. When all the Fourier transforms of all particles have been assigned a position and orientation in the 3D Fourier transform of the object, it can then be inverse Fourier transformed to reveal a model of the 3D object in real space **(Fig. 2.8 A and B (i))**. Given the noisy nature of the data, however, *de novo* 3D model generation isn't straightforward.

2.21   Initial model generation is key for 3D reconstruction of a novel particle

If the sample of interest is of a particle with previously unknown structure then *ab initio* generation of an initial model is required. There are a few strategies available for *ab initio* model generation. The first of these is the random conical tilt method[108]. In this method, images of the same field of view are taken at tilted and untitled angles. Since the angle of tilt is a known quantity, it can be fed into the reconstruction algorithm to obtain a reliable *ab initio* initial model of the particle. This method of initial model generation is robust but it does suffer from some shortcomings such as the missing cone problem[109]. Since one set of the paired image is taken at a tilted angle, an extra data collection step is required. Furthermore, tilted images are usually of poorer quality due to larger specimen movement at higher tilt angles and associated difficulties in properly estimating the CTF[40,41]. While a slightly modified approach known as orthogonal tilt[110] reconstruction overcomes some problems of random conical tilt, both have been supplanted by computational methods for initial model generation.

Computational *ab initio* model generation relies on local search algorithms such as stochastic hill climbing or stochastic gradient descent. Their ease of use and relative success in generating reliable initial models have increased their popularity. Standard refinement techniques or cross-correlation based methods such as hill climbing get stuck at local maxima of likelihood when trying to search for a 3D structure in a space of all possible 3D structures. This means that if a more likely 3D structure that fits the 2D projections better is elsewhere in the data space, the reconstruction algorithm will miss it. Stochastic implementations of hill climbing or stochastic gradient descent get around this problem by sampling the space of 3D structures in a random manner. The *ab initio* modelling algorithms start with a totally random initial model and try to iteratively improve upon it by taking random samplings of the data. Such algorithms have become powerful enough to remove the reliance of cryo-EM on homology based or random conical tilt derived or user guessed initial models.

The *ab initio* generated initial model is a blurry and low resolution representation of the 3D object. The orientations and positions assigned to the particles need to be refined to achieve a clearer and higher resolution 3D reconstruction. Iterative projection matching is used to achieve this **(Fig. 2.8 B)**.

2.22 Iterative projection matching improves orientation assignments to generate a final model

Projection matching starts with a low resolution initial model obtained either from an ab initio reconstruction or from a reference model from a previous experiment. The initial model is first projected in all possible directions to create a set of computationally generated 2D projections **(Fig. 2.8 B ii)**. These 2D projections are then cross-correlated with all available particle images **(Fig. 2.8 B iii)**. When a particle image best matches a 2D projection, the orientation and position of the particle image is updated with those from that particular 2D projection such that it can now be assigned a position in the 3D structure **(Fig. 2.8 B iv)**. Once this process is completed for the entire set of particles, a new model is generated using the particle images with the updated orientation and position assignments. This updating of the assignments is performed in Fourier space **(Fig. 2.8 B (v))**. Using the central slice theorem, an inverse Fourier transform is performed to generate the updated 3D structure. When the updated reconstruction is generated, the steps of re-projecting, cross-correlating the re-projections with the experimental data, updating the angular assignments of the experimental data, and then generating an updated model using those updated images are iterated **(Fig. 2.8 B (ii) to (v))**. These steps are iterated until the angular assignments stop improving. Traditionally, maximum cross-correlation based methods of updating angular assignments were employed and they have been largely supplanted by maximum likelihood based approaches.

In a maximum likelihood based projection matching, a particle is assigned a distribution of positions. What this means is that instead of assigning a specific position in the 3D density to a particle, the particle is assigned a range of probabilities of it occupying every possible position in the 3D density. One of the advantages a maximum likelihood approach is that it takes the noise present in the image into account while assigning the orientation probabilities. Any ambiguously oriented particle thus avoids getting assigned an incorrect position as a result of the noise that surrounds it. This fuzziness of the assignments and the orientations of the particle are then improved with each iteration using an updated probability distribution. An extension to maximum likelihood is Bayesian regularized maximum likelihood approach[111].

In the Bayesian regularized maximum likelihood approach, a prior distribution is implemented on the projection matching algorithm. The prior for this approach is smoothness and is implemented during reconstruction of the 3D density. This prior is based on the knowledge that proteins and molecular densities are smooth. The formulation of this Bayesian approach is called *maximum a posteriori* (MAP)[111] estimation and works as outlined (**Fig. 2.9 A**).

In cases where the sample is homogeneous, 3D reconstruction is relatively straightforward. However, for heterogeneous samples slightly different data processing approach is required.

2.23   Tools such as 3D classification, focused refinement, and multibody refinement help deal with heterogeneity

After a 3D refinement, it is a prudent choice to perform a 3D classification of the obtained density[89]. There is a possibility that despite careful image and particle curation some damaged or low resolution or false positive particles have been carried into the 3D density. When this happens the overall resolution of the density can be impacted. Also, if the sample contains particles from distinct structures, then the sample will be a superposition of two or more states. An example of this is the initial consensus refinement density obtained from a homogeneous reconstruction of an 80S ribosome sample.

While the sample maybe have originated as an 80S sample, there are separate 40S or 60S subunits present in the sample either as contaminants or from splitting of the 80S. Particle picking cannot reliably distinguish between the three states so it picks all of them. Some 40S only particles can be separated out during 2D classification but most of the 60S only particles cannot and they eventually end up in the 3D reconstruction. So a 3D classification is needed in this scenario to get to the 80S particles of interest. The classification can be performed by selecting a number of classes, one each for the three states of the ribosome and at least one each for damaged particles or dirt in the sample. If even after such classification, the density of interest is suspected to be a mix of different states then further 3D classifications can be performed.

Sometimes, for a given 3D density there are parts which are stationary and other which go through some range of motion. If the region of interest is within the mobile part then focused refinement[89,112] can be implemented. Here, the user creates a mask around the region of interest and performs 3D alignment only within the region of the mask. This will result in a sharper reconstruction within the mask at the expense of the unmasked regions. The improved region can then be used to build appropriate models.

A variation to focused refinement is focused refinement with signal subtraction[89,113]. In signal subtraction **(Fig. 2.9 B)**, one mask is created around the region of interest while another encompasses the density outside the first mask. Using the second mask, a set of projections of the 3D density are created. These *in silico* projections are then subtracted from the actual experimental images to create a new set of particles which only correspond to the density inside the first mask. These "signal subtracted" particles are then used to reconstruct the density inside the first mask and resolve the region of interest. Signal subtraction has been used to great success in especially in flexible complexes such as the spliceosome.

An extension of the signal subtraction approach is multibody refinement[114]. In multibody refinement, an automated, iterative, and multi-focused refinement with signal subtraction is implemented on complexes with multiple floppy regions. Multibody refinement in essence is combining multiple signal subtraction based focused refinement jobs in one run. One advantage of using multibody refinement is that it provides a principle component analysis on the orientations of the particles and helps creates molecular motion movies that are grounded on data rather than virtual animation of discrete models. After all the steps the user arrives at a final 3D reconstruction of their sample with a given resolution.

2.24   Resolution in cryo-EM is more of a guideline rather than an absolute metric

The resolution of a cryo-EM density is unlike resolution in optical microscopy or in X-ray crystallography[115]. Resolution usually describes the minimum distance two points need to be separated in an image for them to be clearly distinguished as two different points. This principle is applicable in both optical microscopy and X-ray crystallography. In cryo-EM, however, the resolution of a 3D density is a consistency metric and is arrived at using the Fourier shell correlation (FSC).

To calculate resolution in cryo-EM, two independent half maps created during 3D reconstruction and are correlated with one another in the Fourier space[116]. The 3D Fourier transform of 3D densities is called a Fourier shell. In a perfect dataset with no differences between the two half-maps and no noise present, the Fourier shells at the same spatial frequencies in the two half-maps would exactly correlate and be equal to 1. This would mean that the FSC when plotted against the spatial frequencies will be a straight line. With real world, noisy data, a typical FSC curve assumes a sigmoidal shape which starts with higher correlation between the two half maps at lower spatial frequencies and then degrades to zero at the higher frequencies **(Fig. 2.10)**.

The resolution of the map is then the "maximum spatial frequency at which the information content can be considered reliable.[115]" As we can see form this statement that what is considered to be the resolution of a cryo-EM density map is subjective. Indeed numerous cutoffs of the FSC for what is considered to be reliable information content have been used throughout the history of cryo-EM. Previous version of resolution estimation using the FSC have used 0.5[117] and 0.33[116] as cut-off value for the FSC. The "gold-standard FSC"[116,118] widely used in practice today, uses a 0.143 cut-off for its resolution estimation with the prerequisite that the dataset has been randomly divided into two halves that are independently processed. The search for an objective metric for the resolution of cryo-EM maps is ongoing[119–123] and always generates considerable amount of debate within the community[124,125].

Given the inherent nature of cryo-EM maps, there is always variability in reliable information content throughout the maps. The resolution number of a cryo-EM map generated from the FSC then serves as a guideline for the user. Users should take that number as a starting point and validate whether the features in the map can be explained by such number or not. This should be done both by visual inspection of the map and with available validation tools.

Features in biological molecules characteristically appear at certain resolution thresholds. At low resolutions >10-20 Å, the surface envelope of the complex can be observed. At such resolution it might be possible to estimate the location of certain domains by docking and fitting of X-ray model of the complex or a part of it, if available. Also, any anisotropic resolution because of preferred orientation or overfitting of noise can be easily observed even at low resolutions. At resolutions better than 10 Å, alpha helixes appear as tubes and the overall shape of the complex becomes clearer. At resolutions better than 4.8 Å the individual strands in the beta-sheet become separated. Around 4 Å, most of the bulky aromatic sidechains of the proteins appear distinguishable and the protein backbone becomes traceable. Around 3 Å, most of the amino acid sidechains and nucleotides can be confidently identified and oriented. A user has to look for such features to verify whether the reported global resolution is within range of the features seen in the map. Tools are also available to assess the local variability in resolution throughout a cryo-EM map.

Because of numerous factors such as sample heterogeneity, flexibility, or conformational diversity, the resolution of a cryo-EM map is not always uniform. In a cryo-EM map, outer edges of the macromolecule are usually less resolved than its core. This variability is quite pronounced in molecules such as ribosomes where in the same map the outer edges can be limited to 10+ Å whereas in the inner core amino acid residues and nucleic acid bases are easily distinguishable at better than 3 Å. Such gradient of variation is an inherent property of cryo-EM maps. Various tools such as ResMap[126], Blocres[127], MonoRes[128], and the convolutional neural-net driven DeepRes[129] are available to estimate this variation in local resolution.

Once a map has been deemed of sufficient quality then model building can start. A host of tools and software packages are available for this purpose[130–132]. With high resolution structures at or better than 3 Å, models can be built *de novo* either manually or with the help of an automated tool. At the intermediate range of resolution, the protein backbone and nucleic acid backbone can be traced and docking of available or related models can be performed. Care must be taken when dealing with intermediate and low resolution densities as any newly built or docked models can be over-fitted or improperly fitter into the density. An iterative regime of modeling while enforcing available restraints and validating it as it is being built[133] should be followed to build a model that best represent the density.

2.25   The goal of a cryo-EM experiment is a 3D model of the molecule of interest

The goal of all structural biology techniques is to arrive at a three dimensional model of the molecule of interest. A detailed 3D model provides information about the position of every atom in the molecule of interest. It also provides information about how one atom relates to another in 3D space, how they interact with one another, how they fold into characteristic protein or nucleic acid features, or what isoforms they occupy. How accurate such information in a model can be and whether or not such information can be described at all heavily depends upon the resolution of the 3D reconstruction.

In the early days of electron microscopy in biology, even the best 3D reconstructions were of very low resolution. These reconstructions were only enough to interpret and model global features of the molecule of interest. Still, researchers used such reconstructions to gain valuable insights about biological molecules such as an early model for the tail of a bacteriophage[134] or the number of transmembrane helices in a membrane protein[8]. As the method developed, it became more suited to deciphering the structure of large biological objects or detecting large scale changes within such objects. Cryo-EM became vital in determining the overall structure of large viruses and large conformational changes in megadalton sized molecules such as the ribosome. For the most of cryo-EM's existence, researchers were limited to observing obvious and large scale changes in their large molecules of interest. They had to be satisfied with docking and fitting of models derived from X-ray crystallography into their cryo-EM densities. *De novo* models from cryo-EM densities remained elusive until the resolution revolution at the turn of the past decade.

Now, atomic modeling of the molecule of interest is possible from 3D reconstructions generated using cryo-EM. While some reconstructions are at high enough resolution to distinctly detect and model even water molecules, most of the cryo-EM densities are of either intermediate or close-to-atomic resolution range. Immense care must be taken to generate models from such reconstructions.

While a generalized model building strategy and many tools for validation are available[127,133,135–137], there is no standardized protocol for model building or evaluation. There are variations in model building approaches between labs and sometimes even within the same lab. What some labs might consider reliable density information and build models using such density, others might be more hesitant. Incorrect interpretation of densities will lead to faulty models. As structural biology models are used as a starting point for many types of downstream experiments, any mistake in the model will get amplified and perpetuated throughout different fields of biology. In such a landscape, with the number of new cryo-EM adoptees ever increasing, standardized model building and validation protocols are sorely needed to produce and evaluate models generated using cryo-EM. Until a uniform system of model building and evaluation is implemented in the field, one should take utmost care while building and evaluating their models.

2.26 The future of cryo-EM is obviously atomic but we need to actively work to make it open and democratic

From the early days when it was derisively christened "blob-ology[138]" to its redemption as a Nobel Prize winning technique[139], cryo-EM has come a long way. Since the resolution revolution in the early 2010s[1,2], the developments and advances in cryo-EM have skyrocketed. Newer and better seems to have become the theme in almost every aspect of the technique. Newer and better detectors[79,140], newer and better grids and supports [34,43,45,141], newer and better sample preparation techniques[45–47], newer and better algorithms[86,103,103,142,143], newer and better data collection programs[50,86,144], newer and better data processing programs[107,142], newer and better microscopes[145–147] seem to be announced every month, if not every week. Such advances have led to higher and higher resolution reconstructions.

The field recently broke the 2 Å resolution barrier[148,149] and is well on track[18,150–152] to break the 1 Å resolution barrier within the upcoming months—an astounding feat for a technique that struggled to resolve an amino acid chain just a decade ago. High resolution atomic reconstructions of well-behaved samples will soon become routine and possibly automated[153]. Historically intractable samples or projects that have been gathering dust on the laboratory shelf will become reinvigorated. In its endless and relentless pursuit for the newer and better, however there is a danger that cryo-EM as a field will become more closed and inaccessible.

Currently, cryo-EM is as expensive as it has ever been. With microscopes capable of generating high resolution reconstructions costing north of 7 million US dollars apiece, millions more to prepare a space, and daily running costs in the thousands of US dollars[154], cryo-EM can only be afforded by a small selection of affluent universities and companies. As there are only a couple of manufacturers who make such microscopes and associated DED cameras, prices get exaggerated. The threat of consolidation of such companies driving prices and limiting access further is ever looming[155]. Even if an institute can afford a microscope, there is a long wait to have one delivered. These issues create and exacerbate problems with access. Even a cursory look at the worldwide map[156] of high-end cryo-EM microscopes will show how they are concentrated in the global north. As wait times for access are long and usage charges expensive[154], there is a chance that the global south will be excluded from scientific discoveries and advancement fueled by cryo-EM. Another potential source of inequality is the current trend towards closed-source software and algorithms[91,142,144,157].

While cryo-EM has historically been an open field, there has been a worrying trend of issuing closed-source software or algorithms or issuing them with very restrictive licenses. As cryo-EM gets more and more popular, some have realized that of course there is money that can be made. While sustaining a cryo-EM software or package or algorithm needs a source of revenue, such funds shouldn't be generated by gate-keeping practices such as closed-source software or restrictive licensing. Time and effort that researchers have to spend dealing with ancillary issues caused by such restrictions would be better spent on the bench or on the microscope. Another issue that creates unequal opportunities are the computational demands of cryo-EM.

As cryo-EM has to deal with multiple terabytes of data and has to undergo many millions of calculations before arriving at a usable 3D reconstruction, it has exceptionally high computational demands. Such demands are best handled at large data centers or institutional servers with petabytes of storage, terabytes of memory, and thousands of teraflops of processing power. Such requirements are expensive and out of reach of most labs trying to get started with cryo-EM. It creates another avenue of mismatched resources and unequal access. While the above situations seem bleak, there are numerous projects and opportunities that are democratizing cryo-EM.

One of the most promising is the push to design an affordable cryo-EM microscope at one-tenth of the cost of a high end microscope[158]. It was recently shown that by limiting the accelerating voltage to 100 kV and combining off the shelf-parts including a hybrid pixel detector used for X-ray crystallography, such a microscope can be built[159]. The prototype microscope was able to resolve structures up to 3.7 Å[159] and is designed to work in a regular laboratory space without the need for expensive renovations or ancillary equipment. As an added benefit, it has been shown that using less damaging electrons at 100 keV results in 25% improvement in image quality when compared to images collected on either a 200 kV or a 300 kV microscope[160]. The goal of the 100 kV microscope project is to have an affordable scope in every biology department. For the time being, users can also opt to use the relatively cheaper 200 kV microscopes which are capable of producing high resolution reconstructions[149].

To address issues of access, large facilities such as the NIH National Center for Cryo-EM Access and Training (NCCAT) have been set up by the NIH. The NIH recently established three such national centers throughout the US with a $130 million seed grant[161]. Other such centers, based on the tried and true synchrotron model, are up and running throughout the world[162–165]. Access to such the microscopes and the expertise at such centers are provided at no or minimal cost to the users after a peer-review of their proposal.

On the computational side, a do-it-yourself approach to building data processing machine can significantly reduce costs. By using off-the-shelf computer parts we have in our lab built high-powered desktop machines that have significantly outperformed our server-based solutions. Another approached explored by the field is using cloud-based hosting services to run data processing jobs remotely and at a fraction of the cost of deploying one's own servers.

The issue of restrictive or closed-source software and algorithms in cryo-EM might prove the trickiest to address. Patent holders probably won't invalidate their patents or make their technologies freely available for unrestricted use, so it will be upon us as users to define the path forwards. As users we are always making active choices in what software to use for particular aspect of cryo-EM. While closed-source software might have an easier and user-friendly interfaces, they by their nature cannot be accurately evaluated for the features that they claim to provide. Their algorithms and programming cannot be audited to ensure that they are based on sound mathematical grounds. Users have to take the company's word for it or leave it.

Users can always opt to use open-source software for their processing needs as a way for supporting the research and development and maintenance of such software. With more users, open-source software projects can attract a stable source of funding and thus create newer more advanced feature and support even more users. In a compromise of a situation, hybrid licenses can be implemented so that they are available to academic users at zero or low costs and to commercial users for a fee.

For the future, the field of cryo-EM needs to focus as much on democratizing itself as it focuses on breaking the next resolution barrier. Issues of access and reach remain but with community effort they can be overcome and the field widened to welcome researchers of all levels from all over the globe. There is tremendous potential in the technique and one day we might even be able to see proteins, not in a thin film of ice, but inside the cell itself. With the current trajectory, there is nowhere for cryo-EM to go but up.

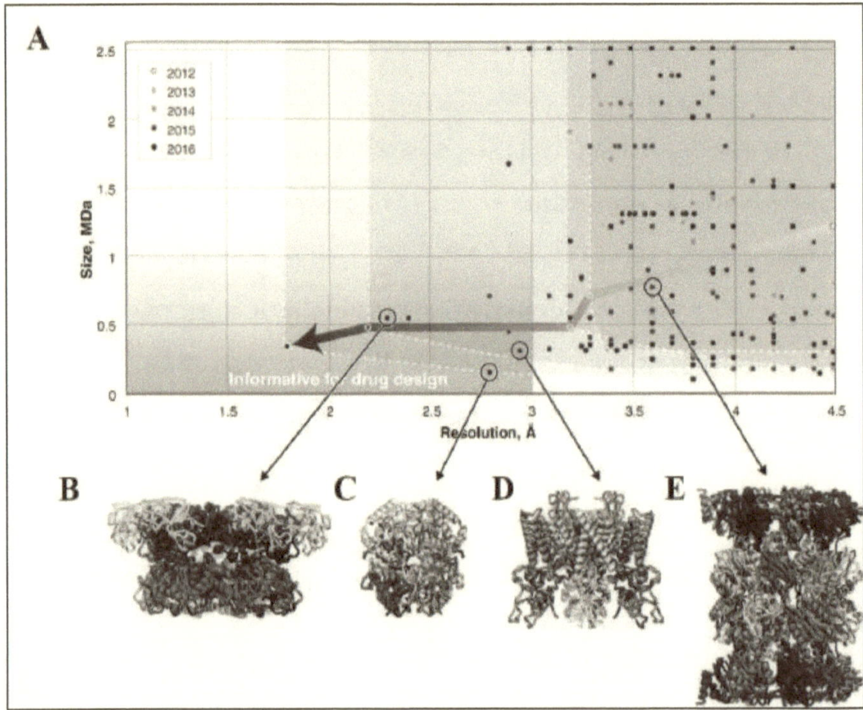

Figure 2.1 | Progress in cryo-EM in the first half of the past decade (Reprinted with permission from Elsevier from Subramaniam et al. [166])

A: Improvements in resolution for cryo-EM derived densities charted for entries deposited in the Electron Microscopy Data Bank (EMDB) for the period from 2012 to 2016. Resolutions better than 4.5 Å for complexes smaller than 2.5 MDa have steadily improved. Select high resolution structures of four complexes obtained using cryo-EM are highlighted.

B: p97 (EMD-3295, PDB-5FTJ) solved at 2.30 Å[167].

C: Lactate dehydrogenase (EMD-8191, PDB-5K0Z) solved at 2.80 Å[168].

D: TRPV1 (EMD-8117, PDB-5IRX) solved at 2.95 Å[169].

E: Plasmodium falciparum 20S proteasome (EMD-3231, PDB-5FMG) solved at 3.60 Å[15].

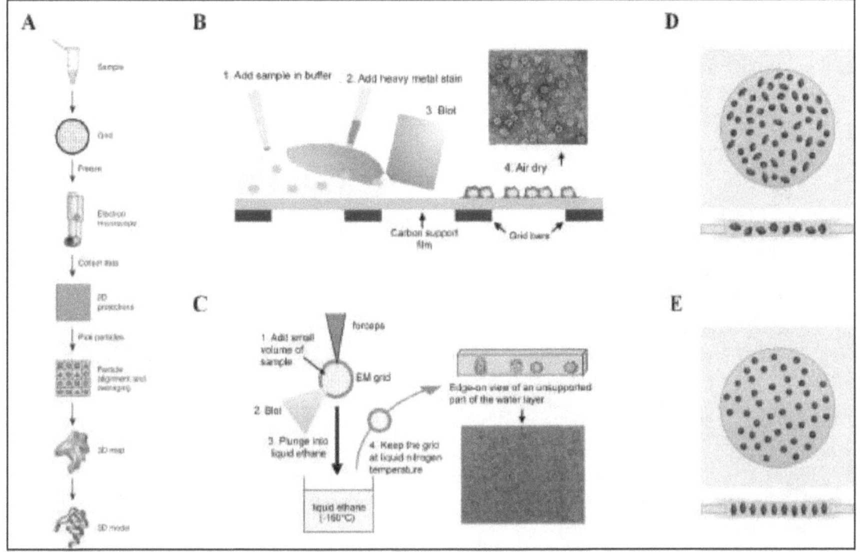

Figure 2.2 | Outline for a cryo-EM workflow, grid preparation methods, and preferred orientation

A: An outline of a general workflow for cryo EM (Reprinted by permission from Nature Springer from Doerr[170]).

B: Steps used for negative staining of samples which is a simple yet vital room temperature method for preliminary screening of samples before making frozen grids (reprinted with permission from ACS, from Orlova and Saibil[12]).

C: Steps involved in plunge freezing of samples. This step is usually done using a plunge freeze device. (modified from Orlova and Saibil[12]).

D: A schematic representation of well behaved, homogeneously distributed, and randomly oriented sample (purple) in a grid (gray). Top down (top) and side on (bottom) view (modified from Drulyte et al. [171]).

E: A schematic representation of a sample (purple) with extreme preferred orientation. Instead of occupying random orientations, the sample is constrained to a single orientation. Top down (top) and side on views (bottom) (modified from Drulyte et al. [171]).

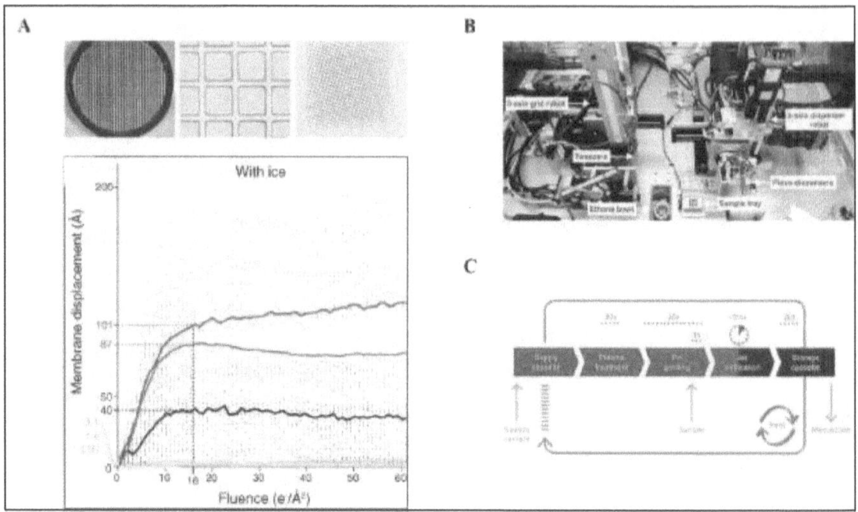

Figure 2.3 | Current and future grid and sample preparation technologies
A: Top: An ultrastable gold grid at different levels of magnification seen using a light microscope. The gold grid (left) is about 3 mm in diameter. Both the grid bars (middle) and the foil mesh (right) with regularly spaced grid holes is made of gold (Reprinted with permission from Science from Passmore and Russo[19]). **Bottom:** Reduction seen in root mean squared vertical motion at multiple holes in multiple squares of a single grid when using gold grids (gold lines) vs. amorphous carbon grids (red, blue, and black lines) (Reprinted with permission from Science from Russo and Passmore[42]).
B: A prototype of the picoliter dispensing grid-making machine called SpotItOn[46] at the New York Structural Biology Center in New York, New York. It is now commercially available as the Chameleon (Reprinted with permission from Elsevier from Jain et al. [46]).
C: Workflow outline of the fully automated, picoliter dispensing, pin-printing grid-making machine known as Vitrojet[45] (modified from Ravelli[45]).

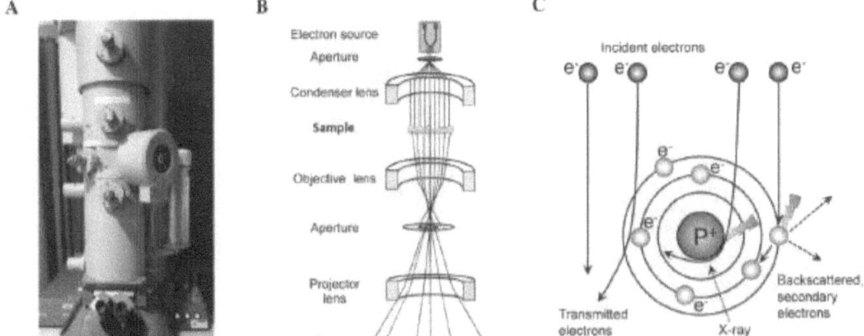

Figure 2.4 | The electron microscope and beam specimen interactions
A: The Tecnai F20 microscope housed at Columbia University was extensively used to screen samples and collect preliminary data for this body of work (personal collection).
B: A schematic of the setup of such an electron microscope showing the path taken by the electron from the source through the sample until it forms an image on a detector (reprinted with permission from ACS, from Orlova and Saibil[12]).
C: A schematic of how the electron beam interacts with atoms of the sample to produce different types of scattering (reprinted with permission from ACS, from Orlova and Saibil[12]).

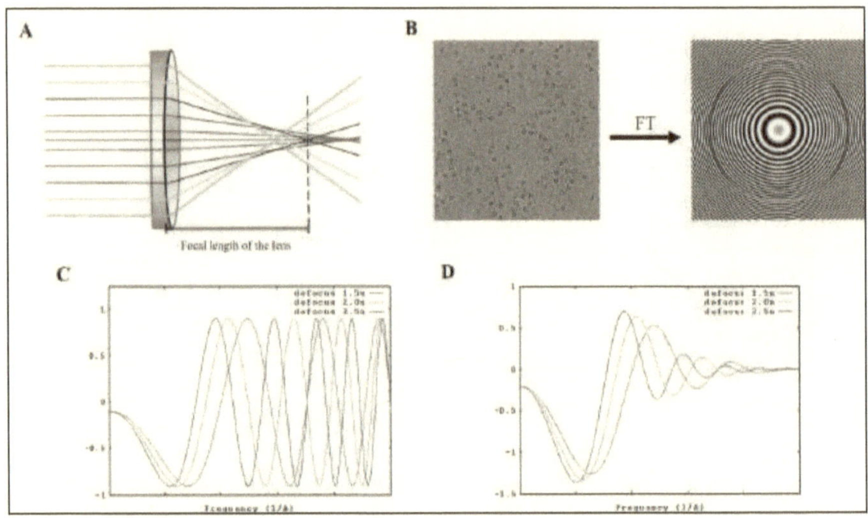

Figure 2.5 | Spherical aberration and the Contrast Transfer Function (CTF)
A: The spherical aberration of a lens is depicted as a schematic. Because of spherical aberration, the incident beam at the periphery (green lines) is focused more strongly than the coaxial beam (purple lines) (reprinted with permission from ACS, from Orlova and Saibil [12]).
B: The Fourier transform of an image from a TEM[172] leads to a characteristic image called the power spectrum (simulated) with alternating and concentric dark and light rings called Thon rings. The power spectrum represents the CTF of an image and is immensely helpful to diagnose and correct beam characteristics, imaging conditions, and ice features. The red circle indicates spatial frequency of 2.5 Å which indicates that high resolution information up to that point is contained in the image.
C: Demonstration of the oscillating nature of the CTF using three different defocus values (retrieved from The Wardsworth Center[173]).
D: Dampening of the CTF at all three defocus values because of the envelope functions that affect the CTF (retrieved from The Wardsworth Center[173]).

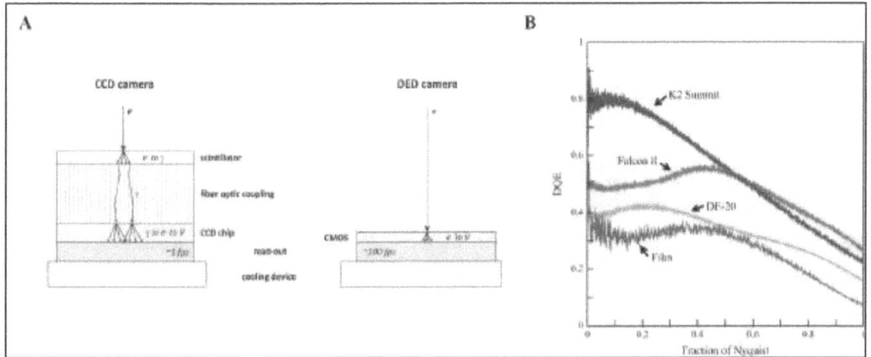

Figure 2.6 | Direct electron detectors (DEDs) have higher detective quantum efficiency (DQE)

A: Comparison between the principles behind charge coupled device (CCD) cameras and direct electron detector (DED) cameras. DED cameras have significantly faster readout rates which is used to collected movies of particles instead of a single static frame (modified from Koning et al[174]).

B: Comparison of the DQE between film (gray) and three commercially available DED cameras showing that all three have higher DQE than film (from Mcmullan et al.[74]). It can also be seen that while some camera such as the Falcon II is worse than the K2 Summit at low spatial frequencies, they can have higher DQEs at high spatial frequencies.

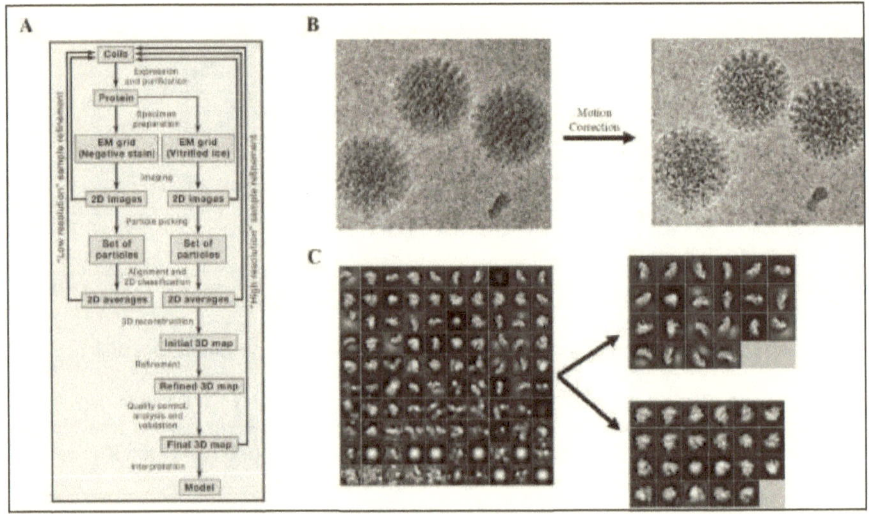

Figure 2.7 | Data processing workflow, motion correction and uses of 2D classification
A: A general outline of the iterative sample preparation, sample screening, data collection, and data processing strategy (Reprinted with permission from Elsevier from Cheng et al.[11]).
B: Motion correction using movies collected from DED cameras significantly improves the quality of the particles which helps generate higher resolution maps. Here, the same field of view before and after motion correction shows how the improvements are visible even to the naked eye (Reprinted with permission from Elsevier from Brilot et al.[77]).
C: 2D classification generates class averages of all the particles in the data (right). We can see that while many classes contain particles of interest, there are classes that contain either mis-picked particles or grid contaminants. Such class averages can be used to iteratively clean up and, if possible, classify a given set of particles (left).

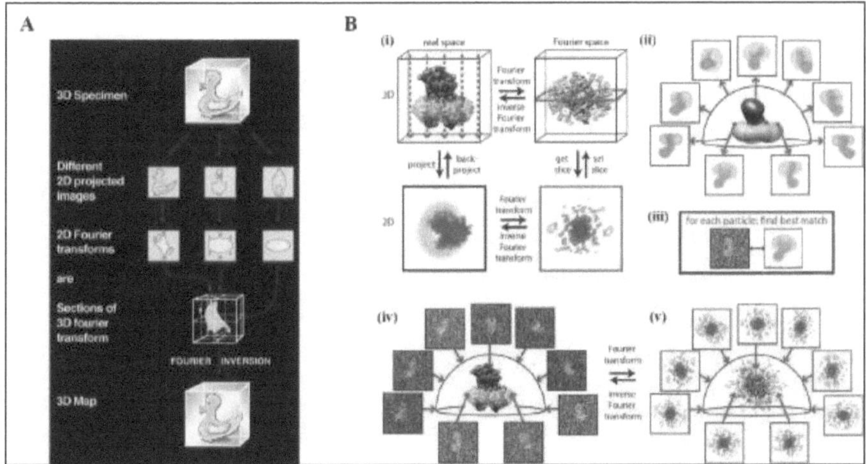

Figure 2.8 | The central slice theorem and the iterative 3D reconstruction strategy
A: A schematic of the central slice theorem and how it can be used to generate 3D maps from 2D project images of a 3D object (Reprinted with permission from John Wiley and Sons from Milne et al[175]).
B: A general outline of the 3D reconstruction strategy used in cryo-EM (Reprinted with permission from Elsevier from Nogales and Scheres[176])
      (i) The central slice theorem as applied to a cryo-EM sample
      (ii-v) Iterative steps of projection matching to generate a 3D density. See text in section
            2.21 for details.

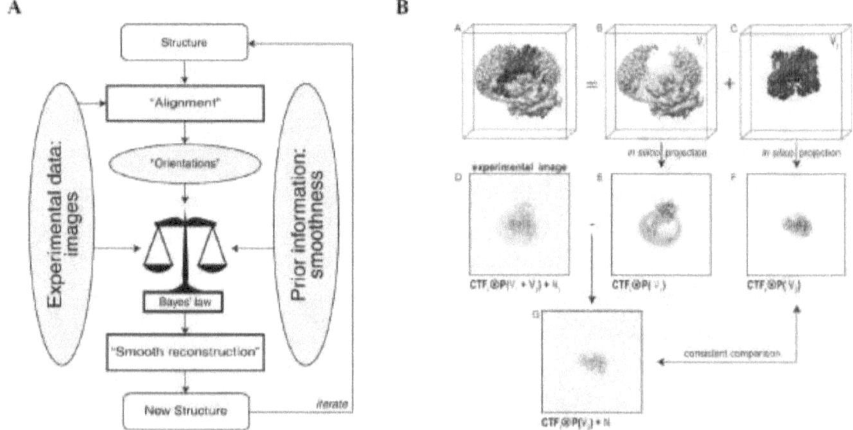

Figure 2.9 | A Bayesian approach to 3D reconstruction as employed by Relion; processing of heterogeneous data with signal subtraction
A: A schematic of the Bayesian approach to 3D reconstruction employed by Relion. Critical is the prior information "smoothness" applied to the algorithm (modified from Scheres[111]).
B: A schematic of the masked classification with signal subtraction strategy that can used to process of heterogeneous data[177] and was extensively used for this body of work.

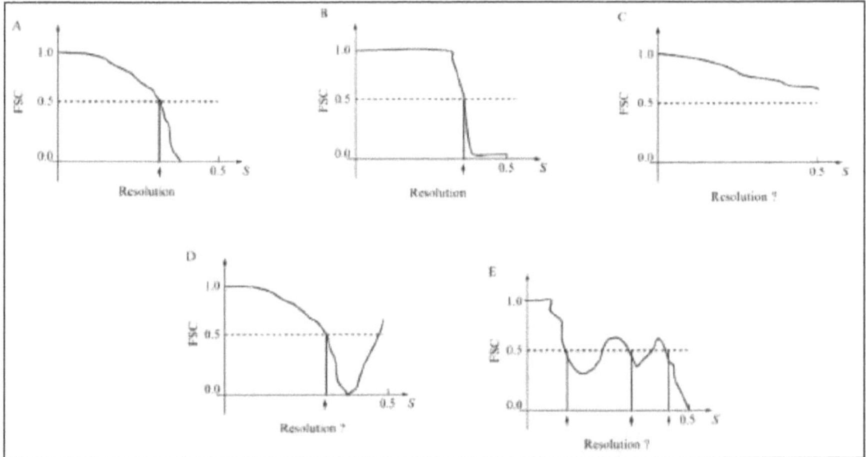

Figure 2.10 | The Fourier shell correlation (FSC) is a consistency metric (modified from Penczek[115])
A: An ideal FSC curve for a cryo-EM reconstruction. Here, the cut-off is set at 0.5 which determines the resolution.
B-E: Various problematic FSC curves that can be used to diagnose issues such as overfitting of noise and too tight a mask during alignment (B), noise was aligned (C), multiple errors during pre-processing or incorrect CTF estimation etc. (D), and data dominated by one defocus subset (E).

2.28   References

1.   Kühlbrandt, W. The Resolution Revolution. *Science* **343**, 1443–1444 (2014).

2.   Callaway, E. The revolution will not be crystallized: a new method sweeps through structural biology. *Nature News* **525**, 172 (2015).

3.   Ameh, E. S. A review of basic crystallography and x-ray diffraction applications. *Int J Adv Manuf Technol* **105**, 3289–3302 (2019).

4.   Marion, D. An Introduction to Biological NMR Spectroscopy. *Molecular & Cellular Proteomics* **12**, 3006–3025 (2013).

5.   Frank, J. *Three-Dimensional Electron Microscopy of Macromolecular Assemblies: Visualization of Biological Molecules in Their Native State. Three-Dimensional Electron Microscopy of Macromolecular Assemblies* (Oxford University Press).

6.   Shi, D., Nannenga, B. L., Iadanza, M. G. & Gonen, T. Three-dimensional electron crystallography of protein microcrystals. *eLife* **2**, e01345 (2013).

7.   Nannenga, B. L. & Gonen, T. The cryo-EM method microcrystal electron diffraction (MicroED). *Nat Methods* **16**, 369–379 (2019).

8.   Henderson, R. & Unwin, P. N. T. Three-dimensional model of purple membrane obtained by electron microscopy. *Nature* **257**, 28–32 (1975).

9.   Gan, L. & Jensen, G. J. Electron tomography of cells. *Quarterly Reviews of Biophysics* **45**, 27–56 (2012).

10.   Marx, V. Calling cell biologists to try cryo-ET. *Nature Methods* **15**, 575–578 (2018).

11.   Cheng, Y., Grigorieff, N., Penczek, P. A. & Walz, T. A Primer to Single-Particle Cryo-Electron Microscopy. *Cell* **161**, 438–449 (2015).

12.   Orlova, E. V. & Saibil, H. R. Structural Analysis of Macromolecular Assemblies by Electron Microscopy. *Chem. Rev.* **111**, 7710–7748 (2011) https://pubs.acs.org/doi/10.1021/cr100353t.

13.   Khoshouei, M., Radjainia, M., Baumeister, W. & Danev, R. Cryo-EM structure of haemoglobin at 3.2 Å determined with the Volta phase plate. *Nature Communications* **8**, 16099 (2017).

14.   Kappel, K. *et al.* Accelerated cryo-EM-guided determination of three-dimensional RNA-only structures. *Nature Methods* **17**, 699–707 (2020).

15.   Huang, G. *et al.* Structure of the cytoplasmic ring of the Xenopus laevis nuclear pore complex by cryo-electron microscopy single particle analysis. *Cell Research* **30**, 520–531 (2020).

16.   Adrian, M., Dubochet, J., Lepault, J. & McDowall, A. W. Cryo-electron microscopy of viruses. 5 (1984).

17.   Ma, M. *et al.* Structure of the Decorated Ciliary Doublet Microtubule. *Cell* **179**, 909-922.e12 (2019).

18.   Nakane, T. *et al.* Single-particle cryo-EM at atomic resolution. *bioRxiv* 2020.05.22.110189 (2020) doi:10.1101/2020.05.22.110189.

19.   Passmore, L. A. & Russo, C. J. Specimen Preparation for High-Resolution Cryo-EM. in *Methods in Enzymology* vol. 579 51–86 (Elsevier, 2016).

20.   Kastner, B. *et al.* GraFix: sample preparation for single-particle electron cryomicroscopy. *Nature Methods* **5**, 53–55 (2008).

21.   Adamus, K., Le, S. N., Elmlund, H., Boudes, M. & Elmlund, D. AgarFix: Simple and accessible stabilization of challenging single-particle cryo-EM specimens through crosslinking in a matrix of agar. *Journal of Structural Biology* **207**, 327–331 (2019).

22.   Shukla, A. K. *et al.* Visualization of arrestin recruitment by a G-protein-coupled receptor. *Nature* **512**, 218–222 (2014).

23.   Brenner, S. & Horne, R. W. A negative staining method for high resolution electron microscopy of viruses. *Biochimica et Biophysica Acta* **34**, 103–110 (1959).

24.   Unwin, P. N. T. Electron microscopy of the stacked disk aggregate of tobacco mosaic virus protein: II. The influence of electron irradiation on the stain distribution. *Journal of Molecular Biology* **87**, 657–670 (1974).

25.   Ohi, M., Li, Y., Cheng, Y. & Walz, T. Negative staining and image classification — powerful tools in modern electron microscopy. *Biol. Proced. Online* **6**, 23–34 (2004).

26.   Gallagher, J. R., Kim, A. J., Gulati, N. M. & Harris, A. K. Negative-Stain Transmission Electron Microscopy of Molecular Complexes for Image Analysis by 2D Class Averaging. *Current Protocols in Microbiology* **54**, e90 (2019).

27.   Dubochet, J. *et al.* Cryo-electron microscopy of vitrified specimens. *Quart. Rev. Biophys.* **21**, 129–228 (1988).

28.   Anderson, M. Vitrobot for Life Sciences. https://www.fei.com/products/vitrobot/#gsc.tab=0 (2018).

29.   Frederik, P. M. & Hubert, D. H. W. Cryoelectron Microscopy of Liposomes. in *Methods in Enzymology* vol. 391 431–448 (Academic Press, 2005).

30.   Noble, A. J. *et al.* Routine single particle CryoEM sample and grid characterization by tomography. *eLife* **7**, e34257 (2018).

31.   Glaeser, R. M. & Han, B.-G. Opinion: hazards faced by macromolecules when confined to thin aqueous films. *Biophys Rep* **3**, 1–7 (2017).

32.   Acosta-Reyes, F., Neupane, R., Frank, J. & Fernández, I. S. The Israeli acute paralysis virus IRES captures host ribosomes by mimicking a ribosomal state with hybrid tRNAs. *EMBO J* **38**, (2019).

33.   Pantelic, R. S. *et al.* Graphene: Substrate preparation and introduction. *Journal of Structural Biology* **174**, 234–238 (2011).

34.   Naydenova, K., Peet, M. J. & Russo, C. J. Multifunctional graphene supports for electron cryomicroscopy. *Proc Natl Acad Sci USA* 201904766 (2019) doi:10.1073/pnas.1904766116.

35.   Palovcak, E. *et al. A simple and robust procedure for preparing graphene-oxide cryo-EM grids.* http://biorxiv.org/lookup/doi/10.1101/290197 (2018) doi:10.1101/290197.

36.   Han, B.-G., Watson, Z., Cate, J. H. D. & Glaeser, R. M. Monolayer-crystal streptavidin support films provide an internal standard of cryo-EM image quality. *Journal of Structural Biology* **200**, 307–313 (2017).

37.   Yu, G., Li, K. & Jiang, W. Antibody-based affinity cryo-EM grid. *Methods* **100**, 16–24 (2016).

38.   Gewering, T., Januliene, D., Ries, A. B. & Moeller, A. Know your detergents: A case study on detergent background in negative stain electron microscopy. *Journal of Structural Biology* **203**, 242–246 (2018).

39.   Chen, J., Noble, A. J., Kang, J. Y. & Darst, S. A. Eliminating effects of particle adsorption to the air/water interface in single-particle cryo-electron microscopy: Bacterial RNA polymerase and CHAPSO. *Journal of Structural Biology: X* **1**, 100005 (2019).

40.   Tan, Y. Z. *et al.* Addressing preferred specimen orientation in single-particle cryo-EM through tilting. *Nature Methods* **14**, 793–796 (2017).

41.   Su, M. goCTF: Geometrically optimized CTF determination for single-particle cryo-EM. *Journal of Structural Biology* **205**, 22–29 (2019).

42.   Russo, C. J. & Passmore, L. A. Ultrastable gold substrates for electron cryomicroscopy. *Science* **346**, 1377–1380 (2014).

43.   Russo, C. J. & Passmore, L. A. Ultrastable gold substrates: Properties of a support for high-resolution electron cryomicroscopy of biological specimens. *Journal of Structural Biology* **193**, 33–44 (2016).

44.   Dobro, M. J., Melanson, L. A., Jensen, G. J. & McDowall, A. W. Chapter Three - Plunge Freezing for Electron Cryomicroscopy. in *Methods in Enzymology* (ed. Jensen, G. J.) vol. 481 63–82 (Academic Press, 2010).

45.   Ravelli, R. B. G. *et al.* Cryo-EM structures from sub-nl volumes using pin-printing and jet vitrification. *Nature Communications* **11**, 2563 (2020).

46.   Jain, T., Sheehan, P., Crum, J., Carragher, B. & Potter, C. S. Spotiton: A prototype for an integrated inkjet dispense and vitrification system for cryo-TEM. *Journal of Structural Biology* **179**, 68–75 (2012).

47.   Rubinstein, J. L. *et al.* Shake-it-off: a simple ultrasonic cryo-EM specimen-preparation device. *Acta Cryst D* **75**, 1063–1070 (2019).

48.   Anderson, M. EPU - Automated Single Particles Acquisition Software for Life Sciences. https://www.fei.com/software/epu-automated-single-particles-software-for-life-sciences/#gsc.tab=0 (2018).

49.   Carragher, B. *et al.* Leginon: An Automated System for Acquisition of Images from Vitreous Ice Specimens. *Journal of Structural Biology* **132**, 33–45 (2000).

50.   Mastronarde, D. N. Automated electron microscope tomography using robust prediction of specimen movements. *Journal of Structural Biology* **152**, 36–51 (2005).

51.   Lander, G. C. *et al.* Appion: An integrated, database-driven pipeline to facilitate EM image processing. *Journal of Structural Biology* **166**, 95–102 (2009).

52.   Thompson, R. F., Iadanza, M. G., Hesketh, E. L., Rawson, S. & Ranson, N. A. Collection, pre-processing and on-the-fly analysis of data for high-resolution, single-particle cryo-electron microscopy. *Nat Protoc* **14**, 100–118 (2019).

53.   Cao, K. *et al.* Imaging an unsupported metal–metal bond in dirhenium molecules at the atomic scale. *Science Advances* **6**, eaay5849 (2020).

54.   Sirohi, D. *et al.* The 3.8 Å resolution cryo-EM structure of Zika virus. *Science* **352**, 467–470 (2016).

55.   Chesler, A. T. & Szczot, M. Portraits of a pressure sensor. *eLife* **7**, e34396 (2018).

56.   Crewe, A. V., Eggenberger, D. N., Wall, J. & Welter, L. M. Electron Gun Using a Field Emission Source. *Review of Scientific Instruments* **39**, 576–583 (1968).

57.   Hewat, E. A. & Neumann, E. Characterization of the performance of a 200-kV field emission gun for cryo-electron microscopy of biological molecules. *Journal of Structural Biology* **139**, 60–64 (2002).

58.   Veesler, D. *et al.* Maximizing the potential of electron cryomicroscopy data collected using direct detectors. *Journal of Structural Biology* **184**, 193–202 (2013).

59.   Hamaguchi, T. *et al.* A new cryo-EM system for single particle analysis. *Journal of Structural Biology* **207**, 40–48 (2019).

60.   Angert, I., Burmester, C., Dinges, C., Rose, H. & Schröder, R. R. Elastic and inelastic scattering cross-sections of amorphous layers of carbon and vitrified ice. *Ultramicroscopy* **63**, 181–192 (1996).

61.   Yonekura, K., Braunfeld, M. B., Maki-Yonekura, S. & Agard, D. A. Electron energy filtering significantly improves amplitude contrast of frozen-hydrated protein at 300kV. *Journal of Structural Biology* **156**, 524–536 (2006).

62.   Tanaka *et al.* A new 200 kV Ω-filter electron microscope. *Journal of Microscopy* **194**, 219–227 (1999).

63.   Gubbens, A. *et al.* The GIF Quantum, a next generation post-column imaging energy filter. *Ultramicroscopy* **110**, 962–970 (2010).

64.   Danev, R. & Baumeister, W. Expanding the boundaries of cryo-EM with phase plates. *Current Opinion in Structural Biology* **46**, 87–94 (2017).

65.   Zernike, F. Phase contrast, a new method for the microscopic observation of transparent objects. *Physica* **9**, 686–698 (1942).

66.   Danev, R., Buijsse, B., Khoshouei, M., Plitzko, J. M. & Baumeister, W. Volta potential phase plate for in-focus phase contrast transmission electron microscopy. *PNAS* **111**, 15635–15640 (2014).

67.   Danev, R., Tegunov, D. & Baumeister, W. Using the Volta phase plate with defocus for cryo-EM single particle analysis. *eLife* **6**, e23006 (2017).

68.   Fan, X. *et al.* Near-Atomic Resolution Structure Determination in Over-Focus with Volta Phase Plate by Cs-Corrected Cryo-EM. *Structure* **25**, 1623-1630.e3 (2017).

69.   Cheng, A. *et al.* Lessons Learned from using a Cs-Corrected, Energy-Filtered, Phase-Plate TEM for Single-Particle CryoEM. *Microscopy and Microanalysis* **23**, 824–825 (2017).

70.   Hamilton, J. F. & Marchant, J. C. Image Recording in Electron Microscopy. *J. Opt. Soc. Am.* **57**, 232 (1967).

71.   Downing, K. H. & Mooney, P. E. A charge coupled device camera with electron decelerator for intermediate voltage electron microscopy. *Review of Scientific Instruments* **79**, 043702 (2008).

72.   McMullan, G., Faruqi, A. R. & Henderson, R. Direct Electron Detectors. in *Methods in Enzymology* vol. 579 1–17 (Elsevier, 2016).

73.   Faruqi, A. R. Principles and prospects of direct high resolution electron image acquisition with CMOS detectors at low energies. *J. Phys.: Condens. Matter* **21**, 314004 (2009).

74.   McMullan, G., Faruqi, A. R., Clare, D. & Henderson, R. Comparison of optimal performance at 300keV of three direct electron detectors for use in low dose electron microscopy. *Ultramicroscopy* **147**, 156–163 (2014).

75.   McMullan, G., Chen, S., Henderson, R. & Faruqi, A. R. Detective quantum efficiency of electron area detectors in electron microscopy. *Ultramicroscopy* **109**, 1126–1143 (2009).

76.   Campbell, M. G. *et al.* Movies of ice-embedded particles enhance resolution in electron cryo-microscopy. *Structure* **20**, 1823–1828 (2012).

77.   Brilot, A. F. *et al.* Beam-induced motion of vitrified specimen on holey carbon film. *Journal of Structural Biology* **177**, 630–637 (2012).

78.   Li, X., Zheng, S. Q., Egami, K., Agard, D. A. & Cheng, Y. Influence of electron dose rate on electron counting images recorded with the K2 camera. *J. Struct. Biol.* **184**, 251–260 (2013).

79.   K3 Camera | Gatan, Inc. https://www.gatan.com/k3-camera.

80.   Sanchez-Garcia, R., Segura, J., Maluenda, D., Sorzano, C. O. S. & Carazo, J. M. MicrographCleaner: A python package for cryo-EM micrograph cleaning using deep learning. *Journal of Structural Biology* **210**, 107498 (2020).

81.   Wagner, T. MPI-Dortmund/sphire-janni. *GitHub* https://github.com/MPI-Dortmund/sphire-janni.

82.   Scheres, S. H. Beam-induced motion correction for sub-megadalton cryo-EM particles. *eLife* **3**, e03665 (2014).

83.   Li, X. *et al.* Electron counting and beam-induced motion correction enable near-atomic-resolution single-particle cryo-EM. *Nat Methods* **10**, 584–590 (2013).

84.   Grant, T. & Grigorieff, N. Measuring the optimal exposure for single particle cryo-EM using a 2.6 Å reconstruction of rotavirus VP6. *eLife* **4**,.

85.   Zheng, S. Q. *et al.* MotionCor2: anisotropic correction of beam-induced motion for improved cryo-electron microscopy. *Nature Methods* **14**, 331–332 (2017).

86.   Tegunov, D. & Cramer, P. Real-time cryo-electron microscopy data preprocessing with Warp. *Nature Methods* **16**, 1146–1152 (2019).

87.   Scheres, S. H. W. RELION: Implementation of a Bayesian approach to cryo-EM structure determination. *Journal of Structural Biology* **180**, 519–530 (2012).

88.   Zivanov, J., Nakane, T. & Scheres, S. H. W. A Bayesian approach to beam-induced motion correction in cryo-EM single-particle analysis. *IUCrJ* **6**, 5–17 (2019).

89.   Scheres, S. H. W. Processing of Structurally Heterogeneous Cryo-EM Data in RELION. in *Methods in Enzymology* vol. 579 125–157 (Elsevier, 2016).

90.   Wade, R. H. A brief look at imaging and contrast transfer. *Ultramicroscopy* **46**, 145–156 (1992).

91.   Zhang, K. Gctf: Real-time CTF determination and correction. *Journal of Structural Biology* **193**, 1–12 (2016).

92.   Rohou, A. & Grigorieff, N. CTFFIND4: Fast and accurate defocus estimation from electron micrographs. *Journal of Structural Biology* **192**, 216–221 (2015).

93.   Huang, Z. & Penczek, P. A. Application of template matching technique to particle detection in electron micrographs. *Journal of Structural Biology* **145**, 29–40 (2004).

94.   Henderson, R. Avoiding the pitfalls of single particle cryo-electron microscopy: Einstein from noise. *Proceedings of the National Academy of Sciences* **110**, 18037–18041 (2013).

95.   Mao, Y. *et al.* Molecular architecture of the uncleaved HIV-1 envelope glycoprotein trimer. *Proceedings of the National Academy of Sciences* **110**, 12438–12443 (2013).

96.   van Heel, M. Finding trimeric HIV-1 envelope glycoproteins in random noise. *Proceedings of the National Academy of Sciences* **110**, E4175–E4177 (2013).

97.   Scheres, S. H. W. Semi-automated selection of cryo-EM particles in RELION-1.3. *Journal of Structural Biology* **189**, 114–122 (2015).

98.   Voss, N. R., Yoshioka, C. K., Radermacher, M., Potter, C. S. & Carragher, B. DoG Picker and TiltPicker: software tools to facilitate particle selection in single particle electron microscopy. *J Struct Biol* **166**, 205–213 (2009).

99.   Valueva, M. V., Nagornov, N. N., Lyakhov, P. A., Valuev, G. V. & Chervyakov, N. I. Application of the residue number system to reduce hardware costs of the convolutional neural network implementation. *Mathematics and Computers in Simulation* **177**, 232–243 (2020).

100.   Aloysius, N. & Geetha, M. A review on deep convolutional neural networks. in *2017 International Conference on Communication and Signal Processing (ICCSP)* 0588–0592 (2017). doi:10.1109/ICCSP.2017.8286426.

101.   Redmon, J., Divvala, S., Girshick, R. & Farhadi, A. You Only Look Once: Unified, Real-Time Object Detection. in 779–788 (2016).

102.   Shafiee, M. J., Chywl, B., Li, F. & Wong, A. Fast YOLO: A Fast You Only Look Once System for Real-time Embedded Object Detection in Video. *arXiv:1709.05943 [cs]* (2017).

103.   Wagner, T. *et al.* SPHIRE-crYOLO is a fast and accurate fully automated particle picker for cryo-EM. *Commun Biol* **2**, 218 (2019).

104.   Bepler, T. *et al.* Positive-unlabeled convolutional neural networks for particle picking in cryo-electron micrographs. *Nat Methods* **16**, 1153–1160 (2019).

105.   The STAR file: a new format for electronic data transfer and archiving | Journal of Chemical Information and Modeling. https://pubs.acs.org/doi/abs/10.1021/ci00002a020.

106.   Grant, T., Rohou, A. & Grigorieff, N. cisTEM, user-friendly software for single-particle image processing. *eLife* **7**,.

107.   Zivanov, J. *et al.* New tools for automated high-resolution cryo-EM structure determination in RELION-3. *eLife* **7**,.

108.   Radermacher, M., Wagenknecht, T., Verschoor, A. & Frank, J. Three-dimensional reconstruction from a single-exposure, random conical tilt series applied to the 50S ribosomal subunit of Escherichia coli. *Journal of Microscopy* **146**, 113–136 (1987).

109.   Boisset, N. *et al.* Overabundant single-particle electron microscope views induce a three-dimensional reconstruction artifact. *Ultramicroscopy* **74**, 201–207 (1998).

110.   Leschziner, A. E. & Nogales, E. The orthogonal tilt reconstruction method: An approach to generating single-class volumes with no missing cone for ab initio reconstruction of asymmetric particles. *Journal of Structural Biology* **153**, 284–299 (2006).

111.   Scheres, S. H. W. A Bayesian View on Cryo-EM Structure Determination. *Journal of Molecular Biology* **415**, 406–418 (2012).

112.   Wong, W. *et al.* Cryo-EM structure of the Plasmodium falciparum 80S ribosome bound to the anti-protozoan drug emetine. *eLife* **3**, e03080 (2014).

113.   Nguyen, T. H. D. *et al.* The architecture of the spliceosomal U4/U6.U5 tri-snRNP. *Nature* **523**, 47–52 (2015).

114.   Nakane, T., Kimanius, D., Lindahl, E. & Scheres, S. H. Characterisation of molecular motions in cryo-EM single-particle data by multi-body refinement in RELION. *eLife* **7**, e36861 (2018).

115.   Penczek, P. A. Resolution Measures in Molecular Electron Microscopy. in *Methods in Enzymology* vol. 482 73–100 (Elsevier, 2010).

116.   Rosenthal, P. B. & Henderson, R. Optimal Determination of Particle Orientation, Absolute Hand, and Contrast Loss in Single-particle Electron Cryomicroscopy. *Journal of Molecular Biology* **333**, 721–745 (2003).

117.   Van Heel, M. Similarity measures between images. *Ultramicroscopy* **21**, 95–100 (1987).

118.   Scheres, S. H. W. & Chen, S. Prevention of overfitting in cryo-EM structure determination. *Nature Methods* **9**, 853–854 (2012).

119.   Penczek, P. A. Reliable cryo-EM resolution estimation with modified Fourier shell correlation. *IUCrJ* **7**, (2020).

120.   Beckers, M. & Sachse, C. Permutation testing of Fourier shell correlation for resolution estimation of cryo-EM maps. *Journal of Structural Biology* **212**, 107579 (2020).

121.   Neumann, P., Dickmanns, A. & Ficner, R. Validating Resolution Revolution. *Structure* **26**, 785-795.e4 (2018).

122.   Beckers, M. & Sachse, C. Adaptive FDR thresholding of Fourier shell correlation for resolution estimation of cryo-EM maps. *bioRxiv* 2020.03.13.990705 (2020) doi:10.1101/2020.03.13.990705.

123.   van Heel, M. & Schatz, M. Information: to Harvest, to Have and to Hold. (2020).

124.   Wlodawer, A. & Dauter, Z. `Atomic resolution': a badly abused term in structural biology. *Acta Crystallogr D Struct Biol* **73**, 379–380 (2017).

125.   Chiu, W. *et al.* Responses to `Atomic resolution': a badly abused term in structural biology. *Acta Cryst D* **73**, 381–383 (2017).

126.   Kucukelbir, A., Sigworth, F. J. & Tagare, H. D. Quantifying the local resolution of cryo-EM density maps. *Nature Methods* **11**, 63–65 (2014).

127.   Heymann, J. B. Guidelines for using Bsoft for high resolution reconstruction and validation of biomolecular structures from electron micrographs. *Protein Science* **27**, 159–171 (2018).

128.   Vilas, J. L. *et al.* MonoRes: Automatic and Accurate Estimation of Local Resolution for Electron Microscopy Maps. *Structure (London, England: 1993)* **26**, 337-344.e4 (2018).

129.   Ramírez-Aportela, E., Mota, J., Conesa, P., Carazo, J. M. & Sorzano, C. O. S. DeepRes: a new deep-learning- and aspect-based local resolution method for electron-microscopy maps. *IUCrJ* **6**, 1054–1063 (2019).

130.   Emsley, P. & Cowtan, K. Coot: model-building tools for molecular graphics. *Acta Cryst D* **60**, 2126–2132 (2004).

131.   DiMaio, F., Leaver-Fay, A., Bradley, P., Baker, D. & André, I. Modeling Symmetric Macromolecular Structures in Rosetta3. *PLOS OPNE* **6**, e20450 (2011).

132. Adams, P. D. *et al.* PHENIX: a comprehensive Python-based system for macromolecular structure solution. *Acta Cryst D* **66**, 213–221 (2010).

133. Chen, V. B. *et al.* MolProbity: all-atom structure validation for macromolecular crystallography. *Acta Cryst D* **66**, 12–21 (2010).

134. De Rosier, D. J. & Klug, A. Reconstruction of Three Dimensional Structures from Electron Micrographs. *Nature* **217**, 130–134 (1968).

135. Brown, A. *et al.* Tools for macromolecular model building and refinement into electron cryo-microscopy reconstructions. *Acta Crystallogr D Biol Crystallogr* **71**, 136–153 (2015).

136. Henderson, R. *et al.* Outcome of the First Electron Microscopy Validation Task Force Meeting. *Structure* **20**, 205–214 (2012).

137. Barad, B. A. *et al.* EMRinger: side chain–directed model and map validation for 3D cryo-electron microscopy. *Nature Methods* **12**, 943–946 (2015).

138. Saibil, H. R. Blob-ology and biology of cryo-EM: an interview with Helen Saibil. *BMC Biology* **15**, 77 (2017).

139. The Nobel Prize in Chemistry 2017. *NobelPrize.org* https://www.nobelprize.org/prizes/chemistry/2017/press-release/.

140. Falcon 4 Electron Detector | Life Sciences EM - US. //www.thermofisher.com/us/en/home/electron-microscopy/products/accessories-em/falcon-4-detector.html.

141. Naydenova, K., Jia, P. & Russo, C. J. Cryo-EM with sub–1 Å specimen movement. *Science* **370**, 223–226 (2020).

142. Punjani, A., Rubinstein, J. L., Fleet, D. J. & Brubaker, M. A. cryoSPARC: algorithms for rapid unsupervised cryo-EM structure determination. *Nature Methods* **14**, 290–296 (2017).

143. Terwilliger, T. C., Ludtke, S. J., Read, R. J., Adams, P. D. & Afonine, P. V. Improvement of cryo-EM maps by density modification. *Nature Methods* **17**, 923–927 (2020).

144. cryoSPARC Live. https://cryosparc.com/live.

145. New Krios Solution: The Highest Productivity in the Most Compact Design. *MediaRoom* https://thermofisher.mediaroom.com/2019-08-05-New-Krios-Solution-The-Highest-Productivity-in-the-Most-Compact-Design.

146. Thermo Fisher Scientific Installs First Glacios Cryo-Electron Microscope for Drug Discovery and Development. *MediaRoom* https://thermofisher.mediaroom.com/2018-05-16-Thermo-Fisher-Scientific-Installs-First-Glacios-Cryo-Electron-Microscope-for-Drug-Discovery-and-Development.

147. JEOL USA CRYO ARM™ 300 TEM. https://www.jeolusa.com/PRODUCTS/Transmission-Electron-Microscopes-TEM/300-kV/CRYO-ARM-300#4372171-resources.

148.	Danev, R., Yanagisawa, H. & Kikkawa, M. Cryo-Electron Microscopy Methodology: Current Aspects and Future Directions. *Trends in Biochemical Sciences* **44**, 837–848 (2019).

149.	Wu, M., Lander, G. C. & Herzik, M. A. Sub-2 Angstrom resolution structure determination using single-particle cryo-EM at 200 keV. *Journal of Structural Biology: X* **4**, 100020 (2020).

150.	Yip, K. M., Fischer, N., Paknia, E., Chari, A. & Stark, H. Breaking the next Cryo-EM resolution barrier – Atomic resolution determination of proteins! *bioRxiv* 2020.05.21.106740 (2020) doi:10.1101/2020.05.21.106740.

151.	Callaway, E. 'It opens up a whole new universe': Revolutionary microscopy technique sees individual atoms for first time. *Nature* **582**, 156–157 (2020).

152.	Kato, T. *et al.* CryoTEM with a Cold Field Emission Gun That Moves Structural Biology into a New Stage. *Microscopy and Microanalysis* **25**, 998–999 (2019).

153.	Li, Y., Cash, J. N., Tesmer, J. J. G. & Cianfrocco, M. A. High-Throughput Cryo-EM Enabled by User-Free Preprocessing Routines. *Structure* **28**, 858-869.e3 (2020).

154.	H, E., Jan. 23, 2020 & Am, 8:00. 'We need a people's cryo-EM.' Scientists hope to bring revolutionary microscope to the masses. *Science | AAAS* https://www.sciencemag.org/news/2020/01/we-need-people-s-cryo-em-scientists-hope-bring-revolutionary-microscope-masses (2020).

155.	Thermo Fisher Scientific Signs Agreement to Acquire Gatan from Roper Technologies. https://ir.thermofisher.com/investors/news-and-events/news-releases/news-release-details/2018/Thermo-Fisher-Scientific-Signs-Agreement-to-Acquire-Gatan-from-Roper-Technologies/default.aspx.

156.	High-End Cryo-EMs Worldwide. *Google My Maps* https://www.google.com/maps/d/viewer?mid=1eQ1r8BiDYfaK7D1S9EeFJEgkLggMyoaT.

157.	Zhang, D. K. Brief Manual of Gautomatch. 10.

158.	Vinothkumar, K. R. & Henderson, R. Single particle electron cryomicroscopy: trends, issues and future perspective. *Quarterly Reviews of Biophysics* **49**, (2016).

159.	Naydenova, K. *et al.* CryoEM at 100 keV: a demonstration and prospects. *IUCrJ* **6**, 1086–1098 (2019).

160.	Peet, M. J., Henderson, R. & Russo, C. J. The energy dependence of contrast and damage in electron cryomicroscopy of biological molecules. *Ultramicroscopy* **203**, 125–131 (2019).

161.	NIH funds three national cryo-EM service centers and training for new microscopists. *National Institutes of Health (NIH)* https://www.nih.gov/news-events/news-releases/nih-funds-three-national-cryo-em-service-centers-training-new-microscopists (2018).

162.	Cryo-EM - - Diamond Light Source. https://www.diamond.ac.uk/Instruments/Techniques/Microscopy/Cryo-EM.html.

163.	CM01 Cryo-electron microscope.
https://www.esrf.eu/home/UsersAndScience/Experiments/MX/About_our_beamlines/CM01.htm
l.

164.	Home - NeCEN | Netherlands Centre for Electron Nanoscopy. https://www.necen.nl/.

165.	LNNano - Brazilian Nanotechnology National Laboratory. *LNNano*
https://lnnano.cnpem.br/.

166.	Subramaniam, S. The cryo-EM revolution: fueling the next phase. *IUCrJ* **6**, 1–2 (2019).

167.	Banerjee, S. *et al.* 2.3 Å resolution cryo-EM structure of human p97 and mechanism of allosteric inhibition. *Science* **351**, 871–875 (2016).

168.	Merk, A. *et al.* Breaking Cryo-EM Resolution Barriers to Facilitate Drug Discovery. *Cell* **165**, 1698–1707 (2016).

169.	Gao, Y., Cao, E., Julius, D. & Cheng, Y. TRPV1 structures in nanodiscs reveal mechanisms of ligand and lipid action. *Nature* **534**, 347–351 (2016).

170.	Doerr, A. Single-particle cryo-electron microscopy. *Nature Methods* **13**, 23–23 (2016).

171.	Drulyte, I. *et al.* Approaches to altering particle distributions in cryo-electron microscopy sample preparation. *Acta Crystallogr D Struct Biol* **74**, 560–571 (2018).

172.	Vossman. *w:Electron microscope image of w:GroEL particle in w:vitreous ice.* (2008).

173.	Contrast transfer function (CTF) correction.
https://spider.wadsworth.org/spider_doc/spider/docs/techs/ctf/ctf.html.

174.	Koning, R. I., Koster, A. J. & Sharp, T. H. Advances in cryo-electron tomography for biology and medicine. *Annals of Anatomy - Anatomischer Anzeiger* **217**, 82–96 (2018).

175.	Milne, J. L. S. *et al.* Cryo-electron microscopy - a primer for the non-microscopist. *FEBS J* **280**, 28–45 (2013).

176.	Nogales, E. & Scheres, S. H. W. Cryo-EM: A Unique Tool for the Visualization of Macromolecular Complexity. *Molecular Cell* **58**, 677–689 (2015).

177.	Bai, X., Rajendra, E., Yang, G., Shi, Y. & Scheres, S. H. Sampling the conformational space of the catalytic subunit of human γ-secretase. *eLife*
https://elifesciences.org/articles/11182/figures (2015) doi:10.7554/eLife.11182.

Chapter 3:

The Israeli acute paralysis virus internal ribosome entry site captures host ribosomes by mimicking a ribosomal state with hybrid tRNAs[*][†]

---

[*] A version of this chapter was published as: Acosta-Reyes, F. [#], Neupane, R. [#], Frank, J. & Fernández, I. S. The Israeli acute paralysis virus IRES captures host ribosomes by mimicking a ribosomal state with hybrid tRNAs. *EMBO J* **38**, (2019).

[#] I and F.A-R. contributed equally as co-first authors.

[†] The final published version of this chapter is included in Appendix A.

3.1   Introduction

Apis mellifera, the common western honey bee, is affected worldwide by an enigmatic

syndrome characterized by a drastic disappearance of the workforce, causing the accelerated

collapse of the hive[1]. Given the essential role bees play in pollination of economically important

crops, the impact of this syndrome, termed colony collapse disorder (CCD), has been estimated

to cost the US economy $15 billion in direct loss of crops and $75 billion in indirect losses[2].

Though the exact etiology of CCD is unknown[3], a group of viruses belonging to the

Dicistroviridae family were found in metagenomic studies of CCD-affected hives[4,5]. Among this

group of viruses, the Israeli acute paralysis virus (IAPV) showed a strong correlation with CCD,

revealing a prominent role in the development of the syndrome[5,6].

The Dicistroviridae family of viruses exhibits a wide environmental distribution,

targeting invertebrates, mainly insects and other arthropods[7]. The genetic architecture of these

viruses is composed of a single positive-stranded RNA molecule which contains two open

reading frames (ORF1 and ORF2)[8,9]. ORF1 encodes non-structural proteins: an RNA helicase, a

cysteine protease, and an RNA-dependent RNA polymerase (RdRP). ORF2 encodes a single

poly-protein that, upon proteolytic digestion, generates the structural proteins that will eventually

compose the viral capsid[10,11].

Both ORFs are preceded by non-coding RNA sequences responsible for the regulation of

the expression of their downstream genes[12,13]. A fine balance between the expression of ORF1

and ORF2 is required for the replication and expansion of the virus[14,15]. This is achieved through

a precise exploitation of host resources, especially the machinery for protein synthesis[10]. The

non-coding RNA regions preceding both ORFs harbor two different internal ribosomal entry

sites (IRESs)[8]. IRESs are structured RNA sequences able to interfere with canonical translation, capturing host ribosomes in order to redirect them toward the production of viral proteins[16]. Eukaryotic ribosomes are operated by a complex collection of cellular factors that regulate the production of proteins in the cell[17]. Specially regulated in eukaryotes is the first step of translation, initiation[18]. During this initiation phase, the small ribosomal subunit (40S), in partnership with many initiation factors, is able to capture an mRNA, localize its AUG initiation codon, deliver the first aminoacyl-tRNA, and finally recruit the large subunit (60S) in order to assemble an elongation competent ribosome (80S) primed with an aminoacyl-tRNA in the P-site and a vacant A-site[19].

The majority of IRES families leverage the complexity of initiation to hijack cellular ribosomes[16,20]. The IAPV-IRES found in the intergenic region of the IAPV virus belongs to the well-characterized type IV family of viral IRESs. IRES sequences from this group dispense with all canonical initiation factors and are able to assemble by themselves an elongation competent ribosome, successfully redirecting the cellular machinery for viral protein production by an RNA-only mechanism[21]. This is accomplished by an elaborate use of intrinsically dynamic elements of the ribosome, naturally involved in translocation[22]. These IRESs are able to induce an artificial state on the ribosome, mimicking a pre-translocation state with tRNAs. Elongation factors eEF2 and eEF1A can then be recruited to effectively by-pass the highly regulated initiation[23–25,25], jumpstarting directly in the elongation phase[26].

The type IV IRES family exhibits a remarkable structural diversity, which remains poorly characterized[21]. Two genera, based on phylogenetic analysis of ORF2 as well as the intergenic region, have been defined: Aparaviruses and Cripaviruses. The cricket paralysis virus IRES (CrPV-IRES), the prototypical Cripavirus, has been extensively studied due to its early discovery and use as model mRNA of early studies in translation[27,28]. Recently, a divergent IRES sequence of a shrimp-infecting virus, the Taura virus syndrome IRES, has been visualized by cryo-EM in complex with yeast ribosomes[23,29].

The IAPV-IRES presents the prototypical features of an Aparavirus, with an additional stem loop (SL-III) nested within the pseudoknot I (PKI) and an extended L1.1 region[30]. Importantly, this IRES can drive translation in two different ORFs, able to produce two different polypeptides from the same mRNA. A frameshift event at the first coding codon is responsible for this multi-coding capacity[31].

Research efforts directed toward finding the cause of CCD and developing strategies to prevent it are underway[5,32]. RNA interference has proved effective in protecting against CCD. Directing double-stranded RNAs complementary to the IRES of the intergenic region of the IAPV virus decreases the probability of hive collapse, preventing effectively the massive death of the workforce, guaranteeing the protection of the queen and thus the survival of the colony[33].

Using single-particle cryo-electron microscopy (cryo-EM), we have characterized how the IAPV-IRES redirects the host machinery for viral protein synthesis exploiting novel ribosomal sites. An early commitment of IRES/ribosome complexes toward global pre-translocation mimicry explains the high efficiency in ribosome hijacking observed for this IRES. These results may inspire structure-based rational designs for the fight against CCD by RNA-interference technology[5].

3.2   Materials and Methods

3.2.1   Plasmids

Expression vector for His-tagged eRF1*(AGQ mutant) has been previously described[34]. A pUC19-based transcription vector for IAPV-IRES-WT was constructed by inserting a T7 promoter sequence upstream of IAPV IGR IRES sequence followed by the first two coding triplets (nucleotides 6,372–6,623 from NC009025). An EcoRI site was included after the second codon. Site-directed mutagenesis was employed to change the first coding codon (GGC) to a stop codon (TAG) to create the IAPV-IRES-STOP construct. Both IRES constructs were transcribed using T7 RNA polymerase. Briefly, 0.1 mg/ml of EcoRI-linearized vector was transcribed using 0.1 mg/ml of homemade T7 RNA polymerase in 10 ml transcription buffer (100 mM HEPES-KOH pH 7.4, 10 mM of each NTP, 22 mM MgCl2, 50 mM DTT, 2 mM Spermidine, 1 μl/ml IPP) for 4 h at 37°C. The RNA was then washed, concentrated, and separated on a 6% UREA-PAGE gel. The IAPV-IRES band was cut from the gel, electro-eluted, buffer exchanged into Buffer A (20 mM Tris–HCl, pH 7.5, 100 mM KCl, 8 mM MgCl2, 2 mM DTT), and snap-frozen in liquid nitrogen.

3.2.2 Purification of translation components

Native 40S and 60S ribosomal subunits and eukaryotic elongation factor 2 (eEF2) were prepared from rabbit reticulocytes as previously described[35]. Recombinant eRF1* was purified according to a previously described protocol[36].

3.2.3 Assembly of ribosomal complexes

To reconstitute ribosomal complexes in the pre-translocation state with IAPV-IRES, we incubated 6 pmol of 40S ribosomal subunits with 60 pmol of IAPV-IRES-WT RNA in a 15 μl reaction mixture in Buffer A for 5 min at 37°C. Then, the reaction mixture was supplemented with 4.5 pmol of 60S ribosomal subunits and incubated for 5 min at 37°C. We maintained an excess of 40S ribosomal subunits to trap both the 40S/IAPV-IRES and the 80S/IAPV-IRES complexes in the same reaction. Assembly of the complexes was verified by running the reaction through an overnight sucrose gradient, phenol extracting the RNA, and running a UREA-PAGE gel using standard methods. The assembly and integrity of the complex was also verified on a F20 screening microscope via negative stain EM and cryo-EM.

To reconstitute ribosomal complexes in a post-translocated state with IAPV-IRES-STOP, we incubated 8.8 pmol of 40S ribosomal subunits with 90 pmol of IAPV-IRES-STOP RNA in a 15 µl reaction mixture containing Buffer A with 1.7 mM GTP for 5 min at 37°C. Then, the reaction mixture was supplemented with 8.7 pmol of 60S ribosomal subunits and incubated for 5 min at 37°C. Next, we added 27 pmol of eEF2 and 90 pmol eRF* and incubated the reaction for an additional 30 min at 37°C.

3.2.4   Cryo-EM sample preparation and data acquisition

For ribosomal complexes in a pre-translocated state: 3 µl aliquots of assembled ribosomal complexes at 240–390 nM concentration were incubated for 15 s either on plasma-treated holey carbon grids (QUANTIFOIL R2/2 with homemade continuous carbon film estimated to be 50 Å thick) or on plasma-treated holey gold grids [UltrAuFoil R1.2/1.3[37]]. Grids were blotted for 2.5– 3.0 s and flash-cooled in liquid ethane using an FEI Vitrobot. Grids were then transferred to a Polara-G2 microscope operated at 300 kV and equipped with a Gatan K2 Summit direct detector. 11,234 movies of 40 frames were collected in counting mode at 8e−/pix/s at a magnification of 31,000 corresponding to a calibrated pixel size of 1.25 Å. Defocus values specified in Leginon[38] ranged from 0.8 to 2.5 µm.

For ribosomal complexes in a post-translocated state: 3 µl aliquots of assembled ribosomal complexes around 600 nM were incubated for 15 s on plasma-treated holey gold grids [UltrAuFoil R1.2/1.3[37]]. Grids were blotted for 2.5 s and flash-cooled in liquid ethane using an FEI Vitrobot. Grids were then transferred to a Titan Krios microscope operated at 300 kV and equipped with an energy filter (slits aperture 20 eV) and a Gatan K2 Summit detector. 6,904 movies of 40 frames were collected in counting mode at 8e−/pix/s at a magnification of 130,000 corresponding to a calibrated pixel size of 1.0605 Å. Defocus values specified in Leginon[38] ranged from 0.5 to 3.0 μm. About half the movies (3,350) were collected at a 35-degree tilt to compensate for preferred orientation that was first identified during screening sessions.

On both microscopes, movies were recorded in automatic mode using the Leginon[38] software and frames were aligned using Motioncor2[39]. Data collection was monitored and checked on the fly using APPION[40].

3.2.5   Image processing and structure determination

For ribosomal complexes in a pre-translocated state, contrast transfer function parameters were estimated using GCTF[41]. For particle picking, a set of templates were generated using density data obtained from a screening session on a F20 microscope that had employed Gaussian picking. Using these templates, particle picking was performed using GAUTOMACH[42] and a particle diameter value of 320 Å. The picked particles were manually screened on the micrographs to remove problematic regions. All 2D and 3D classifications and refinements were performed using RELION[43]. The picked particles were binned four times and subjected to a 2D classification to separate the 40S and 80S particles. We then employed 3D Refine to generate initial consensus models from both the 40S and 80S particle sets **(Fig. 3.2)**. We then used our previous CrPV-IRES model (PDB ID 5IT9)[24] to create appropriate masks for these initial models. The mask for the 40S initial model enclosed the putative IRES binding site. The mask for the 80S initial model enclosed the intersubunit space, the A-site finger, the L1 stalk, and ideal helices for SL-III and SL-VI regions of the IAPV-IRES. Using these masks, we performed a round of 3D classification with signal subtraction to remove 3D classes without the IRES. On the classes with the IRES, we performed focused classification without alignment using a new set of masks **(Fig. 3.2)**. This focused classification step resulted in the final set of classes that were eventually used for modeling.

For ribosomal complexes in a post-translocated state, we employed a similar set of protocols but with a different set of masks that also included regions for eEF2, eRF1*, and the L1 stalk in an extended conformation **(Fig. 3.10)**.

Final refinements with unbinned data for the selected classed yielded high-resolution maps with density features in agreement with the reported resolution. Local resolution was computed with RESMAP[44]. New features implemented in Relion 3.0[45] such as contrast transfer value refinement allowed extending the resolution to close to 3 Å.

3.2.6 Model building and refinement

Models for the mammalian ribosome and eRF1* were docked into the maps using CHIMERA[46] and COOT[47] was used to manually adjust the L1 stalk and build the IAPV-IRES using our CrPV-IRES model as initial step. An initial round of refinement was performed in Phenix using real-space refinement with secondary structure restraints[48]. A final step of reciprocal-space refinement using REFMAC was performed[49] for all complexes. The fit of the model to the map density was quantified using FSCaverage and Cref.

3.3 Results

3.3.1 Biochemical set-up and cryo-EM strategy

Previous biochemical and genetic studies of IAPV-IRES established the secondary structure scheme displayed in **Fig. 3.1 A**[30]. The prototypical architecture of the type IV IRES family consisting of three nested pseudoknots is extended by a 5′ terminal stem loop (SL-VI) proposed to play functional roles in the early positioning of the IAPV-IRES in the [50,51]. Additionally, the genus Aparavirus is characterized by an extended PKI which contains an insertion of a large stem loop **(Fig. 3.1 A, bottom)**. In the IAPV-IRES, SL-III consists of eight Watson–Crick canonical base pairs and a terminal loop of six nucleotides. Notably, two unpaired adenine residues are placed in a strategic position at the core of the three-way helical junction connecting SL-III, the anti-codon stem loop (ASL)-like element of the PKI (residues 6,546–6,574), and the double-helical segment connecting PKI and PKIII. A variable loop region (VLR) bridges the mRNA-like element of PKI (residues 6,613–6,617) with the helical region connecting PKI and PKIII. This single-stranded RNA loop is poorly conserved in sequence; however, even though its role in IRES functioning remains enigmatic, biochemical experiments have proved its integrity is mandatory for productive IRES-driven translation[52].

In order to understand in structural terms how these constituent units of the IAPV-IRES are involved in ribosome hijacking, we produced a full, wild-type IAPV-IRES, including SL-VI and the first two coding codons. Binary complexes with mammalian ribosomes and IAPV-IRES were generated by incubating IRES with ribosomal subunits. We designed a reaction featuring an excess of 40S over 60S in an overall background of IRES excess, to test the ability of the IAPV-IRES to engage both 40S and full 80S ribosomes in a productive and stable binary interaction. The stability of these interactions was tested through a sucrose gradient run overnight **(Fig. 3.1 B)** where the different complexes could be resolved according to their size differences. Each peak was subjected to RNA extraction and UREA-PAGE analysis where the presence of IAPV-IRES bound in the 80S peak as well as in the 40S peak could be confirmed **(Fig. 3.1 B)**.

Leveraging latest cryo-EM maximum-likelihood classification methods implemented in RELION 3.0[43,45,53], we decided to directly image the above reaction without the sucrose gradient step. A large dataset ensuing from this experiment was subjected to an optimized classification scheme combining different in silico classification approaches **(Fig. 3.2)**. This allowed us to identify and refine to high-resolution five distinctive classes from a single dataset **(Fig. 3.1 C–E)**. The nominal resolution of the maps was calculated to be around 3 Å **(Figs. 3.12, 3.13, and 3.14)**. In the best areas, such as the 60S subunit and the body of the 40S subunit, the maps exhibit characteristics in accordance with this resolution, with very well-resolved side chains in proteins and clear base separation in the ribosomal RNA components **(Fig. 3.3 A and B)**. However, the resolution of the IRES density, due to the intrinsic flexibility of this component, deviates from the nominal resolution. Areas of the IRES stabilized by ribosomal components are well-resolved, with local resolution better than 4 Å **(Fig. 3.5 D and E)**, while areas not stabilized by the ribosome or in contact with intrinsically dynamic elements of the ribosome like the L1 stalk exhibit lower local resolution **(Figs. 3.12, 3.13, and 3.14)**. In order to properly visualize the continuity of the maps for the full IRES, we show the unsharpened maps in the figures, especially where large areas of the maps are depicted. In those regions of the maps exhibiting resolution better than 4 Å for the IRES, maps sharpened with B factors reported in **Table 3.1** are shown.

3.3.2 The IAPV-IRES restricts the conformational freedom of the 40S blocking functional sites

The 40S subunit can be roughly divided into two parts: the body, which forms the bulk of the subunit accounting for two-thirds of its mass, and a more mobile part roughly comprising the remaining third, designated as the head **(Fig. 3.4 A)**. The interface between these two components forms the tRNA binding sites of the small subunit. The head of the 40S subunit is a dynamic component, modifying its relative orientation with respect to the body. This dynamics is of critical importance in two aspects of translation: the positioning of the initiator aminoacyl-tRNA and in the concerted movement of mRNA and tRNAs during elongation[54,55]. We identified

two classes of particles showing robust density for IAPV-IRES in the context of a binary

interaction with the 40S **(Fig. 3.4 B and C)**. All elements of the IAPV-IRES included in the

produced construct were identified in the maps except SL-VI, which proved disordered—no

density could be assigned to it even in low-pass filtered maps. The L1.1 region in the context of a

binary interaction with the 40S shows a high degree of mobility and can only be modeled in

maps filtered to 4 Å.

The IAPV-IRES inserts two elements of its structure between the head and the body of

the 40S subunit, effectively restricting the dynamics of the 40S head to specific ranges of

conformations. The ASL/mRNA mimicking part of the PKI is inserted in the decoding site (A-

site) of the small subunit stabilized by the decoding bases of the 18S rRNA A1824-A1825 and

G626 [A1492, A1494, and G530 in Escherichia coli[56]], inducing a decoding event. SL-IV is

deeply inserted in the interface of head and body, in the surroundings of the E-site, clamped by

stacking interactions established with residue A6498 of the IRES and tyrosine 72 from uS7 and

arginine 135 from uS11 **(Fig. 3.5 E)**. In this conformation, the IAPV-IRES fully blocks all three

tRNA binding sites of the 40S subunit, interfering with early steps of canonical initiation **(Fig**

3.4 C)[18].

Density for the full PKI, including SL-III, was clearly visible in the maps **(Figs. 3.4 D**

and E, and 3.13) which allowed accurate modeling of the three-way helical junction

characteristic of Aparavirus IRESs. The VLR was also visible in the maps, partially occupying

the P-site, stabilized by a stacking interaction between A6609 of the IRES and A1085 of the 18S

rRNA **(Fig. 3.4 E)**.

The two classes present a conformation of the IAPV-IRES nearly identical (rmsd = 1.12 Å between the two IRES conformations) but in displaced position with respect to the 40S body **(Fig. 3.5 A and B)**. The IAPV-IRES seems to follow the movement of the 40S head, pivoting around the anchored PKI and SL-IV which are exceptionally stabilized by ribosomal elements from both head and body, effectively "clamping" the IRES to the 40S subunit **(Fig 3.5 D and E)**.

The three-way helical junction modeled in the PKI of the IAPV-IRES resembles a "hammer" shape, with SL-III coaxially stacked on top of the ASL-like stem **(Fig. 3.4 D and E)**. Perpendicular to both and situated in between them, a helical segment connects PKI and PKIII. The coaxially aligned SL-III and ASL-like domain forms a straight unit of shape and dimensions similar to a tRNA, excluding the acceptor stem **(Fig. 3.5 C)**. Alignments of structures containing tRNAs in several configurations [canonical tRNA PDBID:4V5D[57], with A/T-tRNA PDBID:5LZS[58] and hybrid tRNA PDBID:3J7R[59]] with the structure of the IAPV-IRES in complex with the 40S, reveal an interesting positioning of the coaxial unit formed by SL-III and the ASL-like part of PKI **(Fig. 3.5 C)**. Notably, the SL-III/ASL-like unit of IAPV-IRES populates a space more similar to a hybrid A/P-tRNA than a canonical, A/A-tRNA or A/T-tRNA following nomenclature of hybrid states previously proposed[60]. Similarly, PKIII overlaps with the position occupied by a hybrid P/E-tRNA, mimicking its helical components of the elbow region of a tRNA in this intermediate configuration. Overall, the IAPV-IRES is able to manipulate the 40S subunit in isolation, blocking the functional sites where canonical initiation factors eIF1, eIF1A, and eIF5B bind and, at the same time, steering the intrinsic dynamics of the 40S head toward a configuration reminiscent of an early elongation, pre-translocated state.

3.3.3   SL-III interacts with ribosomal protein uL16 stabilizing the 80S in a pre-translocation configuration mimicking hybrid tRNAs

The binary IAPV-IRES/80S complex populates three major conformations, with limited

differences between them **(Fig. 3.1 E)**. The majority of particles populated a class where the 40S

subunit exhibits a small degree of intersubunit rotation (approx. 1°) compared with the unrotated,

canonical configuration **(Fig. 3.7 A)**. No major swiveling or tilt of the 40S head is visible in this

conformation. The IAPV-IRES maintains a similar global conformation as in the binary complex

with 40S, but in the 80S map, both the L1.1 region and the tip of the SL-III show good density as

their dynamics are restricted by specific contacts with elements of the 60S: The L1 stalk

stabilizes the L1.1 region and the A-site finger and the ribosomal protein uL16 the SL-III. The

A-site finger (28S rRNA helix 38) is a flexible component of the 28S rRNA which plays an

important role in translocation of tRNAs from the A- to the P-site[61]. In many structures, it is not

visible due to its intrinsic flexibility, required to perform its role escorting in-transit tRNAs[61,62].

The SL-III of IAPV-IRES contacts the A-site finger, stabilizing it in a fixed conformation, which

allows the apical loop of SL-III to reach deep into the 60S, establishing a novel interaction with

the ribosomal protein uL16 **(Fig. 3.6 A and B)**. The IAPV-IRES positions the apical loop of SL-

III (nucleotides 6,585–6,590) in direct contact with basic residues of uL16, which are in

electrostatic interacting distance with negatively charged phosphates of the RNA backbone of the

IRES **(Fig. 3.6 C)**. The additional anchoring points to the ribosome contributed by SL-III, allows

the IAPV-IRES, in the context of an 80S interaction, to be stabilized in a conformation that

overlaps with the space occupied by a hybrid A/P-tRNA. The coaxial unit SL-III/ASL-like

domain of the IAPV-IRES functionally mimics an A/P-tRNA priming the 80S for eEF2

recruitment, effectively bypassing the initiation stage **(Fig. 3.6 D)**. Additionally, the anchoring

points provided by the IAPV-IRES along the intersubunit space probably contribute to an effective recruitment of the 60S in the absence of the dedicated factor responsible for such function in canonical translation, eIF5B[63].

3.3.4 Aparavirus IRESs restrict the small subunit rotation dynamics in the pre-translocation state

No populations with wide rotations of the small subunit were identified in our large 80S/IAPV-IRES dataset. This suggests that, in contrast with Cripavirus IRESs[29,64], the IAPV-IRES is able to restrict the dynamics of the small subunit, channeling it toward a canonical, non-rotated configuration (**Fig. 3.7 A**). This is accomplished by a solid anchoring of the PKI in the A-site, which not only mimics the ASL of a tRNA interacting with a cognate codon in the A-site, but also, by placing the SL-III in a similar position as a hybrid A/P-tRNA[65,66], mimics the T and D arms of a tRNA (**Figs. 3.6 D, 3.7 B, and 3.8**). In such position, the PKI of the IAPV-IRES establishes a network of interactions with both the large and the small subunits, effectively restricting the rotation of the 40S. Apart from the interactions established by the apical loop of the SL-III with ribosomal protein uL16 (**Fig. 3.6 C**), the decoding event elicited by the placement of the PKI in the decoding center allows the establishment of an interaction with the 28S rRNA base A3760 (A1913 in E. coli), normally involved in decoding (**Fig. 3.7 C**)[67]. This interaction is maintained along the small fluctuations of the 40S, which, locally, are restricted to displacements of a few angströms (**Fig. 3.7 D**). The IAPV-IRES seems to bind very tightly in a binary, pre-translocation complex with the 80S. The recruitment of elongation factors to commit the ribosome to the production of viral proteins seems to be achieved not through a dynamic manipulation of the 40S rotation, but by directly adopting a configuration reminiscent of a ribosome with tRNAs in hybrid configurations.

3.3.5   Remodeling of specific components of the IAPV-IRES allows its translocation through the ribosome

Due to their intrinsic flexibility, IRESs are able to populate multiple conformational states, while maintaining a basic structural framework dictated by their base-pairing scheme. A combination of rigid elements connected via flexible linkers allows these RNAs to tune their interactions with different ribosomal sites as they transit from an early pre-translocated state to a post-translocated one. Along this vectorial movement, these IRESs take advantage of intrinsic dynamic elements of the ribosome, normally involved in translocation of tRNAs and mRNAs[22,68].

In order to visualize the IAPV-IRES in a post-translocated state, we engineered a stop codon in the first coding codon following the IRES sequence[25,69]. By simultaneously incubating a pre-translocated 80S/IAPV-IRES complex with eEF2 and a catalytically inactive version of the eukaryotic release factor 1 (eRF1*)[36] in the presence of GTP, we were able to stabilize the IRES after a single translocation on the ribosome, allowing the visualization of the overall conformation of the IRES in a post-translocated state as well as the specific determinants of the IRES in binding the ribosomal P-site **(Fig. 3.9 A)**.

Globally, the post-translocated IAPV-IRES exhibits an extended conformation with the ASL/SL-III unit deeply inserted in the P-site and the L1.1 region maintaining its original connection with the L1 stalk. The SL-IV and SL-V of the IRES are no longer in contact with the 40S **(Fig. 3.9 B and C, left)**. The SL-III seems to play an important role in orienting the IRES unit formed by the ASL/SL-III to a position that perfectly matches that of a canonical P-site tRNA **(Fig. 3.9D)**. Nucleotides belonging to the SL-III establish contacts with several residues of ribosomal proteins uL5 and uL16 as well as with the 28S rRNA nucleotide A4255, all components of the large ribosomal subunit (60S) **(Fig. 3.9 C, right)**.

In canonical translocation, the movement of tRNAs and mRNA has to be coordinated in order to vacate the ribosomal A-site for the next incoming aminoacyl-tRNA. After peptidyl transfer, the ribosome adopts a rotated configuration of the small ribosomal subunit with tRNAs in hybrid configurations. The movement of the peptidyl-tRNA in the A-site to the P-site has to be coordinated with the movement of the P-site tRNA to the E-site, and the tRNA occupying the E-site has to be ejected from the ribosome with the assistance of the L1 stalk. During this process, it is of capital importance that the correct reading frame on the mRNA be maintained[22,68,70]. This is accomplished by the participation of specific components of the ribosome (mainly RNA bases) which interact with tRNAs and mRNA defining specific checkpoints so as to prevent in-transit tRNAs from slipping or loosing contact with the mRNA on the correct frame[71]. One of the most important checkpoints is defined by the so-called "P-site gate"[71]: a constriction formed by elements of the 18S rRNA (bases 1,054–1,064 and 1,638–1,645, 1,335–1,344, and 785–795 in *E. coli*) that physically block the progression of a translocated peptidyl-tRNA in the P-site from slipping into the E-site. The "closing" of the P-site gate marks the end of a correct translocation cycle, allowing the ribosome to reset to a canonical, non-rotated configuration[70].

By perfectly mimicking a canonical P-site tRNA, the IAPV-IRES in a post-translocated state is able to position the VLR, a flexible element of its structure, in contacting distance with the E-site of the 40S subunit. This is accomplished even with a fully closed P-site gate **(Fig. 3.9 E)**: Sliding through the P-site gate, the single-stranded VLR can contact the ribosomal protein uS7, normally involved in stabilizing E-site tRNAs **(Fig. 3.9 E)**[72]. The VLR thus undergoes a marked remodeling as the IAPV-IRES transitions from the pre-translocation to the post-translocation state. While in the pre-translocation conformation, the VLR establishes interactions with ribosomal components of the P-site, after translocation, it populates a more extended conformation that is able to reach the E-site even with the P-site gate locked. The remodeling of this IRES component seems to be crucial for proper translocation of the IRES, as mutations that alter both its length and its base composition impact negatively on the ability of the IRES to initiate translation[52].

3.4 Discussion

Metagenomic studies of environmental samples have recently underscored the pervasive role RNA viruses exert on the biosphere[7]. With estimates of dozens of RNA viruses infecting a single species, the diversity and impact these molecular entities have in biology and evolution is highly under-appreciated[73]. It has been discovered that insects and arthropods host the highest diversity of RNA viruses[7]. Within that realm, viruses from the *Dicistrovirideae* family have generated special interest given their wide range and distribution of hosts[4,7].

Using cryo-EM, we comprehensively characterized how the intergenic IRES of IAPV, a virus of the *Dicistrovirideae* family and a causing agent of the colony collapse disorder (CCD), binds and manipulates ribosomes to redirect them toward the production of viral proteins. The IAPV-IRES is able to establish a stable binary interaction with the small ribosomal subunit. An early capturing of free 40S subunits, committing them toward viral protein production, may represent a limiting step in a cellular environment where cellular and viral messages contend for ribosomal access. The IAPV-IRES is able to insert its PKI domain in the A-site (decoding site) of the 40S effectively blocking the binding of eIF1/eIF1A, two initiation factors required for canonical initiation **(Fig 3.4 C)**. While blocking the 40S functional sites, the IAPV-IRES is able to steer the intrinsic dynamics of the 40S subunit toward a specific configuration, facilitating the recruitment of the large ribosomal subunit (60S) in the absence of eIF5B, the cellular factor catalyzing this event.

Bypassing the highly regulated initiation stage of translation is a customary requirement for the type IV family of IRES[17]. This is accomplished by inducing an altered ribosome state which is able to recruit elongation factors (eEF2 and eEF1A) directly and without tRNAs. Cripavirus IRESs such as the CrPV-IRES accomplish this by inducing a wide rotation on the 40S, mimicking a pre-translocation state of the ribosome with tRNAs **(Fig. 3.11, top)**. In marked contrast, Aparavirus IRESs such as the IAPV-IRES capture elongation factors not by inducing a rotated state of the 40S, but by directly mimicking a ribosomal state with hybrid tRNAs **(Fig. 3.11, bottom)**. The limited dynamics of 40S subunit rotation observed in our large cryo-EM dataset reflects a solid anchoring of both ribosomal subunits, mediated mainly by the additional contacts contributed by the SL-III characteristic of Aparavirus IRESs. However, a compromise between rigidity and flexibility is required for these IRESs to operate: In order to place the first coding codon in the decoding center of the 40S, the IAPV-IRES has to move (translocate) from the A-site to the P-site. We were able to gain structural information of this transition visualizing the post-translocated state of the IAPV-IRES. In the post-translocated state, the PKI mimics a canonical P-site tRNA and the SL-III establishes new contacts with proteins and rRNA components of the P-site of the 60S **(Fig. 3.9 C)**. Additionally, the extremely dynamics of a single-stranded region of the IAPV-IRES termed the VLR **(Fig. 3.4 D)** is able to modify its configuration in a context-specific manner. In the context of a binary interaction with the 40S subunit and in the pre-translocation state with the 80S ribosome, the VLR exhibits a compact configuration, contacting bases of the 40S subunit's P-site. After translocation, and once

PKI is displaced to the P-site, the VLR is extended, contacting ribosomal protein uS7, a

component of the 40S subunit's E-site (**Fig. 3.9 E**). Biochemical evidence supports a key role of

the VLR in IRES function, as mutations and/or shortening impacts the ability of the IRES to

efficiently translocate[52].

Using high-resolution single-particle cryo-EM analysis, we show how the prototypical

Aparavirus IRES of the intergenic region of the IAPV manipulates the eukaryotic ribosome to

position a viral non-AUG codon in the ribosomal A-site to effectively hijack the host machinery

for protein production[‡]. We have uncovered a new strategy of early pre-translocation mimicry

used by this IRES sub-family as well as visualize the conformational dynamics of a strategic

single-stranded segment of the IRES, the VLR. The structures presented here will allow

structure-based design of new and better RNA-interfering strategies directed toward the vital

intergenic IRES of the IAPV. Ongoing efforts in that direction have already proved successful in

fighting CCD, protecting bee hives from collapse[33,74].

[‡] Link to a movie of the process:
https://www.ncbi.nlm.nih.gov/pmc/articles/PMC6826211/?report=classic#embj2019102226-sup-0002

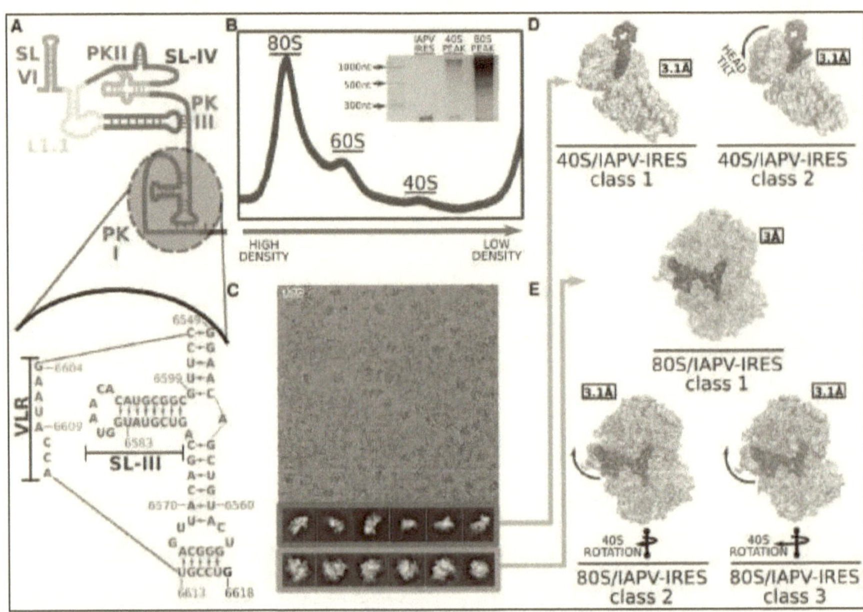

Figure 3.1 | IAPV-IRES secondary structure, experimental set-up, and cryo-EM image processing workflow
A: IAPV-IRES diagram colored according to secondary structure motifs. Bottom, a closer view of the IAPV-IRES PKI highlighting its sequence, with base pairs indicated as well as the variable loop region (VLR) and stem loop III (SL-III).
B: Sucrose gradient UV profile of 60S/40S/IAPV-IRES reaction mixture after an overnight run. The peaks corresponding to 80S and 40S were used for RNA extraction and UREA-PAGE shown in the inset.
C: Representative cryo-EM image where roughly half of the particles correspond to 40S (blue) and the other half to 80S (orange).
D: Two classes with robust density for the IAPV-IRES were found in the 40S group.
E: After classification, three classes with clear IAPV-IRES density and small differences in the conformation of the 40S were found in the 80S group.

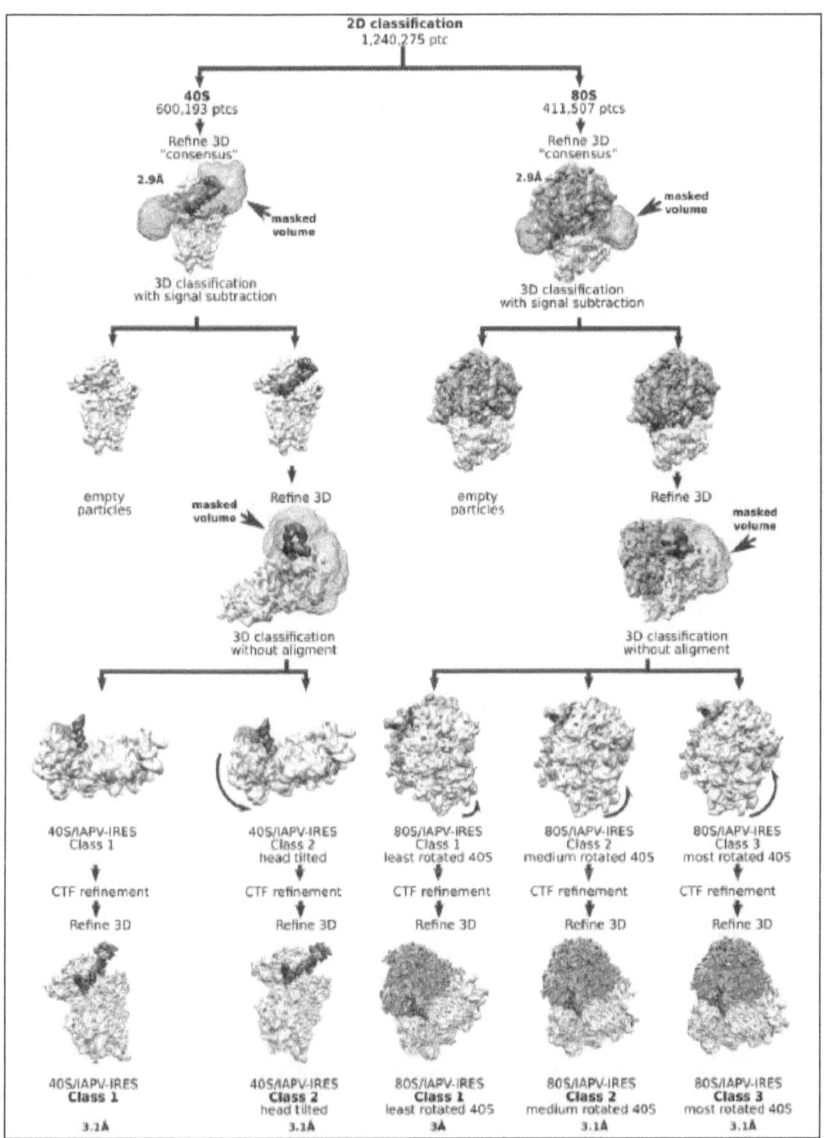

Figure 3.2 | Classification scheme followed for the pre-translocation dataset
Top: 40S and 80S particles were separated by reference-free 2D classification. Homogeneous
subgroups of 40S and 80S particles were subjected to two steps of masked classification

intercalated with refinements to, on a first instance, identified those particles with IAPV-IRES and, in a second instance, distinguish among the groups with IAPV-IRES, different conformations.
Bottom: New features implemented in Relion 3.0[45] such as contrast transfer values refinement allowed extending the resolution to close to 3Å for the five populations.

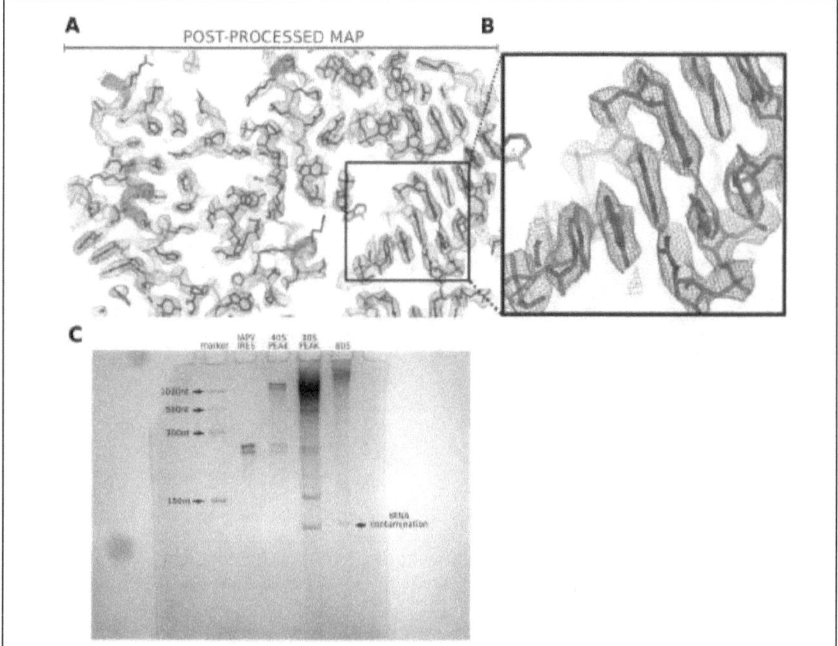

Figure 3.3 | Post-processed map and UREA-PAGE
A and B: Slice through the 80S/IAPV-IRES class 1 post-processed map centered on the 60S. Map features expected for a 3 Å map like protein side chains and base separation for nucleic acids can be appreciated.
C: UREA-PAGE analysis of 40S and 80S peaks resolved in an overnight sucrose gradient run. Bands corresponding to the purified IAPV-IRES RNA can be appreciated in both 40S and 80S peaks.

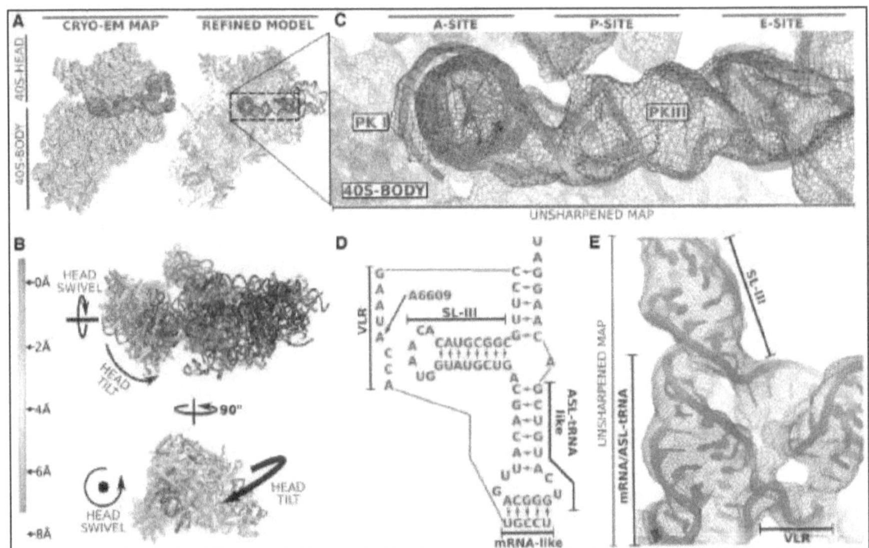

Figure 3.4 | Structure of the IAPV-IRES in complex with the 40S ribosomal subunit
A: Overview of the mammalian 40S in complex with IAPV-IRES. Left, cryo-EM final post-processed map of class 1 with 40S colored yellow and IAPV-IRES maroon. Right, corresponding final refined model with IAPV-IRES colored according to Fig. 3.1 A.
B: Ribbon diagram of the 40S colored by pairwise root-mean-square deviation displacements observed between the two IAPV-IRES/40S classes. The different position of the 40S head between both classes is a composition of swiveling and tilt movements (indicated by arrows in orthogonal views).
C: Close-up view of the ribosomal sites of the 40S for IAPV-IRES/40S class 1 showing cryo-EM unsharpened cryo-EM density.
D: Sequence of the PKI three-way helical junction.
E: Unsharpened cryo-EM density for the PKI region of the IAPV-IRES in class 1 with the SL-III and the tRNA/mRNA mimicking domain indicated.

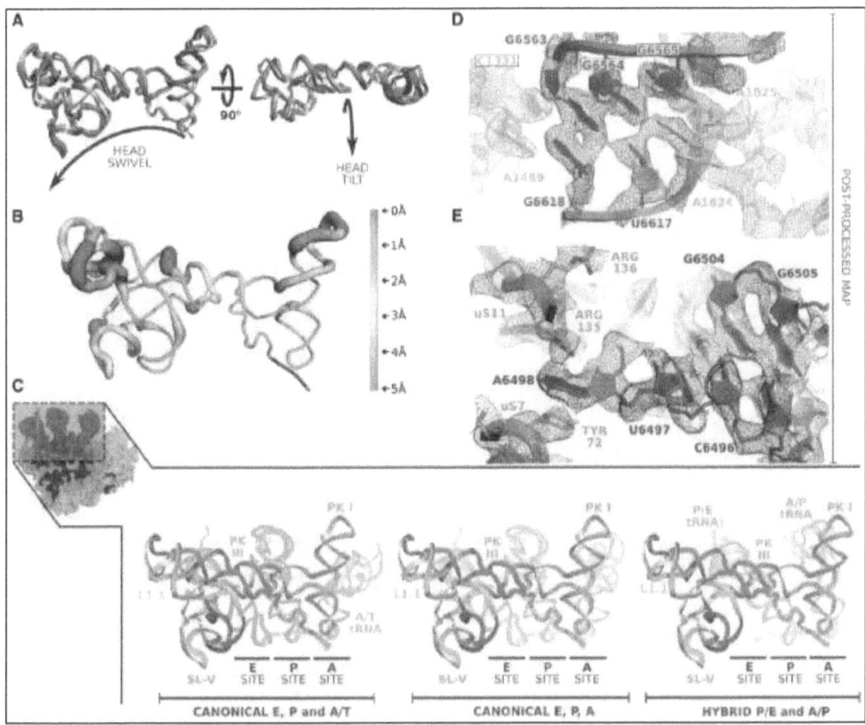

Figure 3.5 | IAPV-IRES conformation in the context of a 40S interaction
A: Superposition of IAPV-IRES models corresponding to IAPV-IRES/40S class 1 and class 2
after alignments excluding the IRES and the 40S head. A similar conformation can be observed
with distinctive relative orientation with respect to the 40S body. The movement characteristic of
the 40S head is indicated by arrows in orthogonal views.
B: Ribbon diagram of the IAPV-IRES colored by pairwise root-mean-square deviation
displacements observed between the two IAPV-IRES/40S classes. The ASL/mRNA-like regions
of the PKI and well as the SL-IV show the lowest degree of displacement (blue), whereas the
apical part of SL-III and the L1.1 the highest (red).
C: Superposition of the IAPV-IRES with tRNAs in different configurations indicated at the
bottom. IAPV-IRES is depicted as ribbons colored according to the secondary structure
elements, and tRNAs are represented as gray ribbons. Alignments of the models were computed
with the 40S body, excluding from the computation the ligands (IRES/tRNAs) and the 40S head.
D: Detailed view of the refined model for IAPV-IRES/40S class 1 inserted in the post-processed
cryo-EM density focused on the decoding center of the 40S. PKI of IAPV-IRES is depicted
green and 18S rRNA yellow.
E: Close-up view of the refined model for IAPV-IRES/40S class 1 inserted in the post-processed
cryo-EM density focused on the SL-IV of the IAPV-IRES (depicted blue).

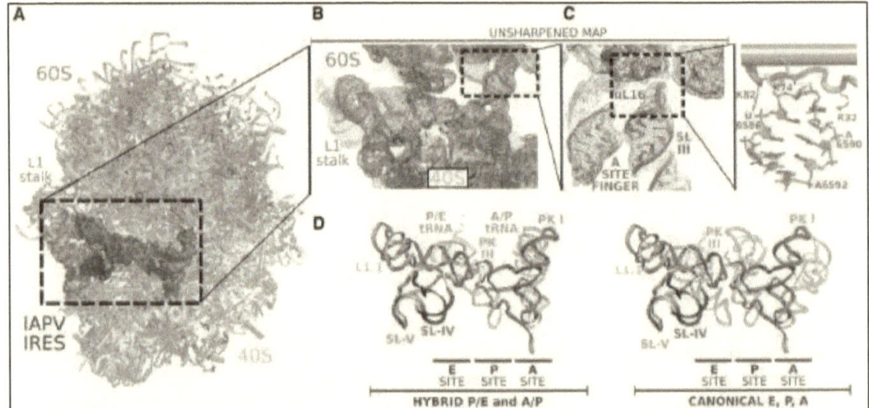

Figure 3.6 | SL-III of IAPV-IRES engages novel sites of the 60S ribosomal subunit
A: Overall view of the IAPV-IRES/80S complex class 1 with 60S represented as cyan ribbons, the 40S as yellow ribbons, and the IAPV-IRES represented as solid Van der Waals surface colored by secondary structure motifs.
B: Close-up view of the intersubunit space with the IAPV-IRES depicted as cartoons colored as in (A) inserted in the unsharpened cryo-EM density.
C: Zoomed view of the A-site finger in interacting distance with the SL-III (green). The apical loop of SL-III reaches deep into the 60S contacting the ribosomal protein uL16.
D: Superposition of the IAPV-IRES in complex with 80S (class 1) with tRNAs in different configurations indicated at the bottom. Alignments of the models were computed with the 40S body, excluding from the computation the ligands (IRES/tRNAs) and the 40S head. IAPV-IRES PKI component SL-III/ASL-like domain populates a space of the intersubunit space similar to a A/P-tRNA. IAPV-IRES PKIII (red) mimics the elbow region of a hybrid P/E-tRNA.

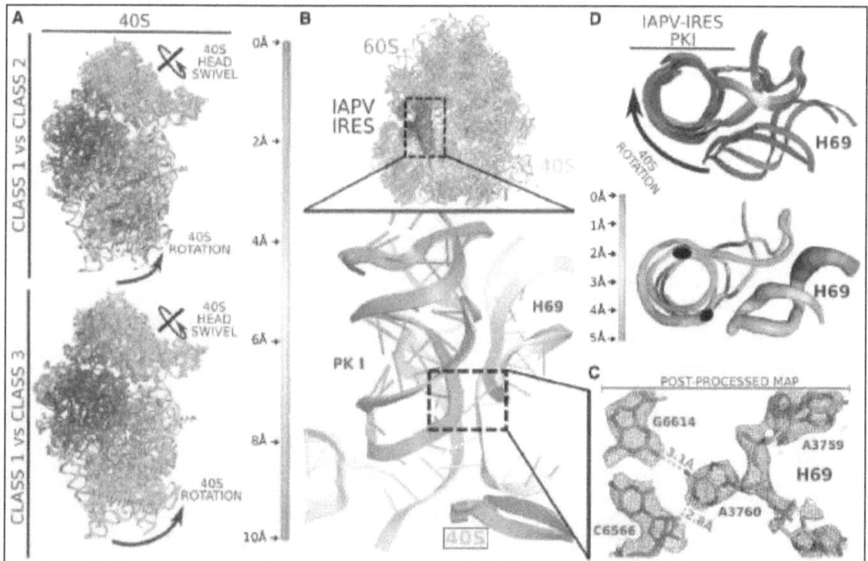

Figure 3.7 | The IAPV-IRES restricts the small subunit rotational dynamics in a pre-translocation complex with 80S

A: Ribbon diagram of IAPV-IRES/80S complex viewed from the 40S colored by pairwise root-mean-square deviation displacements observed between the classes indicated on the left. Class 1 is unrotated, while classes 2 and 3 exhibit a small rotational movement of the 40S.

B: Top, general overview of the non-rotated IAPV-IRES/80S class 1 structure. IAPV-IRES is depicted as solid Van Der Waals surface colored according to the secondary structure motifs. The PKI (green) is solidly anchored to the A-site. Bottom, close-up view of the IAPV-IRES PKI inserted in unsharpened map.

C: Zoomed view of A3760, a nucleotide belonging to the helix 69 (H69) of the 28S rRNA interacting with PKI. Final refined model inserted in the post-processed map is shown.

D: This interaction is not disrupted along the small fluctuations of the 40S. An apical view along the axis of the PKI of a superposition of class 1 versus class 2 shows the IRES displacements are minimal and are followed by the H69 which constantly interacts with the IRES.

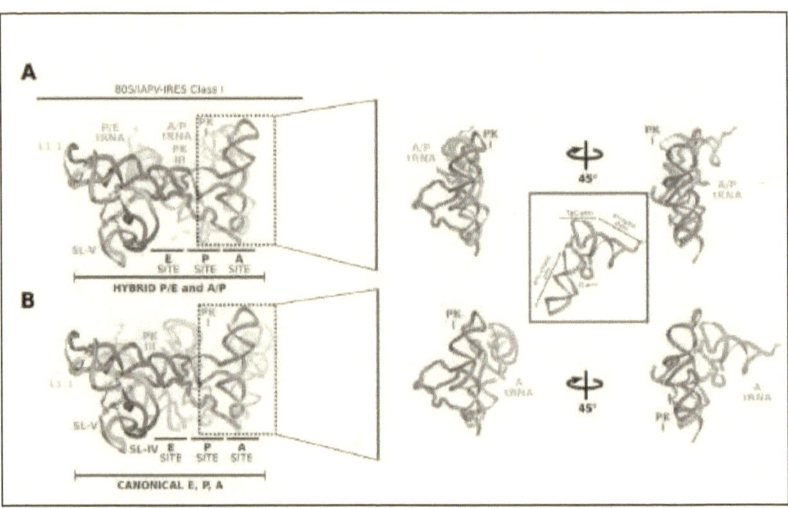

Figure 3.8 | tRNA mimicry in the 80S/IAPV-IRES pre-translocation state
A and B: Superposition of IAPV-IRES in the pre-translocation state class 1 with tRNAs in hybrid-states (A) and with canonical tRNAs (B). On the right, close-up view of the superposition for the PKI domain of the IAPV-IRES in two orientations. It can be appreciated the co-axial unit formed by SL-III and the tRNA/mRNA-like domain of the IRES occupies a space similar to a hybrid A/P tRNA. The tRNA mimicry is however not complete as the acceptor stem of the tRNA is not accounted by any element of the IRES.

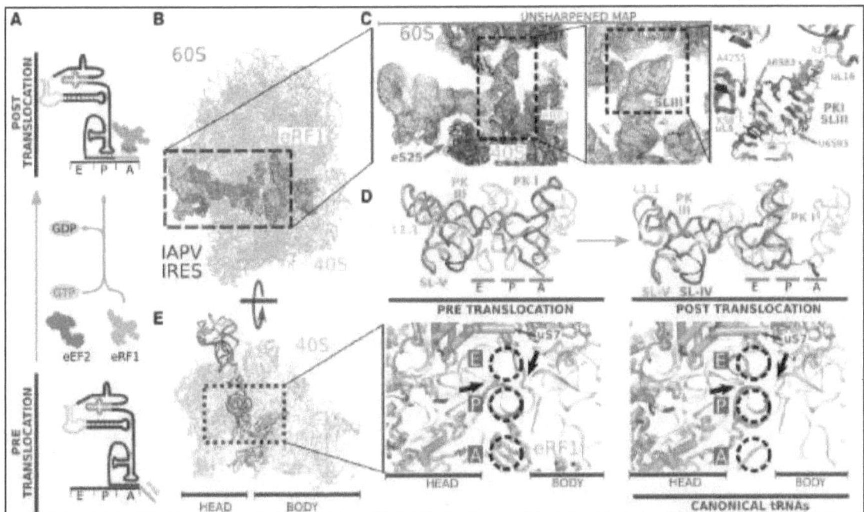

Figure 3.9 | Visualization of IAPV-IRES in a post-translocated state in the ribosome
A: Biochemical strategy employed to trap a post-translocated state of IAPV-IRES in the ribosome.
B: General view of the IAPV-IRES in a post-translocated state on the ribosome: 60S depicted as blue ribbons, 40S as yellow ribbons, IAPV-IRES represented as solid Van Der Waals surface colored according to secondary structure elements described in Fig. 3.1 A, and eRF1* depicted orange.
C: Close-up view of the SL-III inserted in the experimental unsharpened cryo-EM density. On the right, refined model with residues from the 60S (blue) in interacting distance with the SL-III (green) indicated.
D: Comparison of the final refined model for IAPV-IRES colored according to the secondary structure described in Fig. 3.1 A with canonical tRNAs (PDBID: 4V5D) in the pre-translocated state (left) and after translocation (right).
E: Left, overall top view of the intersubunit space of the 40S for the post-translocated state. Inset, close-up view of the 40S tRNA binding sites where it can be appreciated the insertion of PKI of the IAPV-IRES in the P-site, projecting the VLR toward the E-site. The elements of the 18S rRNA forming the "P-site gate" are indicated by solid arrows. On the right, equivalent view for canonical tRNAs.

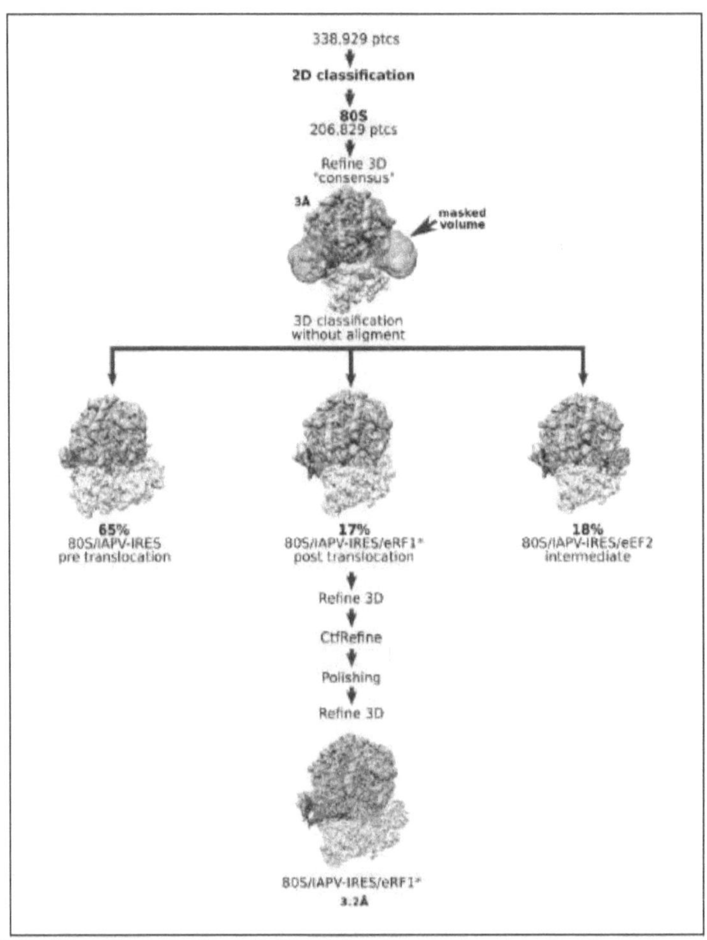

Figure 3.10 | Classification scheme followed for the post-translocation dataset
Top: After 2D and 3D masked classifications, three populations of particles corresponding to pre-translocation, post-translocation and an intermediate state with eEF2 could be identified in the dataset.
Bottom: Contrast transfer values refinement and Bayesian particle polishing implemented in Relion 3.0[45] allowed the extension of the resolution for the post-translocation state to 3.2 Å.

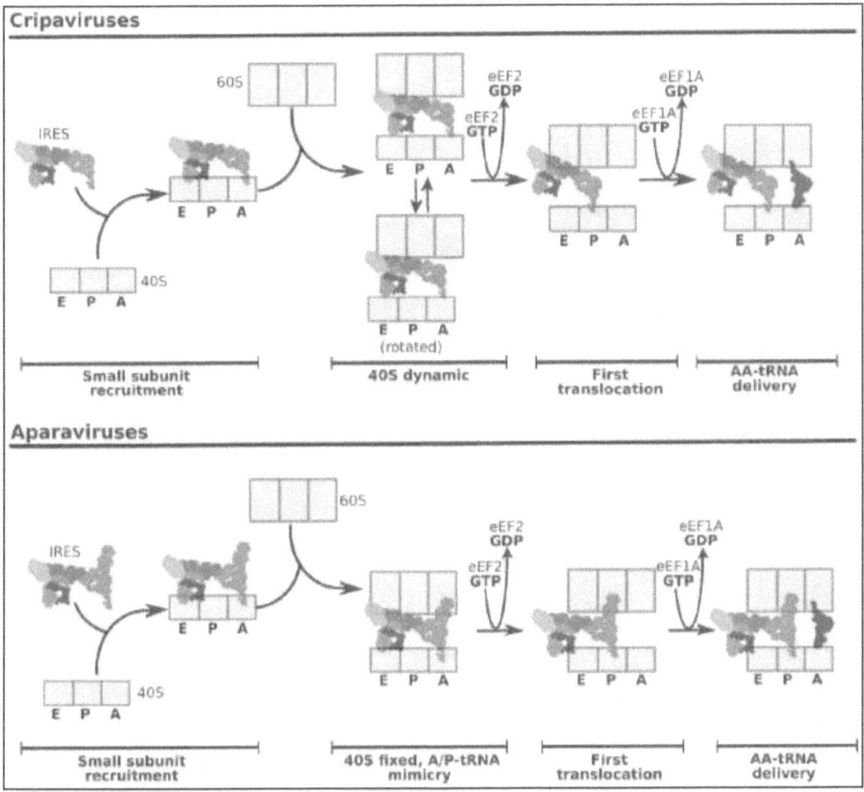

Figure 3.11 | Type IV IRES families exploit different pre-translocation features of canonical translation for ribosome hijacking

Top: The Cripavirus family of type IV IRESs is able to capture free 40S subunits and engage them on a pre-translocation complex by recruiting 60S. This complex is extremely dynamic, with the 40S alternating between non-rotated and rotated configurations with respect to the 60S subunit. IRESs belonging to this family, exemplified by the CrPV-IRES, recruit elongation factors by mimicking a rotated stated of the ribosome with tRNAs.

Bottom: The Aparavirus family of IRESs follows a similar pathway in order to assemble a pre-translocation complex; however, specific structural components of this family allow for additional contacts with the 60S, limiting the rotational freedom of the 40S. Elongation factor engagement and thus effective ribosome hijacking are accomplished by mimicking a ribosome state with tRNAs in hybrid configurations.

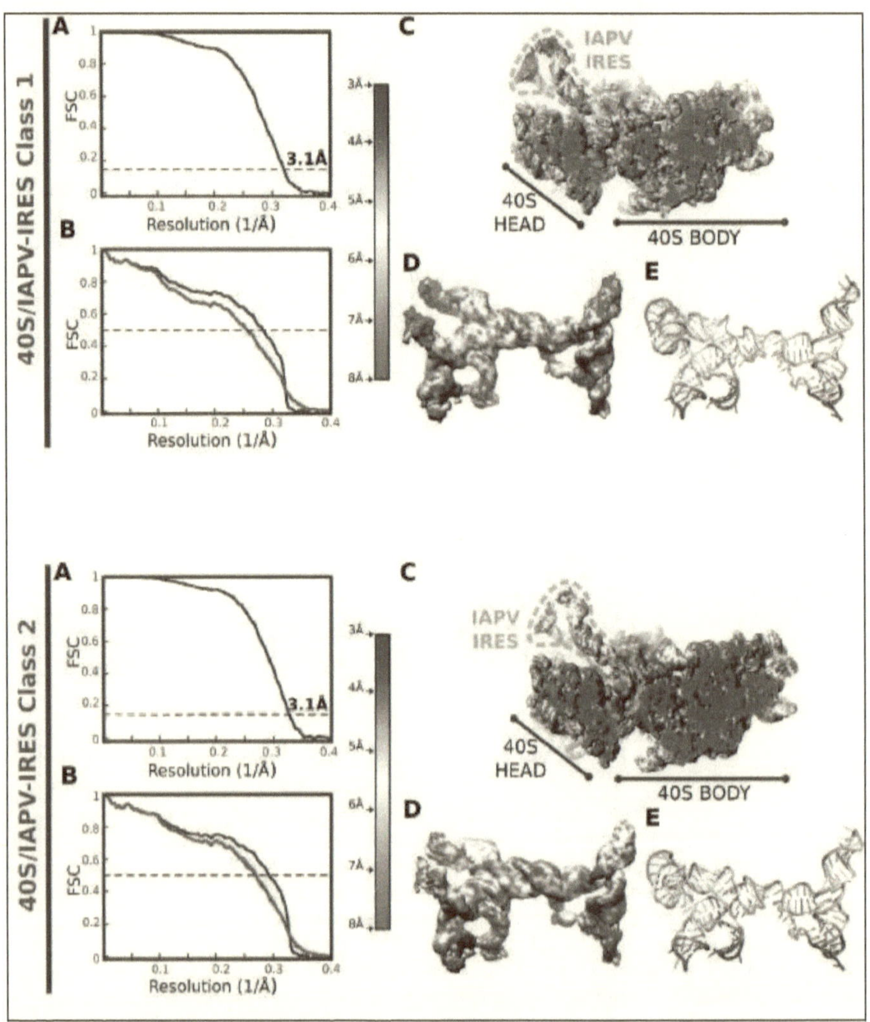

Figure 3.12 | Fourier Shell Correlation curves and local resolution estimation for the 40S/IAPV-IRES complex classes. For each class:

A: Fourier Shell Correlation (FSC) computed for the two half maps of the final subset of particles after classification for the first class identified for the 40S/IAPV-IRES complex. The resolution is estimated to be 3.1 Å using the 0.143 criterion[75].

B: Map-versus-model cross validation FSC. The final model was validated using standard procedures: FSC of the refined model against half map 1 (blue) overlaps with the FSC against half map 2 (red, not included in the refinement). The black curve corresponds to the FSC of the

169

final model against the final map.

C: Slice through the final, unsharpened map colored according to the local resolution as reported by RESMAP[44].

D: Close-up views of the final density colored according to local resolution values as in (C) for the IAPV-IRES.

E: Final refined IAPV-IRES model colored according to the estimated B-factors in $Å^2$ computed by REFMAC[49]. Specific values for estimated resolutions are indicated for each class.

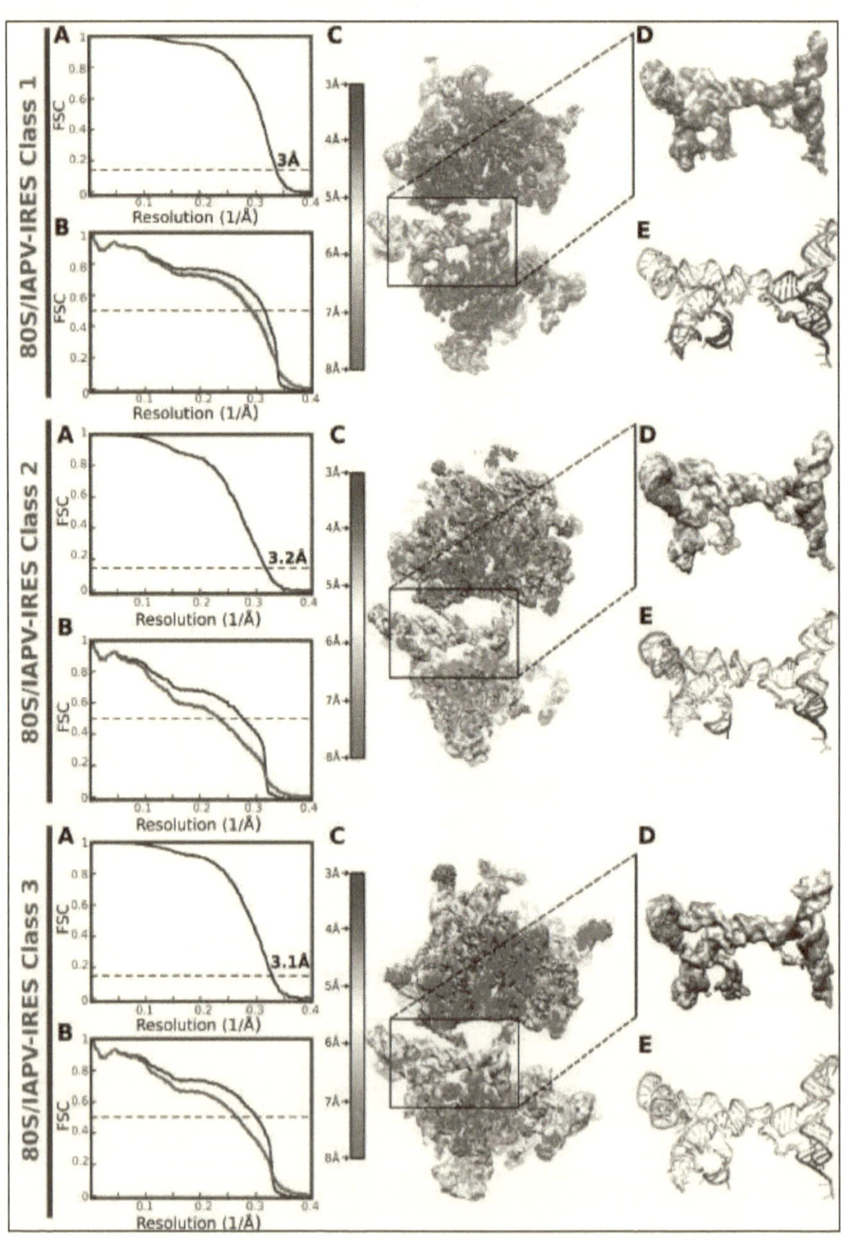

80S/IAPV-IRES Class 1
A
B
FSC
Resolution (1/Å)
3Å
C
3Å
4Å
5Å
6Å
7Å
8Å
D
E
80S/IAPV-IRES Class 2
A
B
FSC
Resolution (1/Å)
3.2Å
C
3Å
4Å
5Å
6Å
7Å
8Å
D
E
80S/IAPV-IRES Class 3
A
B
FSC
Resolution (1/Å)
3.1Å
C
3Å
4Å
5Å
6Å
7Å
8Å
D
E

Figure 3.13 | Fourier Shell Correlation curves and local resolution estimation for the 80S/IAPV-IRES complex classes. For each class:
A: Fourier Shell Correlation (FSC) computed for the two half maps of the final subset of particles after classification for the first class identified for the 80S/IAPV-IRES complex. The resolution are estimated to be 3.0-3.2 Å using the 0.143 criterion[75].
B: Map-versus-model cross validation FSC. The final model was validated using standard procedures: FSC of the refined model against half map 1 (blue) overlaps with the FSC against half map 2 (red, not included in the refinement). The black curve corresponds to the FSC of the final model against the final map.
C: Slice through the final, unsharpened map colored according to the local resolution as reported by RESMAP[44].
D: Close-up views of the final density colored according to local resolution values as in (C) for the IAPV-IRES.
E: Final refined IAPV-IRES model colored according to the estimated B-factors in Å^2 computed by REFMAC[49]. Specific values for estimated resolutions are indicated for each class.

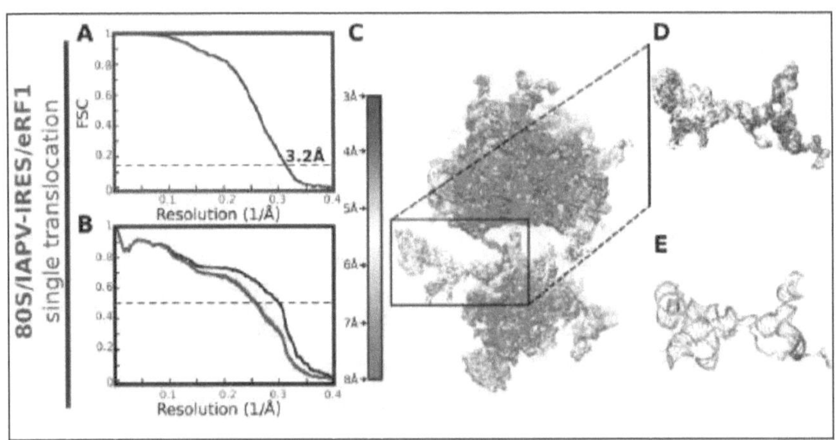

Figure 3.14 | Fourier Shell Correlation curves and local resolution estimation for the post-translocated 80S/IAPV-IRES/eRF1* complex

(A) Fourier Shell Correlation (FSC) computed for the two half maps of the final subset of particles after classification for the first class identified for the 80S/IAPV-IRES/eRF1[*] complex. The resolution is estimated to be 3.2 Å using the 0.143 criterion[75].

(B) Map-versus-model cross validation FSC. The final model was validated using standard procedures: FSC of the refined model against half map 1 (blue) overlaps with the FSC against half map 2 (red, not included in the refinement). The black curve corresponds to the FSC of the final model against the final map.

(C) Slice through the final, unsharpened map colored according to the local resolution as reported by RESMAP[44].

(D) Close-up views of the final density colored according to local resolution values as in (C) for the IAPV-IRES.

(E) Final refined IAPV-IRES model colored according to the estimated B-factors in Å^2 computed by REFMAC[49].

Table 3.1 | Data collection, model refinement and validation statistics.*Defocus range reported by GCTF

Data collection	Collection Polara-F30					Collection Titan-Krios
Micrographs	9973					4887
Picked Particles	1240275					338929
Class 2D particles	1011700					206829
Voltage (KV)	300					300
Defocus range* (µm)	0.4-3.5					1.2-4.9
Defocus mean (µm)	1.5					2.2
Pixel size (Å/pixel)	1.233					1.0605
Frames / Movie	40					40
Electron dose (e-/Å²)	42.1					56.9
Electron dose per frame (e-/Å²)	1.0525					1.4225
Stucture						
Component	40S/IAPV-IRES		80S/IAPV-IRES pre-translocated			80S/IAPV-IRES post-translocated
	Class-1	Class-2	Class-1	Class-2	Class-3	Class-Post T.
Particles	91056	96826	120176	40701	68697	27658
FSC 0.143 (Å)	3.1	3.1	3	3.1	3.1	3.2
Map sharpening (Å²)	-70.3	-66.6	-67.5	-54.1	-65.1	-75.4
Refinement						
Program/Protocol	Refmac5 / Phenix	Refmac5 / Phenix	Refmac5 / Phenix	Refmac5 / Phenix	Refmac5 / Phenix	Refmac5 / Phenix
Used in refinement (Å)	3.1	3.1	3.1	3.1	3.1	3.2
Average B-factors (Å²)	128.29	125.96	77.57	89.34	63.32	123.21
Avg B-fac Prot/RNA (Å²)	129.77/126.89	127.35/124.62	83.85/72.86	93.28/86.39	65.82/61.44	126.50/120.66
R.m.s deviations:						
Bonds (Å)	0.0037	0.0031	0.0031	0.0027	0.0027	0.0032
Angles (deg)	1.1622	1.1208	1.1038	1.0408	1.0511	1.083
Validation						
Molprobity score	1.86	1.65	1.75	1.68	1.74	1.63
Clashcore, all atoms	2.09	2.29	2.23	2.07	2.25	2.11
Rotamer outliers (%)	2.77	1.77	2.16	1.6	2.04	1.5
Ramachandran plot:						
Outliers (%)	0.86	0.67	0.63	0.52	0.62	0.61
Favored (%)	90.36	92.75	91.79	90.01	91.43	91.32
Composition						
Non hydrogen atoms	79372	79379	215972	215966	215976	219296
Protein residues	4862	4862	11491	11490	11490	11892
RNA bases	1900	1900	5766	5766	5766	5768
Ligands	0	0	0	0	0	0
Accession codes						
EMDB	20248	20249	20255	20256	20257	20258
PDB	6P4G	6P4H	6P5I	6P5J	6P5K	6P5N

3.6   References

1.   Ratnieks, F. L. W. & Carreck, N. L. Clarity on Honey Bee Collapse? *SCIENCE* **327**, 152–153 (2010).

2.   Chopra, S. S., Bakshi, B. R. & Khanna, V. Economic Dependence of US Industrial Sectors on Animal-Mediated Pollination Service. *ENVIRONMENTAL SCIENCE & TECHNOLOGY* **49**, 14441–14451 (2015).

3.   Anderson, D. & East, I. J. The latest buzz about colony collapse disorder. *SCIENCE* **319**, 724–725 (2008).

4.   Cox-Foster, D. L. *et al.* A Metagenomic Survey of Microbes in Honey Bee Colony Collapse Disorder. *Science* **318**, 283–287 (2007).

5.   Chen, Y. P. *et al.* Israeli Acute Paralysis Virus: Epidemiology, Pathogenesis and Implications for Honey Bee Health. *PLoS Pathog* **10**, e1004261 (2014).

6.   Doublet, V. *et al.* Unity in defence: honeybee workers exhibit conserved molecular responses to diverse pathogens. *BMC GENOMICS* **18**, (2017).

7.   Shi, M. *et al.* Redefining the invertebrate RNA virosphere. *Nature* **540**, 539–543 (2016).

8.   Wilson, J. E., Powell, M. J., Hoover, S. E. & Sarnow, P. Naturally Occurring Dicistronic Cricket Paralysis Virus RNA Is Regulated by Two Internal Ribosome Entry Sites. *Molecular and Cellular Biology* **20**, 4990–4999 (2000).

9.   Pisarev, A., Shirokikh, N. & Hellen, C. Translation initiation by factor-independent binding of eukaryotic ribosomes to internal ribosomal entry sites. *COMPTES RENDUS BIOLOGIES* **328**, 589–605 (2005).

10.   Kerr, C. H. & Jan, E. Commandeering the Ribosome: Lessons Learned from Dicistroviruses about Translation. *J. Virol.* **90**, 5538–5540 (2016).

11.   Mullapudi, E., Fuezik, T., Pridal, A. & Plevka, P. Cryo-electron Microscopy Study of the Genome Release of the Dicistrovirus Israeli Acute Bee Paralysis Virus. *JOURNAL OF VIROLOGY* **91**, (2017).

12.   Wilson, J. E., Pestova, T. V., Hellen, C. U. T. & Sarnow, P. Initiation of Protein Synthesis from the A Site of the Ribosome. *Cell* **102**, 511–520 (2000).

13.   Gross, L. *et al.* The IRES5′UTR of the dicistrovirus cricket paralysis virus is a type III IRES containing an essential pseudoknot structure. *Nucleic Acids Research* **45**, 8993–9004 (2017).

14.   Carrillo-Tripp, J. *et al.* In vivo and in vitro infection dynamics of honey bee viruses. *Sci Rep* **6**, 22265 (2016).

15.   Khong, A. *et al.* Temporal Regulation of Distinct Internal Ribosome Entry Sites of the Dicistroviridae Cricket Paralysis Virus. *Viruses* **8**, 25 (2016).

16.   Yamamoto, H., Unbehaun, A. & Spahn, C. M. T. Ribosomal Chamber Music: Toward an Understanding of IRES Mechanisms. *Trends in Biochemical Sciences* **42**, 655–668 (2017).

17.   Jackson, R. J., Hellen, C. U. T. & Pestova, T. V. The mechanism of eukaryotic translation initiation and principles of its regulation. *Nat Rev Mol Cell Biol* **11**, 113–127 (2010).

18.   Aylett, C. H. S. & Ban, N. Eukaryotic aspects of translation initiation brought into focus. *PHILOSOPHICAL TRANSACTIONS OF THE ROYAL SOCIETY B-BIOLOGICAL SCIENCES* **372**, (2017).

19.   Hinnebusch, A. G. & Lorsch, J. R. The Mechanism of Eukaryotic Translation Initiation: New Insights and Challenges. *COLD SPRING HARBOR PERSPECTIVES IN BIOLOGY* **4**, (2012).

20.   Jaafar, Z. A. & Kieft, J. S. Viral RNA structure-based strategies to manipulate translation. *Nature Reviews Microbiology* **17**, 110–123 (2019).

21.   Hertz, M. I. & Thompson, S. R. Mechanism of translation initiation by Dicistroviridae IGR IRESs. *Virology* **411**, 355–361 (2011).

22.   Noller, H. F., Lancaster, L., Mohan, S. & Zhou, J. Ribosome structural dynamics in translocation: yet another functional role for ribosomal RNA. *QUARTERLY REVIEWS OF BIOPHYSICS* **50**, (2017).

23.   Abeyrathne, P. D., Koh, C. S., Grant, T., Grigorieff, N. & Korostelev, A. A. Ensemble cryo-EM uncovers inchworm-like translocation of a viral IRES through the ribosome. *ELIFE* **5**, (2016).

24.   Murray, J. *et al.* Structural characterization of ribosome recruitment and translocation by type IV IRES. *ELIFE* **5**, (2016).

25.   Pisareva, V. P., Pisarev, A. V. & Fernández, I. S. Dual tRNA mimicry in the Cricket Paralysis Virus IRES uncovers an unexpected similarity with the Hepatitis C Virus IRES. *eLife* **7**, e34062 (2018).

26.   Johnson, A. G., Grosely, R., Petrov, A. N. & Puglisi, J. D. Dynamics of IRES-mediated translation. *Phil. Trans. R. Soc. B* **372**, 20160177 (2017).

27.   Jan, E. *et al.* Initiator Met-tRNA-independent translation mediated by an internal ribosome entry site element in cricket paralysis virus-like insect viruses. *COLD SPRING HARBOR SYMPOSIA ON QUANTITATIVE BIOLOGY* **66**, 285–292 (2001).

28.   Pestova, T. V., Lomakin, I. B. & Hellen, C. U. T. Position of the CrPV IRES on the 40S subunit and factor dependence of IRES/80S ribosome assembly. *EMBO Rep* **5**, 906–913 (2004).

29.   Koh, C. S., Brilot, A. F., Grigorieff, N. & Korostelev, A. A. Taura syndrome virus IRES initiates translation by binding its tRNA-mRNA-like structural element in the ribosomal decoding center. *Proceedings of the National Academy of Sciences* **111**, 9139–9144 (2014).

30.   Au, H. H. *et al.* Global shape mimicry of tRNA within a viral internal ribosome entry site mediates translational reading frame selection. *Proc Natl Acad Sci USA* **112**, E6446–E6455 (2015).

31.   Wang, Q. S. & Jan, E. Switch from Cap- to Factorless IRES-Dependent 0 and +1 Frame Translation during Cellular Stress and Dicistrovirus Infection. *PLoS ONE* **9**, e103601 (2014).

32.   Hunter, W. *et al.* Large-Scale Field Application of RNAi Technology Reducing Israeli Acute Paralysis Virus Disease in Honey Bees (Apis mellifera, Hymenoptera: Apidae). *PLoS Pathog* **6**, e1001160 (2010).

33.   Maori, E. *et al.* IAPV, a bee-affecting virus associated with Colony Collapse Disorder can be silenced by dsRNA ingestion. *INSECT MOLECULAR BIOLOGY* **18**, 55–60 (2009).

34.   Frolova, L. Y. *et al.* Mutations in the highly conserved GGQ motif of class 1 polypeptide release factors abolish ability of human eRF1 to trigger peptidyl-tRNA hydrolysis. *RNA* **5**, 1014–1020 (1999).

35.   Pisarev, A. V. *et al.* The Role of ABCE1 in Eukaryotic Posttermination Ribosomal Recycling. *Molecular Cell* **37**, 196–210 (2010).

36.   Alkalaeva, E. Z., Pisarev, A. V., Frolova, L. Y., Kisselev, L. L. & Pestova, T. V. In vitro reconstitution of eukaryotic translation reveals cooperativity between release factors eRF1 and eRF3. *CELL* **125**, 1125–1136 (2006).

37.   Russo, C. J. & Passmore, L. A. Ultrastable gold substrates: Properties of a support for high-resolution electron cryomicroscopy of biological specimens. *Journal of Structural Biology* **193**, 33–44 (2016).

38.   Carragher, B. *et al.* Leginon: An Automated System for Acquisition of Images from Vitreous Ice Specimens. *Journal of Structural Biology* **132**, 33–45 (2000).

39.   Zheng, S. Q. *et al.* MotionCor2: anisotropic correction of beam-induced motion for improved cryo-electron microscopy. *Nature Methods* **14**, 331–332 (2017).

40.   Lander, G. C. *et al.* Appion: An integrated, database-driven pipeline to facilitate EM image processing. *Journal of Structural Biology* **166**, 95–102 (2009).

41.   Zhang, K. Gctf: Real-time CTF determination and correction. *Journal of Structural Biology* **193**, 1–12 (2016).

42.   Zhang, D. K. Brief Manual of Gautomatch. 10.

43.   Scheres, S. H. W. RELION: Implementation of a Bayesian approach to cryo-EM structure determination. *Journal of Structural Biology* **180**, 519–530 (2012).

44.   Kucukelbir, A., Sigworth, F. J. & Tagare, H. D. Quantifying the local resolution of cryo-EM density maps. *Nature Methods* **11**, 63–65 (2014).

45.   Zivanov, J. *et al.* New tools for automated high-resolution cryo-EM structure determination in RELION-3. *eLife* **7**,.

46.   Pettersen, E. F. *et al.* UCSF Chimera—A visualization system for exploratory research and analysis. *Journal of Computational Chemistry* **25**, 1605–1612 (2004).

47.   Emsley, P. & Cowtan, K. Coot: model-building tools for molecular graphics. *Acta Cryst D* **60**, 2126–2132 (2004).

48.   Adams, P. D. *et al.* PHENIX: a comprehensive Python-based system for macromolecular structure solution. *Acta Cryst D* **66**, 213–221 (2010).

49.   Murshudov, G. N., Vagin, A. A. & Dodson, E. J. Refinement of Macromolecular Structures by the Maximum-Likelihood Method. *Acta Crystallogr D Biol Crystallogr* **53**, 240–255 (1997).

50.   Schüler, M. *et al.* Structure of the ribosome-bound cricket paralysis virus IRES RNA. *Nat Struct Mol Biol* **13**, 1092–1096 (2006).

51.   Au, H. H. T., Elspass, V. M. & Jan, E. Functional Insights into the Adjacent Stem-Loop in Honey Bee Dicistroviruses That Promotes Internal Ribosome Entry Site-Mediated Translation and Viral Infection. *JOURNAL OF VIROLOGY* **92**, (2018).

52.   Ruehle, M. D. *et al.* A dynamic RNA loop in an IRES affects multiple steps of elongation factor-mediated translation initiation. *eLife* **4**, e08146 (2015).

53.   von Loeffelholz, O. *et al.* Focused classification and refinement in high-resolution cryo-EM structural analysis of ribosome complexes. *Current Opinion in Structural Biology* **46**, 140–148 (2017).

54.   Ramrath, D. J. F. *et al.* Visualization of two transfer RNAs trapped in transit during elongation factor G-mediated translocation. *Proceedings of the National Academy of Sciences* **110**, 20964–20969 (2013).

55.   Llácer, J. L. *et al.* Conformational Differences between Open and Closed States of the Eukaryotic Translation Initiation Complex. *Molecular Cell* **59**, 399–412 (2015).

56.   Ogle, J. M. & Ramakrishnan, V. Structural insights into translational fidelity. *Annu. Rev. Biochem.* **74**, 129–177 (2005).

57.   Voorhees, R. M., Weixlbaumer, A., Loakes, D., Kelley, A. C. & Ramakrishnan, V. Insights into substrate stabilization from snapshots of the peptidyl transferase center of the intact 70S ribosome. *Nature Structural & Molecular Biology* **16**, 528–533 (2009).

58.   Shao, S. *et al.* Decoding Mammalian Ribosome-mRNA States by Translational GTPase Complexes. *CELL* **167**, 1229+ (2016).

59.   Voorhees, R. M., Fernández, I. S., Scheres, S. H. W. & Hegde, R. S. Structure of the Mammalian Ribosome-Sec61 Complex to 3.4 Å Resolution. *Cell* **157**, 1632–1643 (2014).

60.   Ratje, A. H. *et al.* Head swivel on the ribosome facilitates translocation by means of intra-subunit tRNA hybrid sites. *Nature* **468**, 713–716 (2010).

61.   Nguyen, K., Yang, H. & Whitford, P. C. How the Ribosomal A-Site Finger Can Lead to tRNA Species Dependent Dynamics. *JOURNAL OF PHYSICAL CHEMISTRY B* **121**, 2767–2775 (2017).

62.   Brown, A., Fernández, I. S., Gordiyenko, Y. & Ramakrishnan, V. Ribosome-dependent activation of stringent control. *Nature* **advance online publication**, (2016).

63.   Pestova, T. V. *et al.* The joining of ribosomal subunits in eukaryotes requires eIF5B. *Nature* **403**, 332–335 (2000).

64.   Fernandez, I. S., Bai, X.-C., Murshudov, G., Scheres, S. H. W. & Ramakrishnan, V. Initiation of Translation by Cricket Paralysis Virus IRES Requires Its Translocation in the Ribosome. *CELL* **157**, 823–831 (2014).

65.   Moazed, D. & Noller, H. F. Intermediate states in the movement of transfer RNA in the ribosome. **342**, 7 (1989).

66.   Frank, J. Intermediate states during mRNA–tRNA translocation. *Current Opinion in Structural Biology* **22**, 778–785 (2012).

67.   Demeshkina, N., Jenner, L., Yusupova, G. & Yusupov, M. Interactions of the ribosome with mRNA and tRNA. *CURRENT OPINION IN STRUCTURAL BIOLOGY* **20**, 325–332 (2010).

68.   Voorhees, R. M. & Ramakrishnan, V. Structural Basis of the Translational Elongation Cycle*. *Annual Review of Biochemistry* **82**, 203–236 (2013).

69.   Muhs, M. *et al.* Cryo-EM of Ribosomal 80S Complexes with Termination Factors Reveals the Translocated Cricket Paralysis Virus IRES. *Molecular Cell* **57**, 422–432 (2015).

70.   Noller, H. F., Lancaster, L., Zhou, J. & Mohan, S. The ribosome moves: RNA mechanics and translocation. *Nature Structural & Molecular Biology* **24**, 1021–1027 (2017).

71.   Zhou, J., Lancaster, L., Donohue, J. P. & Noller, H. F. How the ribosome hands the A-site tRNA to the P site during EF-G-catalyzed translocation. *Science* **345**, 1188–1191 (2014).

72.   Brown, A., Baird, M. R., Yip, M. C., Murray, J. & Shao, S. Structures of translationally inactive mammalian ribosomes. *eLife* **7**, e40486 (2018).

73.   Koonin, E. V. & Dolja, V. V. A virocentric perspective on the evolution of life. *CURRENT OPINION IN VIROLOGY* **3**, 546–557 (2013).

74.   Piot, N., Snoeck, S., Vanlede, M., Smagghe, G. & Meeus, I. The Effect of Oral Administration of dsRNA on Viral Replication and Mortality in Bombus terrestris. *Viruses* **7**, 3172–3185 (2015).

75.   Rosenthal, P. B. & Henderson, R. Optimal Determination of Particle Orientation, Absolute Hand, and Contrast Loss in Single-particle Electron Cryomicroscopy. *Journal of Molecular Biology* **333**, 721–745 (2003).

Chapter 4:

A complex internal ribosome entry site at the 5′ untranslated region of a viral mRNA assembles a functional 48S complex via an uAUG intermediate[§][**]

[§] A version of this chapter was published as: Neupane, R.[#], Pisareva, V. P.[#], Rodriguez, C. F., Pisarev, A. V. & Fernández, I. S. A complex IRES at the 5′-UTR of a viral mRNA assembles a functional 48S complex via an uAUG intermediate. *eLife* **9**, e54575 (2020).

[#] I and V.P.P. contributed equally as co-first authors.

[**] The final published version of this chapter is included in Appendix B.

4.1 Introduction

Metagenomic studies have uncovered a fascinating diversity of viruses with a pervasive presence in the biosphere[1,2]. Many new viral clades have been discovered, showing an expanded presence compared with previously assumed distributions [3]. Diversity is especially overwhelming in RNA viruses that infect animal hosts[4]. Extensive gene shuffling, horizontal gene transfer events and host switching combined with co-divergence suggest a rich and complex evolutionary scenario within the animal virome, which remains largely unexplored [1,5]. As all viruses need to hijack the host cell's protein production machinery at some point in their life cycle for viral protein production, solutions that viruses have arrived at for this purpose truly reflects their diversity. The diversity of mechanisms is especially pronounced in viruses that infect multicellular eukaryotes and then hijack a highly regulated and complex system.

In multicellular eukaryotes the protein production machinery is complex with numerous large, multi-component protein factors assisting the operation of the significantly expanded and highly decorated eukaryotic ribosomes[6]. Despite all the extra factors and multiple levels of exquisite regulation, protein synthesis in eukaryotes retains the same four key phases present in prokaryotes, namely: initiation, elongation, termination and recycling[7]. While the latter three phases are relatively unchanged or only slightly expanded in eukaryotes, initiation is significantly expanded upon and has become a major hub for transactional regulation[8-11].

Initiation in prokaryotes is streamlined. To load an mRNA onto the small ribosomal subunit, only three initiation factors and base pairing of a ribosomal RNA with the Shine-Dalgarno sequence on an mRNA are required. In contrast, for initiation in eukaryotes docking of mRNAs to the eukaryotic small (40S) ribosomal subunit requires several initiation factors[9] and two GTP-regulated steps are required to correctly position the first aminoacyl-tRNA responsible for setting up the correct reading frame on the messenger RNA (mRNA)[8,9]. The mRNA molecules in eukaryotes are also covalently modified at their 5′ with a methylated nucleotide called the "cap" and do not possess an equivalent of the prokaryotic Shine-Dalgarno sequence [10,11].

Eukaryotic initiation starts when the initiation factors eIF1, eIF1A, eIF3, eIF5 and the Ternary Complex (TC, eIF2/Met-tRNA$_i^{Met}$/GTP) all bind to the 40S subunit to form the 43S Pre-Initiation Complex (43S-PIC) which is ready to recruit mRNA[8]. Eukaryotic mRNAs are then docked to the 43S-PIC at their 5′ ends, forming the 48S complex[12]. This complex is highly dynamic, able to move in the 5′ to 3′ direction along the mRNA in search of an AUG initiation codon in a favorable context, a process called "scanning"[11]. Once the AUG codon is detected, a structural transition in the 48S from an open, scanning-competent conformation to a closed, scanning-arrested conformation occurs[13]. This conformational change is accompanied by the release of eIF1, eIF2 and GDP, leaving the Met-tRNA$_i^{Met}$ at the P site of the 40S base paired with the AUG codon[9]. A second GTP-regulated step, catalyzed by initiation factor eIF5B, is then required for the recruitment of the eukaryotic large (60S) ribosomal subunit [14,15]. A full eukaryotic ribosome (80S) primed with mRNA and Met-tRNA$_i^{Met}$ at the P-site then transitions towards the fast, less regulated elongation phase[16].

The above described pathway is referred to as the canonical 5′-end and cap-dependent translation route of initiation[11]. The bulk of eukaryotic mRNAs transition this route, however, especially in higher eukaryotes, deviations from the canonical route are common and are normally associated with translation under stress conditions[17,18]. Many examples of non-canonical initiation are associate with extended 5′ UnTranslated Regions (5′-UTRs) on mRNAs[19,20]. In higher eukaryotes, 5′-UTRs can be very long and they can harbor short Open

Reading Frames (ORFs) designated upstream ORFs (uORFs)[20,21]. These uORFs are enigmatic as ribosome-profiling experiments clearly show ribosome positioning on them, however, the short peptides encoded could not be unambiguously identified by mass-spectrometry to date[22].

Several uORFs at 5′-UTRs are found in stress regulated genes, like the yeast stress response regulator GCN2 or the mammalian transcription factor ATF4[23,24]. In these stress regulated genes, the presence of several uAUG codons has been shown to be essential for differential translation regimes in homeostasis versus stress conditions[20]. Other examples of translation regulation by uAUG codons are less understood. For example, a uAUG codon immediately followed by a stop codon in what are called start-stop uORFs are also common in 5′-UTRs of mammalian mRNAs[21,25]. Re-initiation events involving translation termination factors have been suggested to play a role in regulating start-stop uORFs, but in general, the function and possible biological activities of the peptides encoded in uORFs and the mechanism that mediate start-stop uORFs regulation remains obscure[24,26].

A better characterized example of non-canonical initiation can be found in viral mRNAs[27]. Viruses exploit the complexity of eukaryotic initiation to gain access to the host machinery for protein production[28]. Strategies like mimicking the cap structure or transferring caps from cellular mRNAs ("cap-snatching") allow viral mRNAs to hijack host ribosomes, redirecting them towards the production of viral proteins[28,29]. A more widespread viral strategy for ribosome hijacking is the use of structured RNA sequences in viral mRNAs[30] called Internal Ribosomal Entry Sites (IRES). A tentative classification based on the degree of their RNA structure and their dependency on canonical initiation factors, divided them in four main types[31,32].

Type I IRESs are mostly unstructured IRESs requiring all canonical initiation factors plus additional proteins not involved in translation (called ITAFs: IRES *Trans* Activator Factors) for efficient initiation[31]. These IRESs also require scanning to locate the initiation codon. Type II IRESs present the same requirements of initiation factors and ITAFs as the type I, but they assemble initiation complexes directly on the AUG codon without the need for scanning. Type III IRES are more structured than either type I or II, they fold in a compact structure able to interact with the 40S and only require initiation factors eIF2 and eIF3. Finally, the type IV IRESs are highly structured IRES with no requirement for any canonical initiation factors to assemble a translating ribosome in viral mRNAs[33].

The *Dicistroviridae* family of (+)-ssRNA viruses employs two types of IRESs to differentially express their regulatory versus the structural genes[34]. The genome architecture of these viruses functionally segregates both kind of genes in two ORFs (**Fig. 4.1 A**)[35]. The first ORF is preceded by a long 5'-UTR which harbors an IRES initially assigned to the type III family due to its dependency on eIF2 and eIF3[35]. *In vitro* characterization of the 5'-UTR-IRES of the Cricket Paralysis Virus (CrPV), a prototypical *Dicistrovirus,* narrowed down the region of the 5'-UTR responsible for the IRES activity and established the strict requirement of eIF3 for this IRES to initiate translation. Interestingly, the AUG codon of the CrPV ORF1 is immediately preceded by a "start-stop uORF"[35]. The combination of these two non-canonical initiation resources, namely a structured IRES and a start-stop uORF at the 5'-UTR, within the same viral mRNA, pose the question of how translation initiation is achieved by this RNA.

We sought to structurally characterize the 5'-UTR-IRES of the CrPV in its ribosome bound configuration to gain insights on the ribosome binding determinants of this peculiar IRES as well as to understand how the delivery Met-tRNA$_i^{Met}$ is accomplished. Two high resolution cryo-EM reconstructions of 40S/5′-UTR-IRES/eIF3 complexes combined with biochemical analysis, allowed us to precisely characterize how this IRES, using an extended structure with a modular, multi-domain architecture, binds and manipulates the 40S. Our results stress the importance of the unstudied diversity of viral IRESs, invite re-examination of the current IRES classification system, and expand our understanding of the role 5′UTR regions play in eukaryotic translation.

4.2   Materials and Methods

4.2.1 5′-UTR-IRES production

A transcription vector for 5′-UTR-IRES (nucleotides 357-720) was constructed inserting a T7 promoter sequence up-stream of 5′-UTR-IRES sequence followed by an EcoRI restriction site, using pUC19 as a scaffold vector. T7 RNA polymerase *in vitro* transcription and UREA-PAGE purification was used to obtain highly purified RNA.

4.2.2 Purification of translation components and ribosomal subunits

Native 40S and 60S subunits, eIF2, eIF3, eIF5B and rabbit aminoacyl-tRNA synthetases were prepared as previously described[36]. Recombinant eIF1 and eIF1A were purified according to a previously described protocol[37]. *In vitro* transcribed Met-tRNA$_i^{Met}$ was aminoacylated with methionine in the presence of rabbit aminoacyl-tRNA synthetases as previously described[38].

4.2.3 Assembly of ribosomal complexes

To reconstitute different ribosomal complexes, we incubated 1.8 pmol 40S ribosomal subunits with 2 pmol 5′-UTR-IRES-RNA and corresponding quantities of eIF1/eIF1A/eIF2 or eIF5B in a 20 uL reaction mixture containing buffer A (20 mM Tris-HCl, pH 7.5, 100 mM KCl,

2.5 mM MgCl2, 0.1 mM EDTA, 1 mM DTT) with 0.4 mM GTP for 5 min at 37∘C. Next, we

added 10 pmol Met-tRNA$_i^{Met}$ and incubated for 5 min at 37∘C. We analyzed the assembled

ribosomal complexes via a toe-printing assay essentially as described[36].

4.2.4 Cryo-EM sample preparation and data acquisition

Aliquots of 3μl of assembled ribosome complexes at concentration range of 250-350 nM

were incubated for 30 seconds on glow-discharged holey gold grids[39] (UltrAuFoil R1.2/1.3).

Grids were blotted for 2.5s and flash cooled in liquid ethane using an FEI Vitrobot. Grids were

transferred to an FEI Titan Krios microscope equipped with an energy filter (slits aperture 20eV)

and a Gatan K2 detector operated at 300 kV. Data was recorded in counting mode at a

magnification of 130,000 corresponding to a calibrated pixel size of 1.08 Å. Defocus values

ranged from 1-3.6 μm. Images were recorded in automatic mode using the Leginon[40] and

APPION[41] software and frames were aligned using the Relion3[42] implementation of the

Motioncor2 algorithm[43].

4.2.5 Image processing and structure determination

Contrast transfer function parameters were estimated using GCTF[44,45] and particle

picking was performed using GAUTOMACH[45] without the use of templates and with a diameter

value of 260 pixels. All 2D and 3D classifications and refinements were performed using

RELION. An initial 2D classification with a 4 times binned dataset identified all ribosome

particles. A second 2D classification step with two times binned data was employed to separate

80S from 40S particles. A consensus reconstruction with all 40S particles was computed using

the AutoRefine tool of RELION. Next, 3D classification without alignment (four classes, T

parameter 4) identified a class with unambiguous density for eIF3. This class was independently refined, and further masked classification allowed the identification of two subclasses distinguishable by different degree of 40S head swiveling and presence or absence of eIF3d density. Final refinements with unbinned data for the classes selected yielded high resolution maps with density features in agreement with the reported resolution. Local resolution was computed with RESMAP[46].

4.2.6 Model building and refinement

Models for the mammalian 40S and eIF3 docked into the maps using CHIMERA[47] and COOT[48] was used to manually adjust these initial models. 5'-UTR-IRES was built manually using COOT. An initial round of refinement was performed in Phenix using real-space refinement[49] with secondary structure restraints and a final step of reciprocal-space refinement with REFMAC[50]. The fit of the model to the map density was quantified using FSCaverage and Cref and model-to-maps over-fitting tests were performed following standard protocols in the field[51,52].

4.3   Results

4.3.1 The 5'-UTR-IRES of the Cricket Paralysis Virus (CrPV) requires eIF3 for a stable interaction with the 40S

A recent study of the IRES located at the 5'-UTR of CrPV (5'-UTR-IRES from hereafter) precisely defined the region in the 5'-UTR responsible for IRES activity (residues 357 to 690). The study also showed that the 5'-UTR-IRES is dependent on initiation factor eIF3 for efficient translation initiation[35]. The 5'-UTR-IRES is grouped together with well described IRESs of the type III family like the Hepatitis C Virus IRES (HCV-IRES) or the Classical Swine Flu Virus IRES (CSFV-IRES) based on its interaction requirement with eIF3. However, sequence alignments of the 5'-UTR region of CrPV with its type III family members and other related *Dicistroviruses* viruses failed to identify any significant similarities[53,54]. This is in contrast with the well characterized type IV family of IRESs found in the InterGenic Region (IGR-IRES) of such *Dicistroviruses*, where strong sequence conservation allows comparisons and identification of structural motifs which can be used to predict how they interact with ribosomes. The 5'-UTRs of these viruses seem to harbor divergent sequences, making structural modelling of the embedded IRESs based on sequence conservation difficult while raising the possibility of a divergent IRES mechanism[33].

In order to address these gaps in our understanding of 5'-UTR IRESs, we sought to obtain structural information of its 40S bound conformation using electron cryo-microscopy (cryo-EM). We first produced a truncated version of the 5'-UTR region of the (+)-ssRNA genome of CrPV containing the IRES (residues 357 to 720, **Fig. 4.1 A**). We then tested *in vitro* whether the 5'-UTR-IRES is able to stably engage purified 40S ribosomal subunits using sucrose density gradients (**Fig. 4.1 B**). If the 5'-UTR-IRES forms a stable complex with the 40S we would observe the two co-migrate in the gradient. Unexpectedly, the 5'-UTR-IRES alone is not able to form a stable complex with the 40S. This is in contrast to the HCV-IRES which forms stable complexes with the 40S and even with elongating (80S) ribosomes all on its own[53]. In the presence of eIF3, however, the 5'-UTR-IRES did co-migrate with purified 40S subunits, demonstrating that eIF3 is required for the formation of a stable complex (**Fig. 4.1 B**). We decided to proceed with the structural determination of this stable 40S/5'-UTR-IRES/eIF3 complex using cryo-EM.

4.3.2 Cryo-EM of the 40S/5′-UTR-IRES/eIF3 complex reveals two classes of particles and a three domain architecture of the 5′-UTR-IRES

The 40S/5′-UTR-IRES/eIF3 complex revealed clear particles in our cryo-EM images (**Fig. 4.2**) and when picked and subjected to 2D classification, they rendered detailed two-dimensional class averages. Density for eIF3 could be identified in these 2D class averages albeit at lower threshold (**Fig. 4.1 C**). Obtaining enough particles to produce a 3D model was challenging because the complex exhibited a delicate behavior under cryo-EM conditions, with a strong tendency to disassemble in thin ice. Thus, extensive screening of the grid for suitable ice areas before launching automated data collection was essential to obtain particles of the fully assembled complex (**Fig. 4.2 A**). The sample also exhibited a high degree of heterogeneity, which we resolved using a series of masked classification steps during image processing in Relion[55,56] (**Fig. 4.1 D and E, 4.2 C, and 4.3**).

After our classification schemes, we found two main classes of particles containing clear densities for 5′-UTR-IRES, 40S and eIF3 in the dataset (**Fig. 4.1 D and E**). Both class-1 and class-2 contain density for the 40S, the 5′-UTR-IRES, and the core subunits of eIF3 (a/c/e/k/l/f/m). Class-2 exhibits the 40S head in a swiveled configuration and presents an extra density attributable to the eIF3 subunit d (**Fig. 4.1 E**, eIF3d). The eIF3d subunit density in class-2 follows the 40S head swivel to establish interactions with a core subunit of eIF3, eIF3a. These interactions seem to stabilize the conformation of eIF3a in this class.

Robust density attributable to the 5′-UTR-IRES could be found in both classes (**Fig. 4.1 D and E**, blue). The ribosome bound 5′-UTR-IRES shows an extended configuration as it encircles the 40S head (**Fig. 4.4 A**). We could identify three distinct domains in the 5′-UTR-

IRES which are connected by flexible linkers: 1) an elongated domain I (DI) at the back of the 40S head that makes novel contacts with the ribosomal proteins uS3 and RACK1 (**Fig. 4.4**), 2) a dual hairpin domain II (DII) located at the back of the 40S body and which interacts with eIF3 (**Fig. 4.5**) and 3) a large helical domain III (DIII) located in close proximity of the 40S E site which makes contacts with ribosomal proteins eS25, uS7 and uS11 (**Fig. 4.6**). We were able to derive a 2D structure diagram from out 3D model (**Fig. 4.7**).

4.3.3 Domain I of the 5′-UTR-IRES makes novel and residue-specific contacts with ribosomal proteins RACK1 and uS3

The 5′ proximal segment of the 5′-UTR-IRES (residues 357 to 486, **Fig 4.7**) form domain I. Domain I (DI) is elongated T-shaped and is anchored to the back of the 40S head (**Fig. 4.4 A and B**). A long double-helical segment in domain I "wraps" around the apical part of ribosomal protein RACK1. Two cytosine residues of this helical segment (C442 and C444) are extruded from the body of the double helix to establish hydrophobic interactions with tyrosine residue 140 (Y140) of RACK1 as if they were pinching the tyrosine residue (**Fig. 4.4 C**). These interactions bend the main helical segment of the 5′-UTR-IRES DI directing the tip of this domains towards ribosomal protein uS3 (**Fig. 4.4 D, left**). A guanine residue (G395) at the apical loop of the helix is inserted deep in a hydrophobic pocket of ribosomal protein uS3 (**Fig. 4.4 D, right**). It establishes contacts with main chain atoms of this ribosomal protein.

These interactions at the back of the 40S head position the apical region of 5′-UTR-IRES domain I close to the mRNA entry channel. This position seems to overlap with the space previously described to be occupied by DHX29, a helicase which is involved in canonical initiation of heavily structured 5′-UTRs[57](**Fig. 4.4 D**). The same apical loop region of domain I also seems to block the binding site of the RRM domain of initiation factor eIF4B, an initiation factor that has been shown to promote helicase activity of the 5′-cap binding eIF4F complex. Docking of eIF4B in low resolution density further supported by cross-linking data places the RRM domain of eIF4B next to uS3. Indeed, one of the main cross-linking interactions between the RRM domain of eIF4B and uS3 is at uS3 residue Lys63—a residue very close to the flipped guanine residue (G395) of domain I (**Fig. 4.4 D, right**). The interaction of domain I with RACK1 and uS3 also disrupts the interaction between the C-terminal tail of uS3 with blade 4 of RACK1, an interaction that is conserved across higher eukaryotes. Interaction of domain I with a regulatory hub protein like RACK1 and it's blocking of binding sites for helicases and helicase simulating proteins might be a strategy employed by the 5′-UTR-IRES to evade detection and stay structured.

4.3.4 5′-UTR-IRES binding to the 40S is compatible with a canonical configuration of eIF3

The second domain of 5′-UTR-IRES (DII) is connected to domain I by a flexible linker that is poorly defined in our maps as it is exposed to the solvent. Domain II is formed by a dual hairpin and is sandwiched in-between the back of the body of the 40S and eIF3 subunits a and c (**Fig. 4.5 A and B, and Fig. 4.7**). Ribosomal protein uS17 contacts domain II at its periphery, establishing contacts with the phosphate backbone of domain II (**Fig. 4.5 B**). A network of interactions involving eIF3 subunits eIF3a, eIF3b, and eIF3c anchors domain II to this position (**Fig. 4.5 C**). These interactions are also established through contacts between negatively charged residues on eIF3 and the phosphate backbone of the IRES. Unlike base specific interactions in domain I, no contacts with specific IRES bases could be observed in domain II, suggesting a structure rather than sequence mediated interaction.

Currently, medium resolution cryo-EM reconstructions of the canonical 48S complex[58] (PDB ID 6FEC) and of the 40S in complex with eIF3 and the CSFV-IRES[54] (PDB ID 4C4Q) are available. Comparisons of these structures with our complex reveal a positioning of eIF3 relative to the 40S that is very similar to the canonical 48S complex and different from the position adopted by eIF3 in the CSFV-IRES–40S complex **(Fig. 4.5 D)**. In the 48S canonical configuration, eIF3 contacts the 40S through helix 1 of eIF3a and helix 22 of eIF3c, as well as through eIF3d, which is isolated in its 40S interaction, away from the core subunits of eIF3(Figure 3D, left). The CSFV-IRES engages the 40S, displacing eIF3 from its position in the canonical48S **(Fig. 4.5 D, right)**. In addition, in the canonical 48S complex, eIF3 interacts with the 40S periphery, allowing the formation of cavities between eIF3 and the back of the 40S. These cavities are exploited by the 5′-UTR-IRES, which inserts its domain II into one of them, adopting a configuration that is compatible with the binding of eIF3 to the 40S in the canonical 48S complex **(Fig. 4.5 D, middle)**. No major rearrangement of eIF3 (compared to its position in the canonical 48S complex) is required for the binding of the 5′-UTR-IRES, so there could be an advantage in hijacking preformed cellular 48S complexes that are ready to transit the cap-dependent route of initiation.

4.3.5 Non-canonical base pairing in the 5′-UTR-IRES domain III places the upstream AUG (uAUG) codon near the P site

A flexible single stranded linker threads through the 40S channel formed by ribosomal proteins uS7 and uS11 to connect domain II of the 5′-UTR-IRES to domain III **(Fig. 4.6 A)**. Domain III forms a prominent, double-helical mass in close proximity of the E site of the small subunit at the inter-subunit side of the 40S. The helical segment is very well defined in our maps as it is stabilized by numerous contacts with ribosomal proteins eS25, uS7, uS11 and 18S ribosomal RNA (rRNA) bases (**Fig. 4.6 and Fig. 4.7**). However, the distal part of this domain forms two short stem loops that given their flexibility could only be modelled at low resolution in our maps.

Inspection of this cryo-EM density reveled a distortion in the canonical double helix of the main segment of this domain as it approaches the E site. The quality of the map in this area allowed *de novo* modelling of these residues which revealed a set of non-canonical interactions between the RNA bases (**Fig. 4.6 A** and **B**). In-plane triple base interactions involving sugar and Hoogsteen edges of the bases as well as purine-purine Hoogsteen base pairs could be found in this stretch of residues of the helical segment of DIII (**Fig. 4 B**)[59]. Overall, these non-canonical base interactions induce a distortion at the base of domain III and help in the positioning of the single stranded segment of the 5′-UTR-IRES harboring the uAUG codon at position 701 (**Fig. 4.1 A**), close to the P site (**Fig. 4.6 C, middle**). The 5′-UTR-IRES thus accesses the 40S P-site through the E-site but what, as a result, gets precluded, in the present configuration, is a concurrent recruitment of the TC (eIF2/Met-tRNA$_i^{Met}$/GTP, **Fig. 4 C**) to the P-site. Interestingly, a similar strategy is followed by the HCV-IRES. A superposition of the structure of the HCV-IRES in complex with the 40S[60,61] (PDB ID 5A2Q) with our structure reveals a very similar positioning of the domain II of HCV-IRES, which accesses the P site through the E site to position the AUG codon in the surroundings of the P site (**Fig. 4.6 C, bottom**). Even though both IRESs differ markedly in their interactions with the back of the 40S and eIF3, both converge to a similar structural solution for the placement of the AUG initial codon close to the 40S P site.

4.3.6 Swiveling of the 40S head locks the 5'-UTR-IRES and induces a compact conformation of eIF3

Initial processing of the cryo-EM data revealed a marked dynamic of the head of the 40S. Masked classification and refinement in Relion3[42] revealed two major populations, distinguished by different degrees of swiveling of the 40S head. The 40S head is attached to the body by a single RNA helix which makes this component of the ribosome extremely [32]. Intrinsic and independent movements of the 40S head are instrumental in tRNA translocation and also in canonical initiation[62,63]. The 5'-UTR-IRES seems to exploit this intrinsic dynamic to, in a first instance, bind to the 40S and, in a second instance, "lock" the IRES in a specific conformation committing the complex towards viral translation (**Fig. 4.8**).

In the class-1 (open conformation) complex, the head of the 40S shows an almost canonical configuration with very little swiveling and no tilt. In this open conformation, the latch of the 40S (an early defined contact between the head and the body of the 40S[64]) is closed. At the other side of the 40S head, the channel formed by ribosomal proteins uS7 and uS11 is exposed and eIF3d density is not well defined, probably due to a high degree of flexibility or low occupancy (**Fig. 4.8 A**, left). In the class-2 (closed conformation) complex, the head of the 40S has experienced a medium range degree of swiveling, compared to the widest displacement reported[62]. The movement of the 40S head is followed by the 5'-UTR-IRES, with DI experimenting the highest displacement compared to its position in class-1 (**Fig. 4.8 A, right and B**). In the swiveled conformation in calss-2, the latch is open, and the channel formed by ribosomal proteins uS7 and uS11 is plugged by eIF3d, which in this class, presents robust density (**Fig. 4.8 C**). These movements are restricted to the 40S-head and the domains of 5'-UTR-IRES, especially domain I. No movement could be detected in the core subunits of eIF3 in the closed class.; an identical configuration of eIF3 with respect the 40S body is maintained between the open and closed classes (**Fig. 4.8 B**). The robust density observed for eIF3d in the closed class results from the stabilization of this subunit due to its contacts with the core subunit eIF3a (**Fig. 4.8 C**). Well defined density in this area could be observed for the interface eIF3a/eIF3d (**Fig. 4.8 C, right**). Thus, in the closed conformation, eIF3 shows a hitherto unknown conformation, which we speculate could reflect a transient state populated at some point during canonical initiation and exploited by the 5′-UTR-IRES in order to gain access to the P-site of the small subunit[65].

4.3.7 The ternary complex (TC) delivers Met-tRNA$_i$^{Met} to uAUG at position 701 and initiation factors eIF1 and eIF1A assist on AUG location

Our structures of the 40S/5′-UTR-IRES/eIF3 complex revealed that the position of domain III of the 5′-UTR-IRES overlaps with the position the ternary complex populates at the E site in canonical initiation[13,58] (**Fig. 4.6 C**). Additionally, in our maps we could only confidently identify density for the single-stranded segment of RNA of the IRES placed close to the P-site until residue 695, whereas the canonical AUG of ORF1 is found at nucleotide 709. These facts prompted us to wonder how the delivery of Met-tRNA$_i$^{Met} to the AUG is accomplished. Using reconstituted initiation assays with native components and toe-printing analysis[37], we could dissect the different steps followed by the 5′-UTR-IRES in order to correctly place Met-tRNA$_i$^{Met} based paired with the correct AUG codon at the P-site (**Fig. 4.9 and Fig. 4.10**). Translation initiation factors in mammals and insects are highly homologous. In particular, eIF2-alpha shares 57% identity and 74% similarity between human and Drosophila, eIF2-beta 74% identity and 83% similarity, eIF2-gamma 82% identity and 88% similarity, eIF5B 71% identity and 85% similarity, eIF3a 46% identity and 63% similarity, and eIF3c 51% identity and 66% similarity.

This high level of homology justifies the utilization of mammalian initiation factors for CrPV analysis, as has been done before for the CrPV IGR-IRES.

Toe-print assays identify the location of functional ribosomal complexes assembled on mRNAs using reverse transcription of a primer annealed to the mRNA. The length of the resulting cDNA fragment provides information about the position of the ribosome on the mRNA. Due to its large size, the paused ribosome protects a segment of the mRNA, inhibiting primer extension and generating toe-print signals approximately 15–17 nucleotides downstream of the P-site of 40S. Cognate aminoacyl or peptidyl-tRNAs in the P-site or post-termination complexes with eRF1 in the A-site yield robust toe-print signals[66]. The 40S–5'-UTR-IRES–eIF3 complex produces signal that is ascribable to the secondary structure of the 5'-UTR-IRES **(Fig. 4.10, Lanes 1–3)**, indicating no measurable pausing of the ribosome on the mRNA around any of the AUG codons. By itself, the TC (eIF2– Met-tRNA$_i^{Met}$ –GTP) is able to load Met-tRNA$_i^{Met}$ onto the P-site of the 40S–5'-UTR-IRES–eIF3 complex, producing a robust toe-print **(Fig. 4.9 A, lane 2, labelled as 48S–uAUG)** 15–17 nucleotides away from the uAUG located at 701. A similar uAUG delivery of Met-tRNA$_i^{Met}$ can be accomplished by eIF5B which, under stress conditions, has been described as substituting for eIF2 in Met-tRNA$_i^{Met}$ delivery[67–70], with eukaryotic initiation then following a 'bacterial-like' mode **(Fig. 4.9 A, Lane 4)**. Transitioning to the correct AUG could only be detected in the presence of eIF1and eIF1A, and only when the TC was present, and not for eIF5B **(Fig. 4.9 A, Lanes 3 and 5)**. Notably, the presence of eIF1and eIF1A seems to be detrimental to uAUG Met-tRNA$_i^{Met}$ loading by eIF5B, as their presence significantly reduces the toe-print signal that can be observed for eIF5B in isolation. However, no concomitant increase in toe-print signal for the canonical AUG could be observed for the eIF1/eIF1A/eIF5B reaction.

Only eIF2 as part of the TC and assisted by eIF1 and eIF1A can properly locate the *bona fide* AUG of ORF1. The role of the uAUG located at nucleotide 701 is not clear, but the fact that eIF5B can deliver Met-tRNA$_i^{Met}$ only to this codon points towards an important role for this uAUG in initiation when eIF2 is unavailable.

4.4 Discussion

The development of ribosome profiling and its use in dissecting translation in higher eukaryotes, especially mammals, has expanded our once simplified perception of translational landscapes[22,71]. There is now direct evidence that translation regulation can be accomplished via numerous alternative pathways that involve not only the direct regulation of ribosomes but also sequences adjacent to the coding regions of mRNAs[72]. As such, 5′-UTRs are of special interest, given their pervasive presence in mammalian transcriptomes[19,73]. Ribosome profiling has revealed the presence of translating ribosomes paused on 5′-UTRs, specially under stress conditions, showing that such regions play critical role in regulating translation[72,74]. Given the proficiency and diversity viruses exhibit in exploiting the eukaryotic protein production machinery, it would be surprising if they didn't circumvent the plethora of regulatory mechanisms implemented at the level of 5′-UTRs in one way or another for viral protein synthesis. An IRES found in the 5′-UTR of the (+)-ssRNA of the *Dicistovirus* Cricket Paralysis Virus seems to have evolved one such mechanism.

The CrPV 5′-UTR-IRES directs initiation towards ORF1 in the early phase of viral infection[75,76]. Expression of ORF1 is instrumental for virus replication, as the RNA-dependent RNA polymerase (RdRp) and the protease used for the proteolytic digestion of the structural polyprotein are encoded in ORF1[77]. Given the poor sequence homology between 5′-UTR regions within the family of viruses that CrPV belongs to, it was challenging even to propose a secondary structure diagram for this IRES. Applying cryo-EM techniques combined with toe-print analysis of initiation reactions assembled with native components, we were able to determine the three-dimensional structure of the 5'-UTR-IRES in its ribosome bound configuration **(Fig. 4.7)** and define the initiation route followed by this IRES to assemble an elongation competent initiation complex.

The 5'-UTR-IRES features a novel multi-domain, extended architecture that it uses to encircle three quarters of the 40S head, exploiting binding sites not previously described for any IRESs **(Figs. 4.4, 4.5, and 4.6)**. The IRES uses residue specific binding with ribosomal proteins uS3 and RACK1 and these interactions are used by the IRES to anchor its domain I to the back of the 40S head **(Fig. 4.4)**. The structure thus rationalizes previous data showing a preeminent role of RACK1 in CrPV and related viruses that infect *Drosophila*[78]. The interaction of domain I with RACK1 is also instrumental to position domain II at the back of the 40S body, sandwiched in between ribosomal protein uS17 and eIF3 **(Fig. 4.5)**. Interestingly and in contrast with the HCV-IRES and the CSFV-IRES, the conformation observed for eIF3 in the complex with 5'-UTR-IRES is very similar to the conformation observed for eIF3 in the canonical/native 48S complex, with the IRES "filling up" cavities present between the 40S and eIF3 in this complex[58]. The HCV-IRES and related IRESs like the CSFV-IRES displace eIF3 from its canonical location, using a very different mechanism for docking the IRES to the 40S.

In order to place the uAUG of ORF1 in the proximity of the P site, the 5'-UTR-IRES access the P site through the E site, in a similar manner as the HCV-IRES **(Fig. 4.6, C)**[60]. One difference even within this mechanism, however, is that HCV uses a trans-element (domain II) while 5'-UTR-IRES uses a cis-element (domain III). In this aspect, the 5'-UTR-IRES recapitulates binding strategies known for other IRESs like the IGR-CrPV-IRES that make use of ribosomal protein eS25 for its movement through the ribosome or the HCV-IRES which places its domains II and IV in close proximity of the P site, sliding the elongated domain II from the back of the 40S to the P site through the E site[79].

The placement of the AUG of ORF1 in the surroundings of the P-site seems to be exerted by a mechanism involving the intrinsic dynamics of the 40S head[32] (**Fig. 4.8**). The 5′-UTR-IRES seems to exploit the intrinsic dynamics of the 40S head to place the AUG of ORF1 in close proximity of the P-site[32]. It exploits the characteristic swiveling movement of the 40S head to bind and progress towards a conformation that "locks" the IRES on the 40S. At the same time, this locked conformation further induces a compact conformation of eIF3 where the eIF3d subunit comes in close contact with the core subunits of eIF3[65]. These dynamics are probably instrumental for the ability of the 5′-UTR-IRES–40S complex to localize the annotated AUG, in a genomic context where a uAUG-stop is physically close. The capacity of the 40S to scan an mRNA bi-directionally upon termination on a stop codon has been previously reported[66]. It is thus plausible that the peculiar genetic configuration of the CrPV around the annotated AUG of ORF1 **(Figure 4.1 A)** evolved to leverage these re-initiation mechanisms already present in the translation of cellular messengers. However, these considerations are highly speculative, as the particular role that the uAUG exerts in Met-tRNA$_i^{Met}$ recruitment, or more generally its involvement in initiation of viral messengers, remains enigmatic. A comprehensive understanding of the role of uAUG and the start-stop configuration will demand further studies, ideally in vivo.

Taking all these observations into account, we propose the following integrated model for how the CrPV 5′-UTR-IRES operates: immediately after the (+)-ssRNA genome of the CrPV is injected in the cytoplasm of the host cell, the 5′-UTR-IRES captures 40S subunits using

Domain I **(Fig. 4.11, bottom right)**. Next, recruitment of eIF3 is mediated by Domain II, allowing the sliding of the flexible linker connecting domain II and domain III through the cannon formed between the head and the platform of the 40S to place domain III in the surroundings of the E site **(Fig. 4.11, bottom left)**. A swivel movement of the 40S head close this channel "locking" the 5′-UTR-IRES in the 40S, inducing a compact conformation of eIF3 with eIF3d subunit in interacting distance with eIF3′s core subunit a **(Fig. 4.11, top left)**. With this configuration, eIF2 as part of the TC can deliver Met-tRNA$_i$^{Met} to the uAUG located at nucleotide 701 and further assistance by initiation factors eIF1/eIF1A allow for a downstream location of the AUG codon of ORF1 at nucleotide 709. Large subunit recruitment will allow transitioning towards elongation, committing the ribosome to the production of viral proteins located in ORF1 **(Fig. 4.11, top right)**.

The extent and diversity of viral populations contrast with our limited understanding of viral strategies for translation competency[3]. Little is known about the structural diversity of viral IRESs. We also know little about how 5′-UTR regions and accessory mRNA sequences can be leveraged by viruses for ribosome hijacking[80]. The study of IRESs has been limited to a handful of model viral IRESs because they offer advantageous traits like compact conformations and structural stability[30]. A preliminary classification of viral IRES sequences sorts known IRESs into four main families[31]. Though this classification has proved useful in guiding research into IRESs, such a simplified system may be insufficient in describing the extensive diversity of IRES mechanism found in nature. For example, the 5′-UTR-IRES of the CrPV described here was assigned to type III family of IRES given its dependency on eIF3 and eIF2. This family of IRES is also defined by the direct assembly of initiation complexes in AUG initiation codons, without the need for "scanning"[11]. However, we have shown that this 5'-UTR-IRES assembles an initiation complex via an intermediate at a uAUG codon, and then, transition to the *bona fide* AUG codon. Even though we cannot call this transition a proper "scanning" event, it isn't either a direct assembly on the AUG codon, which challenges the definition for type III IRESs and the inclusion of this IRES in that class.

Given the rich diversity of viral sequences in the animal virome, new IRESs exploiting different aspects of translation by employing intriguing secondary structures will probably be discovered paving the way for a new system of sorting and, ultimately, understanding the fascinating world of IRESs.

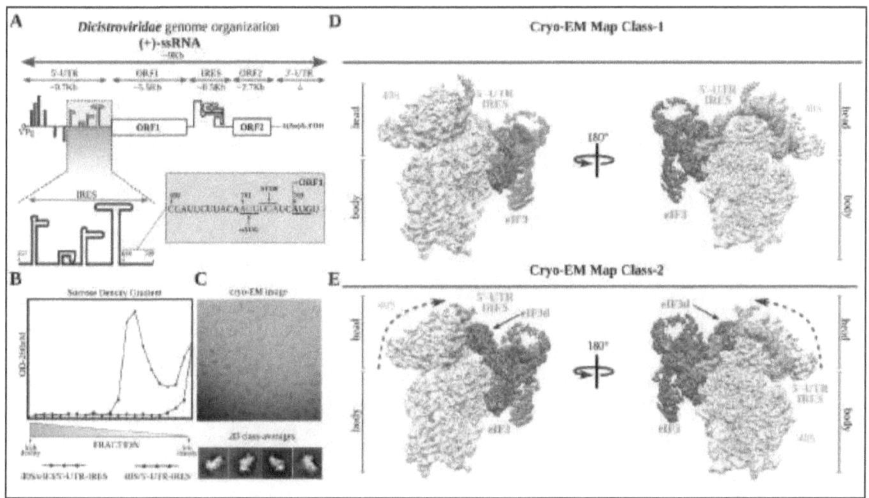

Figure 4.1 | *Dicistroviridae* genome organization, *in vitro* complex formation and cryo-EM maps

A: Top, schematic describing the genome organization of *Dicistroviruses*. Indicated with arrows are approximate genomic lengths for the different components. Bottom, detailed view of the region described to harbor the IRES activity of the 5'-UTR of the CrPV. On the right, sequence adjacent to the initiation AUG codon of ORF1located at nucleotide 701 and preceded by a "star-stop uORF" indicated in red.

B: Sucrose-gradient analysis showed the dependency on eIF3 of the 5'-UTR-IRES in order to form a stable complex with the 40S. Only in the presence of eIF3, 5'-UTR-IRES co-migrates with the 40S.

C: Representative cryo-EM micrograph of the 40S/5'-UTR-IRES/eIF3 complex. Bottom, representative reference-free 2D class averages used for further image processing.

D: After 3D classifications, two classes showing density for 40S (yellow), eIF3 (red) and 5'-UTR-IRES (blue) could be found in the data set. Class 1 (top) present a non-swiveled configuration of the 40S head and density for eIF3d is absent. Class 2, bottom, shows a swiveled configuration of the 40S head (arrows) with eIF3d (indicated) contacting eIF3's core subunit a.

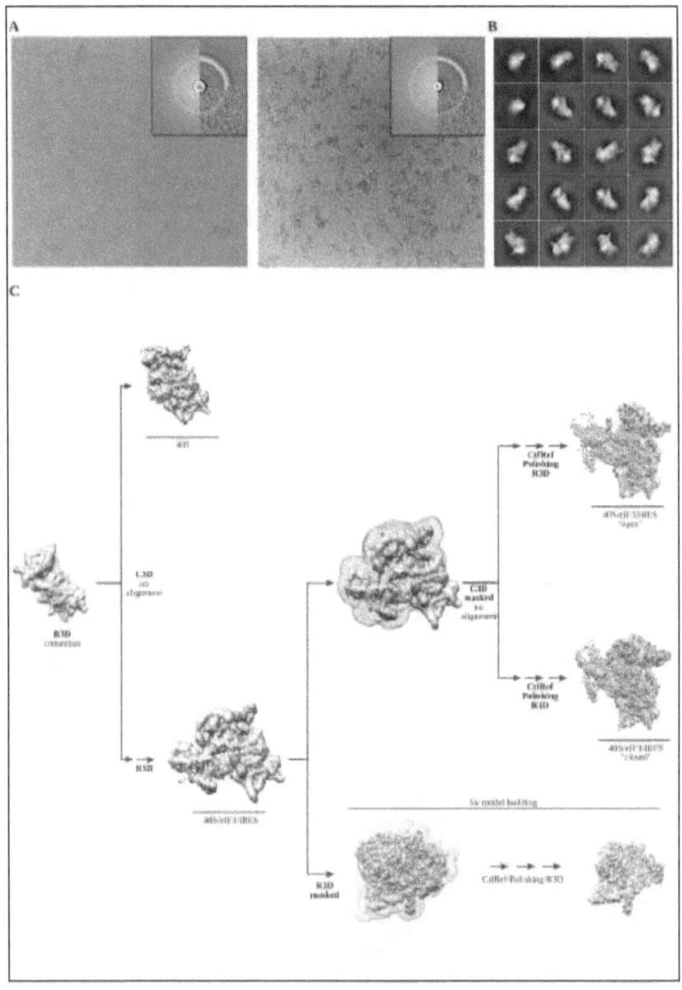

Figure 4.2 | Cryo-EM representative images and classification workflow
A: Two examples of aligned micrographs used for image processing. Data collection in thick ice was instrumental to avoid complex disassembly and preferential orientation.
B: Representative reference-free 2D averages.
C: Classification scheme followed to identify the two classes described.

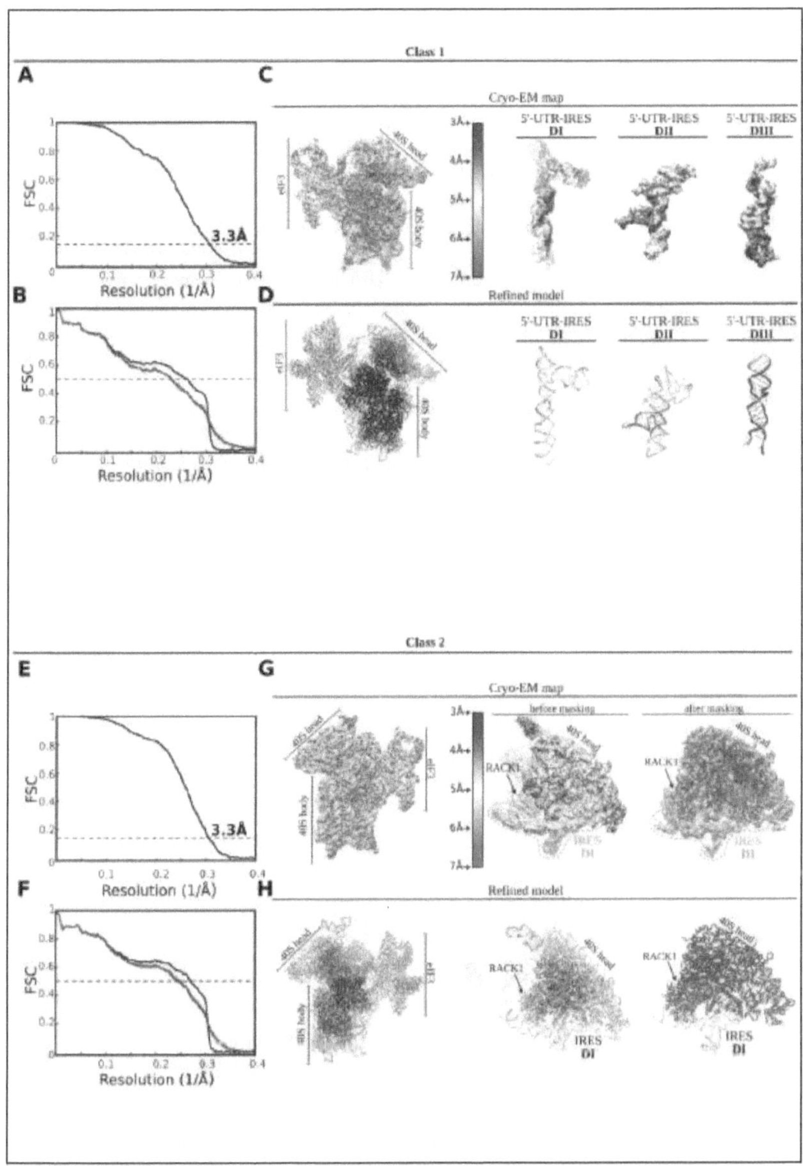

Figure 4.3 | FSC correlation curves, local resolution and model validation

204

Top, class-1:
A: Gold-Standard Fourier Shelf Correlation (FSC) curves between half-maps independently refined in Relion3. Global resolution by the 0.143 cutoff criterion was estimated to be 3.3Å.
B: FSC between the final refined model and final map (black curve). Absence of model overfitting is demonstrated by the overlapping of FSC curves between half-map 1 (included in the refinement, blue) and the map and half-map 2 (not included in the refinement, red).
C: Unsharpened map colored according to local resolution as computed by ResMap. On the right, detailed views for the three 5'-UTR-IRES domains.
D: Refined model colored according to estimated B-factors computed by Refmac.
Bottom, class-2:
E: Gold-Standard Fourier Shelf Correlation (FSC) curves between half-maps independently refined in Relion3. Global resolution by the 0.143 cutoff criterion was estimated to be 3.3Å.
F: FSC between the final refined model and final map (black curve). Absence of model overfitting is demonstrated by the overlapping of FSC curves between half-map 1 (included in the refinement, blue) and the map and half-map 2 (not included in the refinement, red).
G: Unsharpened map colored according to local resolution as computed by ResMap. On the right, detailed views for the 40S head before and after masked refinements.
H: Refined model colored according to estimated B-factors computed by Refmac.

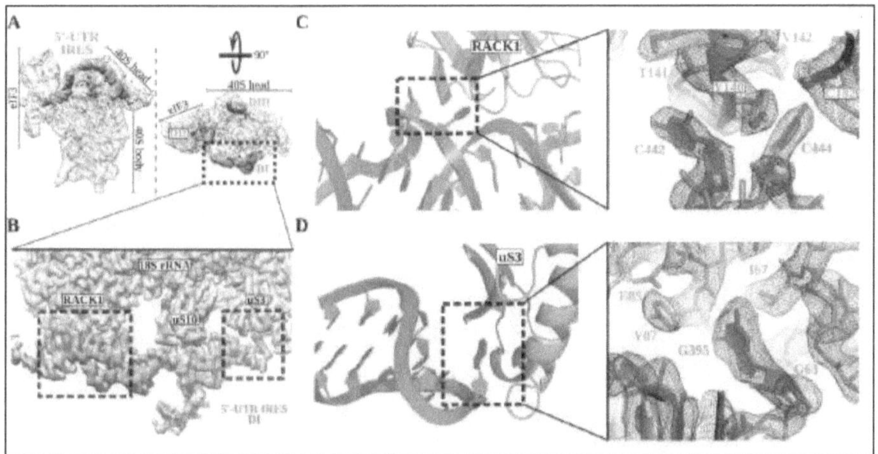

Figure 4.4 | 5′-UTR-IRES domain I engages ribosomal proteins RACK1 and uS3
A: Overview of the 40S/5′-UTR-IRES/eIF3 map with 40S and eIF3 depicted grey and 5′-UTR-IRES blue.
B: Detailed view of the cryo-EM density of the 40S/5′-UTR-IRES/eIF3 map centered around 5′-UTR-IRES domain I (DI). Ribosomal proteins are colored dark yellow, 18S rRNA yellow and 5′-UTR-IRES blue.
C and D: Contacts between 5′-UTR-IRES domain I and ribosomal proteins RACK1 (**C**) and uS3 (**D**) could be defined due to excellent local cryo-EM densities.

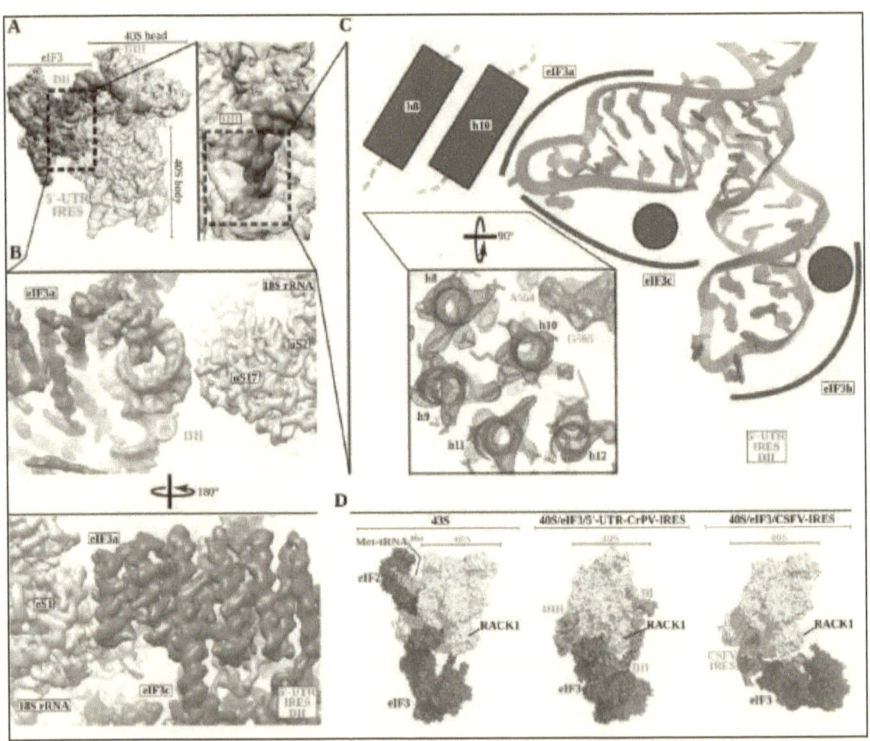

Figure 4.5 | 5′-UTR-IRES domain II is formed by a dual hairpin mediating eIF3 recruitment

A, Overview of the 40S/5′-UTR-IRES/eIF3 cryo-EM map with 40S colored grey, eIF3 red and 5′-UTR-IRES blue. On the right a zoomed view centered around 5′-UTR-IRES domain II. **B,** Detailed view of the cryo-EM map for the region occupied by 5′-UTR-IRES domain II with 40S components colored yellow, eIF3 red and 5′-UTR-IRES blue. Domain II is sandwiched between ribosomal protein uS17 located at the back of the 40S body and eIF3 core subunits a and c. **C,** 5′-UTR-IRES domain II is formed by a dual hairpin that stablishes interactions with α-helices 8 and 10 from eIF3a. These contacts are mediated mainly by basic residues of eIF3 and the phosphate backbone of the IRES. **D,** superposition of the 40S/5′-UTR-IRES/eIF3 complex with canonical 43S complex (left, PDB ID 6FEC) and with the CSFV-IRES/40S complex (right, PDB ID 4c4q). The 5′-UTR-IRES binds to the 40S with a conformation compatible with the canonical position described for eIF3 in the 43S complex.

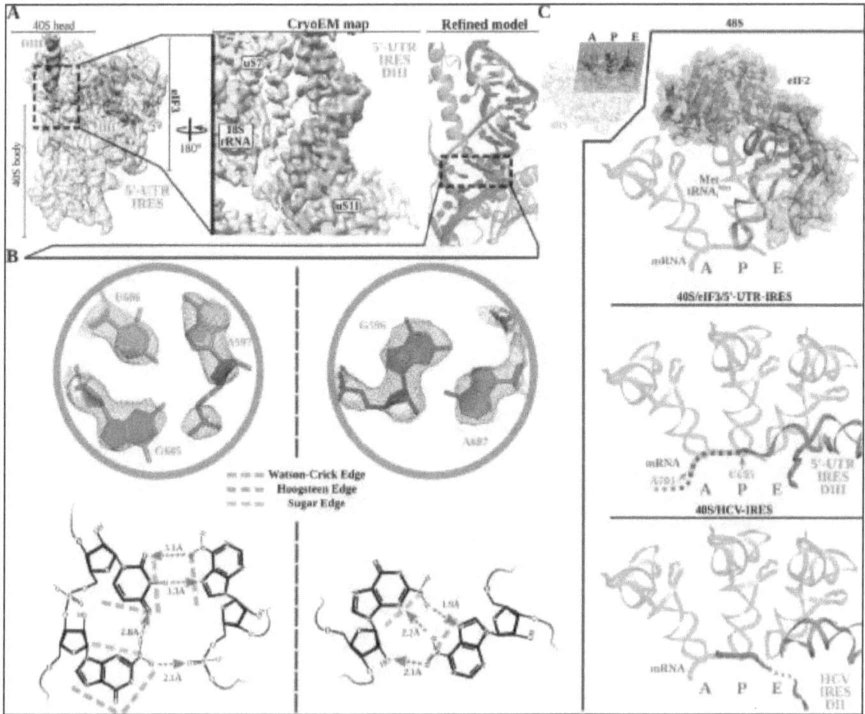

Figure 4.6 | Non-canonical base pairing in 5′-UTR-IRES domain III assist on P site access
A: Overview of the 40S/5′-UTR-IRES/eIF3 cryo-EM map with 40S and eIF3 colored grey and
5′-UTR-IRES blue. On the right, detailed view of the E site where 5′-UTR-IRES domain III is
placed. Shown is the cryo-EM map with the 5′-UTR-IRES colored blue and 40S components
yellow and, on the left, the final refined model colored following the same color scheme.
Ribosomal proteins uS7 and uS11 as well as several 18S rRNA bases contact 5′-UTR-IRES
domain III.
B: Non-canonical base pairs found in 5′-UTR-IRES domain III induce a distortion of the double
helix near the E site. Two examples are shown with the refined model inserted in the
experimental cryo-EM density at the top, and corresponding chemical diagrams of the bases
involved with the base edges and hydrogen bonds at the bottom.
C: Superpositions of canonical 43S complex (PDB ID 6FEC, top), HCV-IRES/40S complex
(PDB ID 5A2Q, bottom) with the 40S/5′-UTR-IRES/eIF3 (middle) model focused on the tRNA
A, P and E binding sites. 5′-UTR-IRES domain III and HCV-IRES occupies a space on the E site
that overlaps with the position occupied by eIF2 in the canonical 43S complex.

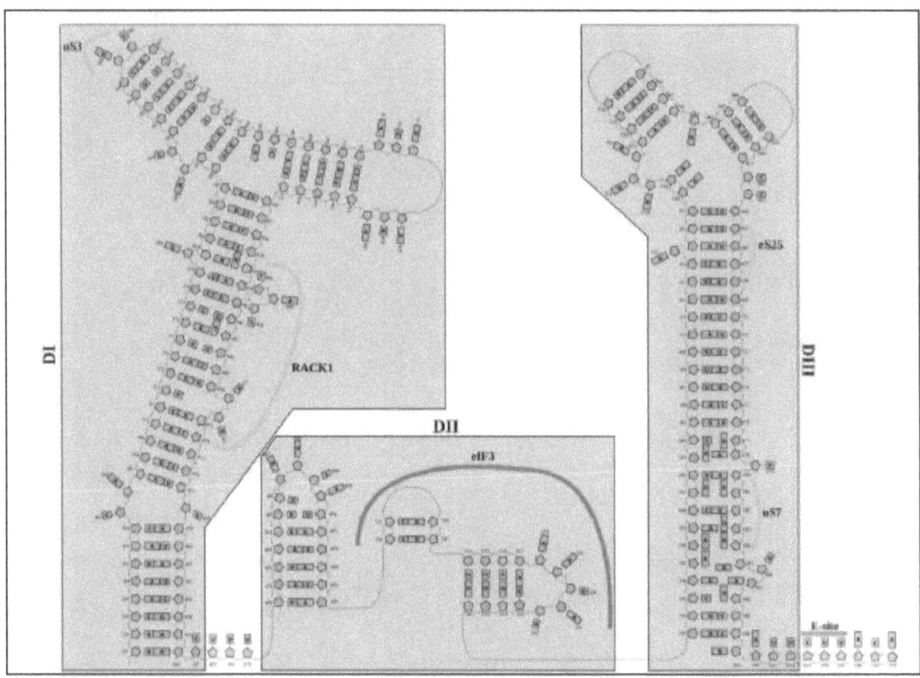

Figure 4.7 | Structurally derived secondary structure diagram for 5′-UTR-IRES
Secondary structure diagram for the 5′-UTR-CrPV IRES derived from the cryo-EM structure.

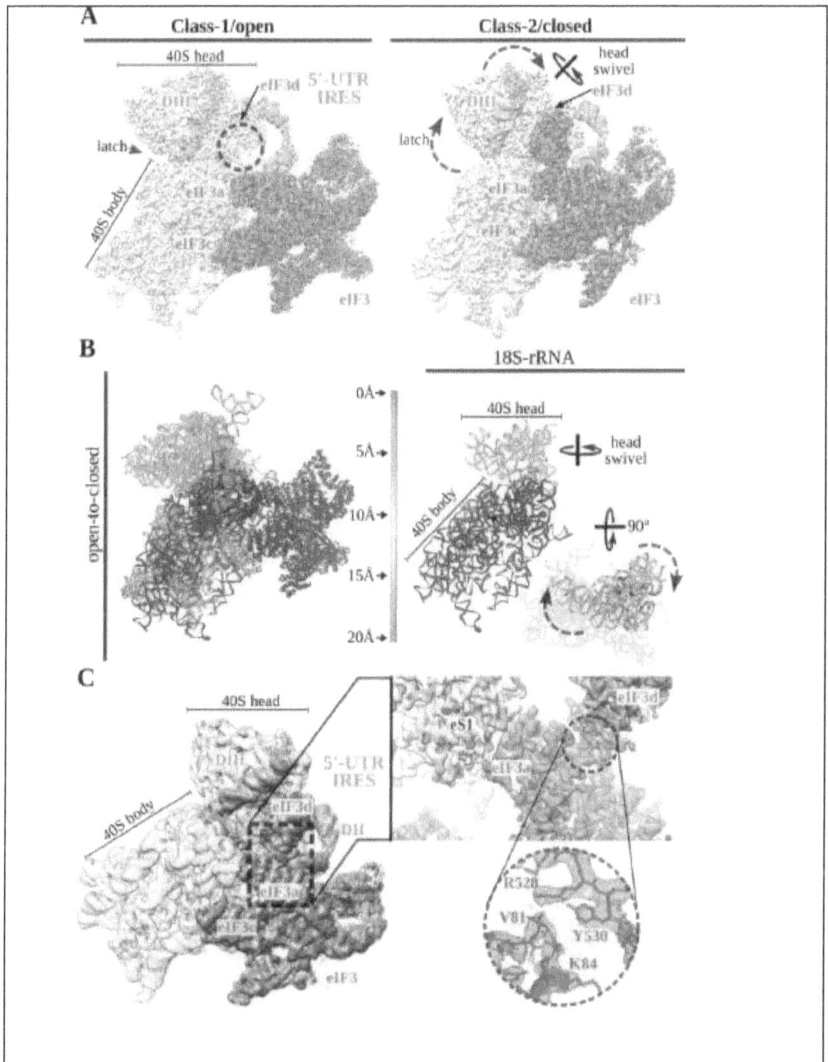

Figure 4.8 | A 40S head swiveling movement "locks" 5′-UTR-IRES on the 40S inducing a compact eIF3 configuration

A: Cryo-EM maps obtained for the two classes present in the 40S/5′-UTR-IRES/eIF3 dataset with 40S colored yellow, eIF3 red and 5′-UTR-IRES blue. Indicated, the position of the latch, eIF3d and the swiveling rotation axis. A 40S head swiveling movement in class 2 bring closer

210

eIF3d to the core subunit of eIF3, eIF3a, establishing interactions that stabilize its conformation.
B: Left, ribbon diagram colored according to pairwise root mean square deviation (rmsd) displacements for the open to close transition with displacements scale at the center. On the right, simplified diagram showing only the 18S rRNA colored with the same scale as on the left. Two orthogonal view are shown where it can be appreciated the main displacement is localized on the 40S head.
C: Overview of the closed class with 40S colored grey, eIF3 red and 5'-UTR-IRES blue. Inset, detail of the experimental density obtained for the eIF3a/eIF3d interface for this class. Clear side chains information was present in the maps, allowing proper building and model refinement.

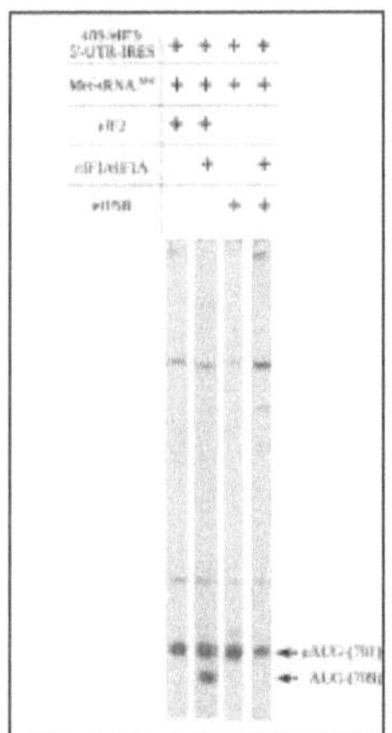

Figure 4.9 | 5′-UTR-IRES requires TC, eIF1 and eIF1A to assemble a functional initiation complex via an uAUG intermediate

Toe-print analysis for initiation reactions with components indicated on the top, all of them performed in the presence of GTP. eIF2 is able to deliver Met-tRNA$_i^{Met}$ to the uAUG (left line, arrow 701) and requires the presence of eIF1/eIF1A to transition to the *bona fide* AUG codon of ORF1 (second line from left, arrow 709). Initiation factor eIF5B, under some conditions, can substitute eIF2 in Met-tRNA$_i^{Met}$ delivery. A robust toe-print signal could be detected in the presence of eIF5B (third line from left, arrow 701) however, Met-tRNA$_i^{Met}$ was deliver to the uAUG at nucleotide 701. Initiation factor eIF5B is unable to find the "correct" AUG even in the presence of eIF1/eIF1A (fourth line from left, arrow 709).

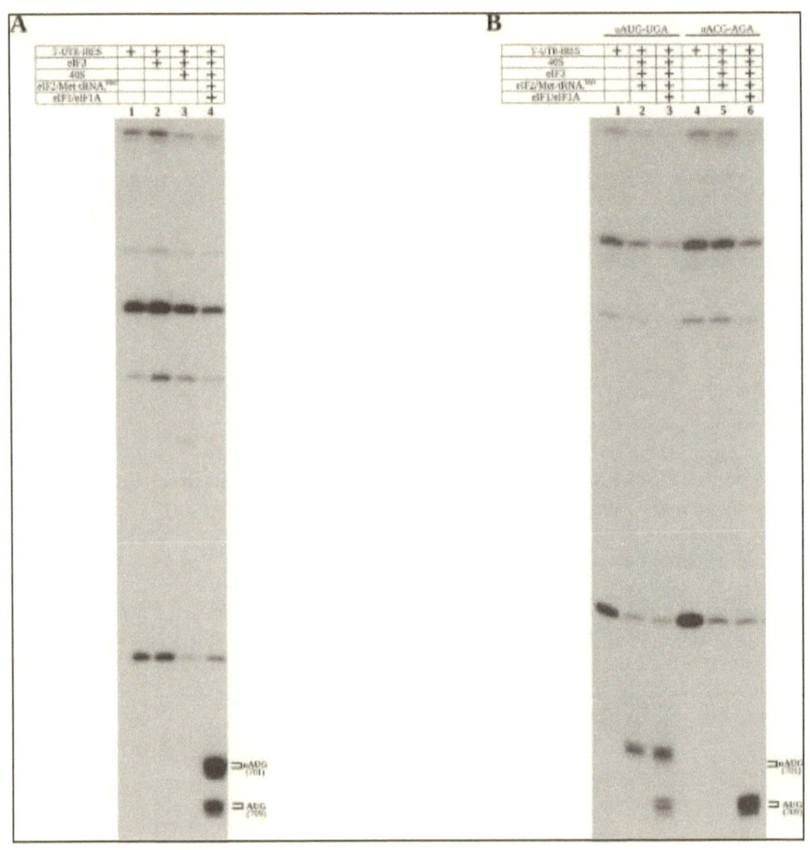

Figure 4.10 | Control toe-print experiments
A: Toe-print analysis for 5′-UTR-IRES, 5′-UTR-IRES–eIF3 and 40S/5′-UTR-IRES–eIF3 complexes (lanes 1, 2 and 3, respectively). No toe-print signal was detected that is ascribable to uAUG or annotated AUG in any of these reactions. Robust toe-print signal could be detected 15–17 nucleotides downstream of uAUG and AUG only for 40S–5′-UTR-IRES–eIF3, and in the presence of TC (eIF2–Met-tRNAi^{Met}–GTP) and eIF1 or eIF1A.
B: The uAUG toe-print signal can be abolished by mutation of the uAUG to ACG, redirecting the Met-tRNAi^{Met} loading event exclusively to the annotated AUG in the presence of TC and eIF1 or eIF1A.

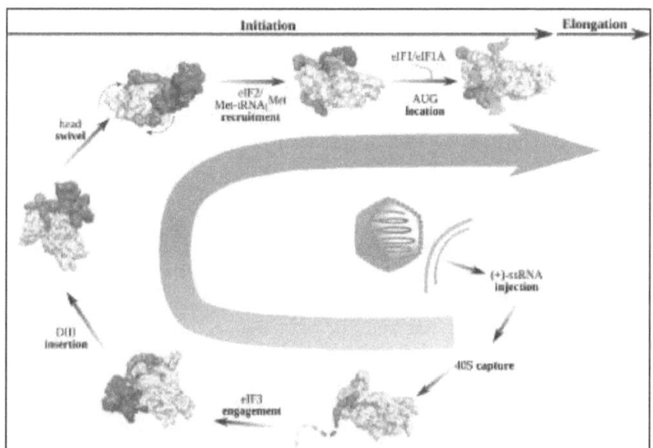

Figure 4.11 | 5′-UTR-IRES requires TC, eIF1 and eIF1A to assemble a functional initiation complex via an uAUG intermediate

A model for 5′-UTR-IRES mediated translation initiation, from the bottom right: injection of the genomic (+)-ssRNA of the CrPV into the cytosol allows the pre-folded 5′-UTR-IRES DI to capture free 40S subunits. Via interactions with DII, eIF3 is recruited to the complex with an initial canonical conformation of the 40S head without tilt or swivel. Insertion of 5′-UTR-IRES DIII in the vicinity of the E site and a swiveling movement of the 40S head, induce a "locked" conformation of the complex with the uAUG at 701 in the vicinity of the P-site. Delivery of Met-tRNA$_i$^{Met} and location of ORF1 AUG is achieved by the concerted action of eIF2, eIF1 and eIF1A via an uAUG intermediate. Large subunit recruitment mediated by eIF5B will allow transitioning into elongation.

Table 4.1 | Data collection, model refinement and validation statistics

Cryo-EM data collection, refinement and validation statistics		
	Class-1 (open) **(EMDB-21529) (PDB 6W2S)**	**Class-2 (closed)** **(EMDB-21530) (PDB 6W2T)**
Data collection and processing		
Magnification	130,000	
Voltage (kV)	300	
Electron exposure (e–/Å^2)	59.55	
Defocus range (μm)	−1 to − 3	
Pixel size (Å)	1.06	
Symmetry imposed	C1	
Initial particle images (no.)	915,647	
Final particle images (no.)	14,257	23,444
Map resolution (Å)	3.3	3.3
FSC threshold	0.143	0.143
Map resolution range (Å)	3–8	3–8
Refinement		
Initial model used (PDB code)	5A2Q	5A2Q
Model resolution (Å)	3.6	3.6
FSC threshold	0.5	0.5
Model resolution range (Å)	3.3–8	3.3–8
Map sharpening B factor (Å^2)	−31.94	−43.41
Model composition		
Non-hydrogen atoms	106,817	109,684
Ligands	-	-
B factors (Å^2)		
Protein	92.47	96.5
RNA	114.1	117.4
R.m.s. deviations		
Bond lengths (Å)	0.014	0.014
Bond angles (°)	1.77	1.78
Validation		
MolProbity score	2.12	1.99
Clashscore	6.13	4.92
Poor rotamers (%)	1.62	1.39
Ramachandran plot		
Favored (%)		
Allowed (%)	88.92	90.25
Disallowed (%)	98.37	98.50
RNA validation	1.63	1.50
Angles outliers (%)	0.18	0.17
Sugar puckers outliers (%)	2.35	2.25
Average suit	0.442	0.428

4.6 References

1.   Zhang, Y.-Z., Wu, W.-C., Shi, M. & Holmes, E. C. The diversity, evolution and origins of vertebrate RNA viruses. *Current Opinion in Virology* **31**, 9–16 (2018).

2.   Greninger, A. L. A decade of RNA virus metagenomics is (not) enough. *Virus Research* **244**, 218–229 (2018).

3.   Zhang, Y.-Z., Chen, Y.-M., Wang, W., Qin, X.-C. & Holmes, E. C. Expanding the RNA Virosphere by Unbiased Metagenomics. *Annu. Rev. Virol.* **6**, 119–139 (2019).

4.   Shi, M. *et al.* Redefining the invertebrate RNA virosphere. *Nature* **540**, 539–543 (2016).

5.   Dolja, V. V. & Koonin, E. V. Metagenomics reshapes the concepts of RNA virus evolution by revealing extensive horizontal virus transfer. *Virus Research* **244**, 36–52 (2018).

6.   Hashem, Y. & Frank, J. The Jigsaw Puzzle of mRNA Translation Initiation in Eukaryotes: A Decade of Structures Unraveling the Mechanics of the Process. *Annu. Rev. Biophys.* **47**, 125–151 (2018).

7.   Schmeing, T. M. & Ramakrishnan, V. What recent ribosome structures have revealed about the mechanism of translation. *Nature* **461**, 1234–1242 (2009).

8.   Jackson, R. J., Hellen, C. U. T. & Pestova, T. V. The mechanism of eukaryotic translation initiation and principles of its regulation. *Nat Rev Mol Cell Biol* **11**, 113–127 (2010).

9.   Aitken, C. E. & Lorsch, J. R. A mechanistic overview of translation initiation in eukaryotes. *Nature Structural & Molecular Biology* **19**, 568–576 (2012).

10.   Myasnikov, A. G., Simonetti, A., Marzi, S. & Klaholz, B. P. Structure–function insights into prokaryotic and eukaryotic translation initiation. *Current Opinion in Structural Biology* **19**, 300–309 (2009).

11.   Hinnebusch, A. G. The Scanning Mechanism of Eukaryotic Translation Initiation. *Annu. Rev. Biochem.* **83**, 779–812 (2014).

12.   Hinnebusch, A. G. Structural Insights into the Mechanism of Scanning and Start Codon Recognition in Eukaryotic Translation Initiation. *Trends in Biochemical Sciences* **42**, 589–611 (2017).

13.   Hussain, T. *et al.* Structural Changes Enable Start Codon Recognition by the Eukaryotic Translation Initiation Complex. *Cell* **159**, 597–607 (2014).

14.   Pestova, T. V. *et al.* The joining of ribosomal subunits in eukaryotes requires eIF5B. *Nature* **403**, 332–335 (2000).

15.   Lee, J. H. *et al.* Initiation factor eIF5B catalyzes second GTP-dependent step in eukaryotic translation initiation. *Proceedings of the National Academy of Sciences* **99**, 16689–16694 (2002).

16.   Voorhees, R. M. & Ramakrishnan, V. Structural Basis of the Translational Elongation Cycle*. *Annual Review of Biochemistry* **82**, 203–236 (2013).

17.   Starck, S. R. *et al.* Translation from the 5′ untranslated region shapes the integrated stress response. *Science* **351**, aad3867–aad3867 (2016).

18.   Shatsky, I. N., Dmitriev, S. E., Terenin, I. M. & Andreev, D. E. Cap- and IRES-independent scanning mechanism of translation initiation as an alternative to the concept of cellular IRESs. *Mol Cells* **30**, 285–293 (2010).

19.   Sendoel, A. *et al.* Translation from unconventional 5′ start sites drives tumour initiation. *Nature* **541**, 494–499 (2017).

20.   Young, S. K. & Wek, R. C. Upstream Open Reading Frames Differentially Regulate Gene-specific Translation in the Integrated Stress Response. *J. Biol. Chem.* **291**, 16927–16935 (2016).

21.   Wethmar, K. The regulatory potential of upstream open reading frames in eukaryotic gene expression: uORF-mediated translational control. *WIREs RNA* **5**, 765–768 (2014).

22.   Brar, G. A. & Weissman, J. S. Ribosome profiling reveals the what, when, where and how of protein synthesis. *Nat. Rev. Mol. Cell Biol.* **16**, 651–664 (2015).

23.   Hinnebusch, A. G. Gene-specific translational control of the yeast GCN4 gene by phosphorylation of eukaryotic initiation factor 2. *Molecular Microbiology* **10**, 215–223 (1993).

24.   Vattem, K. M. & Wek, R. C. Reinitiation involving upstream ORFs regulates ATF4 mRNA translation in mammalian cells. *Proceedings of the National Academy of Sciences* **101**, 11269–11274 (2004).

25.   Gunišová, S., Hronová, V., Mohammad, M. P., Hinnebusch, A. G. & Valášek, L. S. Please do not recycle! Translation reinitiation in microbes and higher eukaryotes. *FEMS Microbiol Rev* **42**, 165–192 (2018).

26.   Weisser, M. *et al.* Structural and Functional Insights into Human Re-initiation Complexes. *Molecular Cell* **67**, 447-456.e7 (2017).

27.   Lozano, G. & Martínez-Salas, E. Structural insights into viral IRES-dependent translation mechanisms. *Current Opinion in Virology* **12**, 113–120 (2015).

28.   Jaafar, Z. A. & Kieft, J. S. Viral RNA structure-based strategies to manipulate translation. *Nature Reviews Microbiology* **17**, 110–123 (2019).

29.   Jan, E., Mohr, I. & Walsh, D. A Cap-to-Tail Guide to mRNA Translation Strategies in Virus-Infected Cells. *Annu. Rev. Virol.* **3**, 283–307 (2016).

30.   Yamamoto, H., Unbehaun, A. & Spahn, C. M. T. Ribosomal Chamber Music: Toward an Understanding of IRES Mechanisms. *Trends in Biochemical Sciences* **42**, 655–668 (2017).

31.   Filbin, M. E. & Kieft, J. S. Toward a structural understanding of IRES RNA function. *Current Opinion in Structural Biology* **19**, 267–276 (2009).

32.   Johnson, A. G., Grosely, R., Petrov, A. N. & Puglisi, J. D. Dynamics of IRES-mediated translation. *Phil. Trans. R. Soc. B* **372**, 20160177 (2017).

33.   Kieft, J. S. Comparing the three-dimensional structures of Dicistroviridae IGR IRES RNAs with other viral RNA structures. *Virus Research* **139**, 148–156 (2009).

34.   Nakashima, N. & Uchiumi, T. Functional analysis of structural motifs in dicistroviruses. *Virus Research* **139**, 137–147 (2009).

35.   Gross, L. *et al.* The IRES5′UTR of the dicistrovirus cricket paralysis virus is a type III IRES containing an essential pseudoknot structure. *Nucleic Acids Research* **45**, 8993–9004 (2017).

36.   Pestova, T. V. & Hellen, C. U. T. Reconstitution of eukaryotic translation elongation in vitro following initiation by internal ribosomal entry. *Methods* **36**, 261–269 (2005).

37.   Kolupaeva, V. G., Breyne, S. de, Pestova, T. V. & Hellen, C. U. T. In Vitro Reconstitution and Biochemical Characterization of Translation Initiation by Internal Ribosomal Entry. in *Methods in Enzymology* vol. 430 409–439 (Elsevier, 2007).

38.   Pisarev, A. V. *et al.* The Role of ABCE1 in Eukaryotic Posttermination Ribosomal Recycling. *Molecular Cell* **37**, 196–210 (2010).

39.   Russo, C. J. & Passmore, L. A. Controlling protein adsorption on graphene for cryo-EM using low-energy hydrogen plasmas. *Nature Methods* **11**, 649–652 (2014).

40.   Carragher, B. *et al.* Leginon: An Automated System for Acquisition of Images from Vitreous Ice Specimens. *Journal of Structural Biology* **132**, 33–45 (2000).

41.   Lander, G. C. *et al.* Appion: An integrated, database-driven pipeline to facilitate EM image processing. *Journal of Structural Biology* **166**, 95–102 (2009).

42.   Zivanov, J. *et al.* New tools for automated high-resolution cryo-EM structure determination in RELION-3. *eLife* **7**,.

43.   Zheng, S. Q. *et al.* MotionCor2: anisotropic correction of beam-induced motion for improved cryo-electron microscopy. *Nature Methods* **14**, 331–332 (2017).

44.   Zhang, K. Gctf: Real-time CTF determination and correction. *Journal of Structural Biology* **193**, 1–12 (2016).

45.   Biology, ©2020 MRC Laboratory of Molecular, Avenue, F. C., Campus, C. B., CB2 0QH, C. & Uk. 01223 267000. Zhang Software. *MRC Laboratory of Molecular Biology* https://www2.mrc-lmb.cam.ac.uk/research/locally-developed-software/zhang-software/.

46.   Kucukelbir, A., Sigworth, F. J. & Tagare, H. D. Quantifying the local resolution of cryo-EM density maps. *Nature Methods* **11**, 63–65 (2014).

47.   Pettersen, E. F. *et al.* UCSF Chimera—A visualization system for exploratory research and analysis. *Journal of Computational Chemistry* **25**, 1605–1612 (2004).

48.   Emsley, P. & Cowtan, K. Coot: model-building tools for molecular graphics. *Acta Cryst D* **60**, 2126–2132 (2004).

49.   Afonine, P. V. *et al.* Real-space refinement in PHENIX for cryo-EM and crystallography. *Acta Cryst D* **74**, 531–544 (2018).

50.   Murshudov, G. N., Vagin, A. A. & Dodson, E. J. Refinement of Macromolecular Structures by the Maximum-Likelihood Method. *Acta Crystallogr D Biol Crystallogr* **53**, 240–255 (1997).

51.   Brown, A. *et al.* Tools for macromolecular model building and refinement into electron cryo-microscopy reconstructions. *Acta Crystallogr D Biol Crystallogr* **71**, 136–153 (2015).

52.   Amunts, A. *et al.* Structure of the Yeast Mitochondrial Large Ribosomal Subunit. *Science* **343**, 1485–1489 (2014).

53.   Yokoyama, T. *et al.* HCV IRES Captures an Actively Translating 80S Ribosome. *Molecular Cell* **74**, 1205-1214.e8 (2019).

54.   Hashem, Y. *et al.* Hepatitis-C-virus-like internal ribosome entry sites displace eIF3 to gain access to the 40S subunit. *Nature* **503**, 539–543 (2013).

55.   Scheres, S. H. W. RELION: Implementation of a Bayesian approach to cryo-EM structure determination. *Journal of Structural Biology* **180**, 519–530 (2012).

56.   Scheres, S. H. W. Processing of Structurally Heterogeneous Cryo-EM Data in RELION. in *Methods in Enzymology* vol. 579 125–157 (Elsevier, 2016).

57.   Hashem, Y. *et al.* Structure of the Mammalian Ribosomal 43S Preinitiation Complex Bound to the Scanning Factor DHX29. *Cell* **153**, 1108–1119 (2013).

58.   Eliseev, B. *et al.* Structure of a human cap-dependent 48S translation pre-initiation complex. *Nucleic Acids Research* **46**, 2678–2689 (2018).

59.   Leontis, N. B. & Westhof, E. Conserved geometrical base-pairing patterns in RNA. *Quart. Rev. Biophys.* **31**, 399–455 (1998).

60.   Yamamoto, H. *et al.* Molecular architecture of the ribosome-bound Hepatitis C Virus internal ribosomal entry site RNA. *EMBO J* **34**, 3042–3058 (2015).

61.   Quade, N., Boehringer, D., Leibundgut, M., van den Heuvel, J. & Ban, N. Cryo-EM structure of Hepatitis C virus IRES bound to the human ribosome at 3.9-Å resolution. *Nat Commun* **6**, 7646 (2015).

62.   Ratje, A. H. *et al.* Head swivel on the ribosome facilitates translocation by means of intra-subunit tRNA hybrid sites. *Nature* **468**, 713–716 (2010).

63.   Roy, C. R. & Cherfils, J. Structure and function of Fic proteins. *Nature Reviews Microbiology* **13**, 631–640 (2015).

64.   Frank, J. *et al.* A model of protein synthesis based on cryo-electron microscopy of the E. coli ribosome. *Nature* **376**, 441–444 (1995).

65.   Lee, A. S. Y., Kranzusch, P. J., Doudna, J. A. & Cate, J. H. D. eIF3d is an mRNA cap-binding protein that is required for specialized translation initiation. *Nature* **536**, 96–99 (2016).

66.   Skabkin, M. A., Skabkina, O. V., Hellen, C. U. T. & Pestova, T. V. Reinitiation and Other Unconventional Posttermination Events during Eukaryotic Translation. *Molecular Cell* **51**, 249–264 (2013).

67.   Terenin, I. M., Dmitriev, S. E., Andreev, D. E. & Shatsky, I. N. Eukaryotic translation initiation machinery can operate in a bacterial-like mode without eIF2. *Nat Struct Mol Biol* **15**, 836–841 (2008).

68.   Pestova, T. V., de Breyne, S., Pisarev, A. V., Abaeva, I. S. & Hellen, C. U. T. eIF2-dependent and eIF2-independent modes of initiation on the CSFV IRES: a common role of domain II. *EMBO J* **27**, 1060–1072 (2008).

69.   Yamamoto, H. *et al.* Structure of the mammalian 80S initiation complex with initiation factor 5B on HCV-IRES RNA. *Nat Struct Mol Biol* **21**, 721–727 (2014).

70.   Kenner, L. R. *et al.* eIF2B-catalyzed nucleotide exchange and phosphoregulation by the integrated stress response. *Science* **364**, 491–495 (2019).

71.   Ingolia, N. T., Ghaemmaghami, S., Newman, J. R. S. & Weissman, J. S. Genome-Wide Analysis in Vivo of Translation with Nucleotide Resolution Using Ribosome Profiling. *Science* **324**, 218–223 (2009).

72.   Andreev, D. E. *et al.* Insights into the mechanisms of eukaryotic translation gained with ribosome profiling. *Nucleic Acids Res* **45**, 513–526 (2017).

73.   Resch, A. M., Ogurtsov, A. Y., Rogozin, I. B., Shabalina, S. A. & Koonin, E. V. Evolution of alternative and constitutive regions of mammalian 5′UTRs. *BMC Genomics* **10**, 162 (2009).

74.   Archer, S. K., Shirokikh, N. E., Beilharz, T. H. & Preiss, T. Dynamics of ribosome scanning and recycling revealed by translation complex profiling. *Nature* **535**, 570–574 (2016).

75.   Garrey, J. L., Lee, Y.-Y., Au, H. H. T., Bushell, M. & Jan, E. Host and Viral Translational Mechanisms during Cricket Paralysis Virus Infection. *Journal of Virology* **84**, 1124–1138 (2010).

76.   Hertz, M. I. & Thompson, S. R. In vivo functional analysis of the Dicistroviridae intergenic region internal ribosome entry sites. *Nucleic Acids Research* **39**, 7276–7288 (2011).

77.   Jan, E., Kinzy, T. G. & Sarnow, P. Divergent tRNA-like element supports initiation, elongation, and termination of protein biosynthesis. *Proceedings of the National Academy of Sciences* **100**, 15410–15415 (2003).

78.   Majzoub, K. *et al.* RACK1 Controls IRES-Mediated Translation of Viruses. *Cell* **159**, 1086–1095 (2014).

79.   Pisareva, V. P., Pisarev, A. V. & Fernández, I. S. Dual tRNA mimicry in the Cricket Paralysis Virus IRES uncovers an unexpected similarity with the Hepatitis C Virus IRES. *eLife* **7**, e34062 (2018).

80.   Terenin, I. M., Smirnova, V. V., Andreev, D. E., Dmitriev, S. E. & Shatsky, I. N. A researcher's guide to the galaxy of IRESs. *Cell. Mol. Life Sci.* **74**, 1431–1455 (2017).

Chapter 5:

Discussion

5.1 Structural biology using cryo-EM informs us of a different mechanism of action for a type IV IRES

When this body of work was initiated little was known about the structural diversity of the type IV IRESs. Structural study of type IV IRESs from the *Cripavirus* family of IRESs was extensive but limited to the prototypical Cricket Paralysis Virus IRES (CrPV-IRES)[1–5]. Structural understanding of the *Aparavirus* family of IRESs was unavailable. While the Taura Syndrome Virus IRES (TSV-IRES) was structurally characterized using cryo-EM[6], the Taura Syndrome Virus doesn't neatly fall into either of the two families[7] and is postulated to represent a new family viruses[8]. This meant that the structural diversity of type IV IRESs from the *Aparavirus* family remained unexplored. Since members of the *Aparavirus* family cluster tightly within a phylogenetic tree[8] and since some of them infect agriculturally important organisms[9,10], structural characterization of a prototypical IRES from this family was warranted. With these considerations in mind we sought to structurally characterize the Israeli Acute Paralysis Virus IRES (IAPV-IRES)[11,12].

As one goal of this body of work, we selected to structurally characterize the type IV IAPV-IRES motivated by intriguing RNA features unique to this IRES and by its enhanced coding capacity [13–16]. To this end, as detailed in chapter 3, we have generated accurate models from high resolution cryo-EM reconstructions of the IAPV-IRES as it binds to and transits through the ribosome[17]. From these models are able to deduce the trajectory of the IRES from its initial binding to the small subunit of the ribosome, to its pre-translocated state on the A-site of the 80S ribosome, and, finally to its post-translocated state on the P-site of the ribosome. Once the IAPV-IRES is in the P-site and the first viral coding codon in a vacant A-site is ready to accept the first elongator aminoacyl-tRNA, the peculiar translation initiation route these IRESs follow can be considered finished.

We gained structural insights on how the unique RNA features of the IAPV-IRES interact with the ribosome which helped explain their effect on initial IRES binding and their role in facilitating translocation of the IRES through the ribosome. These interactions mimic particular conformations of endogenous tRNAs which the IRES uses to hijack and steal host cell ribosomes for viral protein production. We saw how different such a mechanism is from previously characterized mechanism of a different type IV IRES. Key difference of the type IV class II *Aparavirus* IRESs seems to be their usage of a stem-loop structure (SLIII) to interact with the large subunit of the ribosome and restrict rotational freedom of the small subunit and use that to their advantage. Since the type IV class I *Cripavirus* IRESs lack such a feature, they follow a different regime where the small subunit of the ribosome is rotating uninhibited[1,2]. Thus, there seems to be two different strategies within the type IV IRESs for stealing ribosomes. These insights and the structural models produced from this body of work lay the ground work for further studies of the IAPV-IRES.

A still unanswered question about the IAPV-IRES is how it achieves +1 frameshifting[13]. Before this body of work was initiated, it was discovered that the IAPV-IRES undergoes a +1 frameshifting event to produce a short peptide from a hidden gene labelled ORFx[13]. ORFx was shown to be translated during infection of the host by the virus[14] but how the IRES switches from the 0 frame to the +1 frame and thus expand its coding capacity was unknown. It was hypothesized that base pairing of the +1 base of the first coding codon with another base on PKI unlocks +1 frame[14] but this could not be supported by our data[17]. Even though the structural underpinnings of the +1 frameshifting by the IAPV-IRES remains elusive, the various RNA bases throughout the IRES that either couple or decouple the 0 and +1 frames when mutated can now be contextualized on a detailed structural model of the IRES bound to the ribosome. This can help explain how such mutations might affect frame selection when the IRES is bound to the ribosome. The models should also help in rational design of other mutations to study this particular type of +1 frameshifting.

This body of work also lays excellent groundwork for visualizing the exit of an IRES from the ribosome. From our studies we have found that the IAPV-IRES binds particularly well to the ribosomes and is able to better tolerate the harsh conditions required for cryo-EM grid making and imaging[17]. Studies exploring what this IRES looks like in the E-site of the ribosome and what it looks like outside the ribosome can now be pursued using our work as a starting point. The structural contribution of the elusive SLVI found at the 5′-end of the IAPV-IRES in +1 frameshifting, if any, might also be finally elucidated from such studies.

Another avenue this body of work has opened is related to structural determination of RNA only samples using cryo-EM. Since we know that the IAPV-IRES binds to the ribosome as a highly structured and compact unit, we hypothesized that it exists in solution as a preformed unit and can be imaged using cryo-EM. Indeed, in a small set of cryo-EM images of only the IAPV-IRES containing sample, we were able to identify the IRES by eye and pick the particles without using any reference templates **(Appendix C, Fig. C1 A)**. Using these particles we were ultimately able to obtain low resolution densities that showed the characteristic shape of the IAPV-IRES **(Appendix C, Fig. C1 C, D)**. Our preliminary work identifies the IAPV-IRES as an excellent test sample for a high resolution structure determination of an RNA-only sample using cryo-EM[18,19].

5.2 Structure of a novel 5′-UTR IRES identifies a divergent mechanism within the type III IRESs

The second goal of this body of work was to determine the structure of the type III 5′-UTR-IRES found in the genome of CrPV. To this end, as detailed in chapter 4, we have generated a model of the 5′-UTR IRES when it is bound to the small subunit of the ribosome[20]. From this *de novo* built structural model of the IRES we were able to see the complex shape of the IRES as it wraps itself around the head of the small subunit. We also observed how the different domains of the IRES interact with the small subunit at novel binding sites. The model derived from high resolution density data, allowed us to visualize individual RNA bases at some key regions of the IRES that either extruded from the IRES body to interact with the ribosome or formed non-canonical base pairs or triplets.

We also observed a previously unknown compact conformation of a key initiation factor induced by the IRES. We also saw how in two different IRESs using two different RNA modules, evolution has arrived at a similar solutions for placing the start codon at the P-site. The HCV-IRES, another member of the type III IRES family, uses a 5′-end stem-loop (domain II) to interact with the small subunit from the E-site to gain access to the P-site[21–23]. The 5′-UTR-IRES, on the other hand, uses a complex 3′-end structure (domain III) to do the same. The end result of which seems to be access to the P-site via the E-site and a possible restriction on concurrent recruitment of the ternary complex delivering the initiator tRNA.

We also observed how the 5′-UTR-IRES binds to the small subunit with a canonical-like configuration of the pre initiation complex. The configuration adopted by eIF3 on the small subunit when it is bound to the 5′-UTR-IRES is more similar to the configuration of eIF3 when bound to a canonical 48S complex rather than to a configuration of eIF3 when bound to another type III IRES. Toeprinting assays performed using the 5′-UTR-IRES indicated that while an alternative "bacterial-like" initiation pathway can deliver the initiator tRNA to an upstream start codon of the IRES, only the conventional set of eukaryotic initiation factors can locate and deliver the initiator tRNA to the genuine start codon of the IRES.

We believe that the above observations, the reagents generated, the data collection strategies, and the processing insights we learned while undertaking this goal provide us with a firm launchpad to further explore this IRES. Our initial attempts to resolve a state with the initiator tRNA delivered to the start codon of the IRES were unsuccessful. But we believe that careful control of the reaction conditions using newly available grid making, reagent mixing, grid screening, and real-time data processing technologies in cryo-EM will make such state more tractable. Such technologies will also help resolve other states of the IRES such as the 5′-UTR-IRES bound to the 80S ribosome. Given that the enormous multi-domain initiation factor eIF3 was better resolved in our structure than any before it[24–26], the 5′-UTR-IRES can potentially be used as a system to further resolve and study this multifaceted factor.

The two different IRESs studied in this body of work have thus yielded secrets that have advanced our understanding of IRESs. As discussed above and detailed in chapters 3 and 4—and in the published papers resulting from this body of work[17,20]—the features and mechanisms that the two IRESs deploy are unique. In some aspects, such features and mechanisms are novel and distinct even from other IRESs that they are supposedly similar to. While they are indeed unique and different in many respects, looking at them a bit closer, we can see that these two IRESs share some commonalities.

5.3   Two different IRESs share some commonalities

One of most obvious commonalities between the two IRESs is their complex RNA structure. Complex RNA structure is a theme that runs deep through all IRESs families[27]. 2D structure probing assays can inform about the 3D folds and organization of an RNA molecule but only so much. Only after looking at high resolution maps and models of the IRESs can one truly appreciate the marvel of the RNA molecule in IRESs. In our high resolution maps and the subsequently derived models, we see that both the IAPV-IRES and the 5′-UTR-IRES have complex folds, interactions, and geometries within their RNA structure[17,20]. We can also appreciate how modular these IRESs are. Both have stretches of RNA that form distinct modules which are usually tasked with one particular job. For example, a pseudoknot containing module in IAPV-IRES is tasked with mimicking the codon-anticodon interactions while a dual hairpin module recruits an initiation factor in the 5′-UTR-IRES. Such modules can be mixed-and-matched within IRES types indicating wide versatility and potential for rapid evolution.

Another commonality we can appreciate is how these two IRESs restrict key movements of the small subunit. The small subunit, as discussed in chapter 1, is highly dynamic with many distinct movement patterns[28–30]. Key among them is head tilt and head swivel, both of which are important in translocation during elongation and in loading of the initiator tRNA during initiation[31,32]. Both IRESs seem to heavily dampen and subsequently exploit these movements to their advantage. This restriction of movement most likely helps anchor the IRES into place and prevents binding sites for canonical initiation site from becoming available. In the case of the IAPV-IRES, even the rotation of the small subunit during ratcheting is severely restricted and consequently exploited by the IRES to adopt a particular conformation of endogenous tRNAs[17].

While the 5′-UTR-IRES exploits head swiveling[20], it is yet to be seen if it also restricts subunit rotation.

Another key commonality between the two IRESs is that a subset of the ribosomal proteins that they interact with are the same. Despite being remarkably different in both sequence and structure, the IAPV-IRES and the 5′-UTR-IRES both interact with the universally conserved ribosomal proteins uS7 and uS11. Both IRESs also interact with the eukaryotic specific protein eS25[33,34]. This binding to shared set of evolutionarily conserved proteins, their common tactic of exploiting universal ribosomal dynamics, and their RNA biased mechanism point to a common origin of these two IRESs in an RNA based world.

5.4 IRESs are possibly the missing link between cap-independent and cap-dependent translation initiation systems

The RNA world hypothesis posits that life emerged from RNA[35,36]. What started as self-replicating RNA molecules in the primordial soup of the ancient earth has resulted in life as we know it today. When the prototypical molecules of life—RNA, DNA, and protein—got enclosed in water filled lipid sacs by chance, the very first cells came into existence. As the unrelenting forces of evolution acted on these proto-cells, life sprang forth into the incredible diversity of forms that can be seen in the fossil records of yesterday and in the branches and leaves of today's tree of life[37]. On this tree of life, one can also see the most critical juncture in the evolution of life: the acquisition of a nucleus.

The acquisition of a nucleus has been one of the most successful accidental ventures that life has ever undertaken. Instead of staying as unicellular flotsam in tide pools of water, nucleated cells have evolved into a myriad of both unicellular and multicellular forms which have successfully made almost every niche on earth their home. Acquisition of nucleus was probably a singular event in the entire history of life and it happened when an archaeal cell merged with an alphaproteobacterial endosymbiont[38–40]. This common ancestor of all nucleated cells was the first eukaryote.

Eukaryogenesis presented with its own set of challenges for the ancestral eukaryotic cell and its descendants. It is hypothesized that an intron invasion concurrent with eukaryogenesis led to interruptions in and breaking up of the polycistronic messages that were prevalent in the archaeal host cell[41]. The newly acquired nuclear membrane also meant that transcription and translation were separated in both time and space[42]. These challenges led the ancestral eukaryote to evolve transcriptionally coupled systems of protecting and correcting the intron interrupted mRNAs before exporting it to the cytoplasm for translation by ribosomes[41,42]. It is likely that a version of the Shine-Dalgarno based ribosome initiation system found in currently extant prokaryotes still existed in the archaeal host cell ancestor of eukaryotes[43,44]. The ancestral eukaryotes and its direct descendants most likely used such a system to make proteins. This system however was most likely severely damaged or lost because of the aforementioned intron invasion during eukaryogenesis. As evolutionary pressures mounted yet again, ancestral eukaryotes are thought to have adapted by evolving a translation initiation system which eventually led to the highly expanded cap-dependent but nimble and responsive initiation system found in modern day eukaryotes. How did eukaryotes get there? IRESs provide some clues.

The primary goal of an IRES, a goal identical to that of either the 5′-cap based system or the Shine-Dalgarno based system, is to recruit and properly place the small subunit of the ribosome at the start codon of a message. The 5′-cap based system achieves this using a complex set of dynamics and interplay between a host of different initiation factors[45]. The Shine-Dalgarno based system, on the other hand, uses relatively simple geometries of RNA base pairing to achieve the same goal[46]. IRESs lie somewhere in between the two.

Some IRESs, as shown previously by others[1–6] and as we saw in chapter 3[17], can first recruit the small subunit and then assemble an entire elongation competent ribosome with total independence from any eukaryotic initiation factors. Still other IRESs[21–23,26,47], as we saw in chapter 4[20], can recruit the small subunit with the help of just a single factor and then assemble an elongation competent ribosome using just a bare minimum set of initiation factors. While these IRESs are modern day IRESs found in modern day viruses, one can imagine an evolutionary past when similar IRESs with minimal to zero factor requirements would have come in handy.

Indeed, it has been hypothesized that IRESs could have acted as an intermediate step in the evolution of protein synthesis[43,44,48]. In early eukaryotes, it is thought that unspliced introns located at the 5'-UTR acted as proto-IRESs[48]. Splicing factors bound to such proto-IRESs and transported them to the cytoplasm. The splicing factors also played a critical role in recruiting the small subunit of the ribosome. As RNA binding proteins were probably already evolved during this evolutionary period, it is thought that such proteins bound to the proto-IRES and facilitated protein synthesis. Such protein eventually evolved into the modern day initiation factors along with the incorporation of eukaryote specific features such as scanning and 5'-cap dependence.

If we look at the different types of modern day IRESs, we can imagine a hypothetical pathway for evolution of cap-dependent initiation from IRESs. In an early eukaryote, a type IV-like IRES could have handled all the protein synthesis needs of the cell[27]. Given its zero factor requirement and RNA-only mechanism, it wouldn't have needed any of the yet to evolve eukaryotic specific initiation factors to function. Lending credence to this idea is the fact that type IV IRESs, though are found in viruses that infect eukaryotes, bind to ribosomes across kingdoms[1,2,6,49–51]. As they bind to a highly conserved region of the ribosome they have been shown to bind to yeast[6] or mammalian[2] or human[50] or even bacterial[49] ribosomes. In the evolutionary past, such binding was probably possible with the archaeal host cell ribosomes too.

Next, as more IRES-binding proteins and initiation factors evolved, a situation similar to that found in type III IRESs could have been possible[27]. A case where a few initiation factors evolved to assist the IRES, similar to one found in the IRES we saw in chapter 4[20], is plausible. Such a scenario would still be largely dependent on the structure of the RNA. Then, with the evolution of large scaffolding proteins such as eIF4G, helicases such as eIF4A, and energy generating factors, IRESs could have moved away from a RNA structure dependent model to a protein factor reliant model. This is similar to the mechanism found in type II IRESs.

Addition of even more factors and gradual relaxation of the RNA structure dependence could have created a system similar to the type I IRESs[27]. Here, early versions of scanning[43,44,48] could have been explored before the eukaryotic cell switched to a completely cap-dependent initiation system similar to the one found in eukaryotes of today.

Four types of IRESs exist in the world today but they aren't molecular fossils preserved in the amber of time since the first eukaryotes. Even if the first eukaryotes used an IRES-like system such early IRESs have most likely evolved out of existence. Yes, an IRES-like system could have evolved into the cap-dependent initiation found in eukaryotes of today but such an event happened so long ago that any evidence of its existence is also lost to time. The modern day IRESs are found in a small number of viruses that infect a narrow range of hosts[52]. As viruses more often than not evolve from their own hosts, it is most likely that the IRESs in such viruses evolved after the eukaryotes had established themselves. The distant set of hosts of such

IRESs containing viruses suggests that the different types of IRESs evolved independently from

one another.

Whatever their origins, modern day IRESs help shed light on an evolutionary past where

both cap-dependent and IRES-dependent initiation systems could have co-existed in the same

cell.

5.5   References

1.   Fernandez, I. S., Bai, X.-C., Murshudov, G., Scheres, S. H. W. & Ramakrishnan, V. Initiation of Translation by Cricket Paralysis Virus IRES Requires Its Translocation in the Ribosome. *CELL* **157**, 823–831 (2014).

2.   Muhs, M. *et al.* Cryo-EM of Ribosomal 80S Complexes with Termination Factors Reveals the Translocated Cricket Paralysis Virus IRES. *Molecular Cell* **57**, 422–432 (2015).

3.   Murray, J. *et al.* Structural characterization of ribosome recruitment and translocation by type IV IRES. *ELIFE* **5**, (2016).

4.   Abeyrathne, P. D., Koh, C. S., Grant, T., Grigorieff, N. & Korostelev, A. A. Ensemble cryo-EM uncovers inchworm-like translocation of a viral IRES through the ribosome. *ELIFE* **5**, (2016).

5.   Pisareva, V. P., Pisarev, A. V. & Fernández, I. S. Dual tRNA mimicry in the Cricket Paralysis Virus IRES uncovers an unexpected similarity with the Hepatitis C Virus IRES. *eLife* **7**, e34062 (2018).

6.   Koh, C. S., Brilot, A. F., Grigorieff, N. & Korostelev, A. A. Taura syndrome virus IRES initiates translation by binding its tRNA-mRNA-like structural element in the ribosomal decoding center. *Proceedings of the National Academy of Sciences* **111**, 9139–9144 (2014).

7.   Bonning, B. C. & Miller, W. A. Dicistroviruses. *Annual Review of Entomology* **55**, 129–150 (2010).

8.   Hertz, M. I. & Thompson, S. R. In vivo functional analysis of the Dicistroviridae intergenic region internal ribosome entry sites. *Nucleic Acids Research* **39**, 7276–7288 (2011).

9.   Maori, E. *et al.* Isolation and characterization of Israeli acute paralysis virus, a dicistrovirus affecting honeybees in Israel: evidence for diversity due to intra- and inter-species recombination. *Journal of General Virology* **88**, 3428–3438 (2007).

10.   Cox-Foster, D. L. *et al.* A Metagenomic Survey of Microbes in Honey Bee Colony Collapse Disorder. *Science* **318**, 283–287 (2007).

11.   Firth, A. E., Wang, Q. S., Jan, E. & Atkins, J. F. Bioinformatic evidence for a stem-loop structure 5 '-adjacent to the IGR-IRES and for an overlapping gene in the bee paralysis dicistroviruses. *Virology Journal* **6**, 193 (2009).

12.   Sabath, N., Price, N. & Graur, D. A potentially novel overlapping gene in the genomes of Israeli acute paralysis virus and its relatives. *Virol J* **6**, 144 (2009).

13.   Ren, Q. *et al.* Alternative reading frame selection mediated by a tRNA-like domain of an internal ribosome entry site. *Proceedings of the National Academy of Sciences* **109**, E630–E639 (2012).

14.   Au, H. H. *et al.* Global shape mimicry of tRNA within a viral internal ribosome entry site mediates translational reading frame selection. *Proc Natl Acad Sci USA* **112**, E6446–E6455 (2015).

15.	Ren, Q., Au, H. H. T., Wang, Q. S., Lee, S. & Jan, E. Structural determinants of an internal ribosome entry site that direct translational reading frame selection. *Nucleic Acids Res* **42**, 9366–9382 (2014).

16.	Au, H. H. T., Elspass, V. M. & Jan, E. Functional Insights into the Adjacent Stem-Loop in Honey Bee Dicistroviruses That Promotes Internal Ribosome Entry Site-Mediated Translation and Viral Infection. *JOURNAL OF VIROLOGY* **92**, (2018).

17.	Acosta-Reyes, F., Neupane, R., Frank, J. & Fernández, I. S. The Israeli acute paralysis virus IRES captures host ribosomes by mimicking a ribosomal state with hybrid tRNAs. *EMBO J* **38**, (2019).

18.	Kladwang, W., VanLang, C. C., Cordero, P. & Das, R. A two-dimensional mutate-and-map strategy for non-coding RNA structure. *Nature Chemistry* **3**, 954–962 (2011).

19.	Zhang, K. *et al.* Cryo-EM structure of a 40 kDa SAM-IV riboswitch RNA at 3.7 Å resolution. *Nature Communications* **10**, 5511 (2019).

20.	Neupane, R., Pisareva, V. P., Rodriguez, C. F., Pisarev, A. V. & Fernández, I. S. A complex IRES at the 5'-UTR of a viral mRNA assembles a functional 48S complex via an uAUG intermediate. *eLife* **9**, e54575 (2020).

21.	Quade, N., Boehringer, D., Leibundgut, M., van den Heuvel, J. & Ban, N. Cryo-EM structure of Hepatitis C virus IRES bound to the human ribosome at 3.9-Å resolution. *Nat Commun* **6**, 7646 (2015).

22.	Yamamoto, H. *et al.* Structure of the mammalian 80S initiation complex with initiation factor 5B on HCV-IRES RNA. *Nat Struct Mol Biol* **21**, 721–727 (2014).

23.	Yamamoto, H. *et al.* Molecular architecture of the ribosome-bound Hepatitis C Virus internal ribosomal entry site RNA. *EMBO J* **34**, 3042–3058 (2015).

24.	Cate, J. H. D. Human eIF3: from 'blobology' to biological insight. *Phil. Trans. R. Soc. B* **372**, 20160176 (2017).

25.	Eliseev, B. *et al.* Structure of a human cap-dependent 48S translation pre-initiation complex. *Nucleic Acids Research* **46**, 2678–2689 (2018).

26.	Hashem, Y. *et al.* Structure of the Mammalian Ribosomal 43S Preinitiation Complex Bound to the Scanning Factor DHX29. *Cell* **153**, 1108–1119 (2013).

27.	Filbin, M. E. & Kieft, J. S. Toward a structural understanding of IRES RNA function. *Current Opinion in Structural Biology* **19**, 267–276 (2009).

28.	Guo, Z. & Noller, H. F. Rotation of the head of the 30S ribosomal subunit during mRNA translocation. *PNAS* **109**, 20391–20394 (2012).

29.	Mohan, S., Donohue, J. P. & Noller, H. F. Molecular mechanics of 30S subunit head rotation. *Proc Natl Acad Sci U S A* **111**, 13325–13330 (2014).

30.	Ratje, A. H. *et al.* Head swivel on the ribosome facilitates translocation by means of intra-subunit tRNA hybrid sites. *Nature* **468**, 713–716 (2010).

31.   Noller, H. F., Lancaster, L., Zhou, J. & Mohan, S. The ribosome moves: RNA mechanics and translocation. *Nature Structural & Molecular Biology* **24**, 1021–1027 (2017).

32.   Zhou, J., Lancaster, L., Donohue, J. P. & Noller, H. F. How the ribosome hands the A-site tRNA to the P site during EF-G-catalyzed translocation. *Science* **345**, 1188–1191 (2014).

33.   Hertz, M. I., Landry, D. M., Willis, A. E., Luo, G. & Thompson, S. R. Ribosomal Protein S25 Dependency Reveals a Common Mechanism for Diverse Internal Ribosome Entry Sites and Ribosome Shunting. *Molecular and Cellular Biology* **33**, 1016–1026 (2013).

34.   Landry, D. M., Hertz, M. I. & Thompson, S. R. RPS25 is essential for translation initiation by the Dicistroviridae and hepatitis C viral IRESs. *Genes & Development* **23**, 2753–2764 (2009).

35.   Cech, T. R. The RNA Worlds in Context. *Cold Spring Harb Perspect Biol* **4**, a006742 (2012).

36.   Neveu, M., Kim, H.-J. & Benner, S. A. The "Strong" RNA World Hypothesis: Fifty Years Old. *Astrobiology* **13**, 391–403 (2013).

37.   Hug, L. A. *et al.* A new view of the tree of life. *Nat Microbiol* **1**, 16048 (2016).

38.   Cox, C. J., Foster, P. G., Hirt, R. P., Harris, S. R. & Embley, T. M. The archaebacterial origin of eukaryotes. *Proc Natl Acad Sci U S A* **105**, 20356–20361 (2008).

39.   Koonin, E. V. Origin of eukaryotes from within archaea, archaeal eukaryome and bursts of gene gain: eukaryogenesis just made easier? *Philosophical Transactions of the Royal Society B: Biological Sciences* **370**, 20140333 (2015).

40.   Martin, W. F., Garg, S. & Zimorski, V. Endosymbiotic theories for eukaryote origin. *Philos Trans R Soc Lond B Biol Sci* **370**, 20140330 (2015).

41.   Koonin, E. V. The origin of introns and their role in eukaryogenesis: a compromise solution to the introns-early versus introns-late debate? *Biol Direct* **1**, 22 (2006).

42.   Martin, W. & Koonin, E. V. Introns and the origin of nucleus–cytosol compartmentalization. *Nature* **440**, 41–45 (2006).

43.   Hernández, G. On the origin of the cap-dependent initiation of translation in eukaryotes. *Trends in Biochemical Sciences* **34**, 166–175 (2009).

44.   Londei, P. Evolution of translational initiation: new insights from the archaea. *FEMS Microbiol Rev* **29**, 185–200 (2005).

45.   Hinnebusch, A. G. The Scanning Mechanism of Eukaryotic Translation Initiation. *Annu. Rev. Biochem.* **83**, 779–812 (2014).

46.   Shine, J. & Dalgarno, L. The 3′-Terminal Sequence of Escherichia coli 16S Ribosomal RNA: Complementarity to Nonsense Triplets and Ribosome Binding Sites. *PNAS* **71**, 1342–1346 (1974).

47.   Gross, L. *et al.* The IRES5′UTR of the dicistrovirus cricket paralysis virus is a type III IRES containing an essential pseudoknot structure. *Nucleic Acids Research* **45**, 8993–9004 (2017).

48.   Hernández, G. Was the initiation of translation in early eukaryotes IRES-driven? *Trends in Biochemical Sciences* **33**, 58–64 (2008).

49.   Colussi, T. M. *et al.* Initiation of translation in bacteria by a structured eukaryotic IRES RNA. *Nature* **519**, 110–113 (2015).

50.   Spahn, C. M. T. *et al.* Cryo-EM Visualization of a Viral Internal Ribosome Entry Site Bound to Human Ribosomes: The IRES Functions as an RNA-Based Translation Factor. *Cell* **118**, 465–475 (2004).

51.   Schüler, M. *et al.* Structure of the ribosome-bound cricket paralysis virus IRES RNA. *Nat Struct Mol Biol* **13**, 1092–1096 (2006).

52.   Mailliot, J. & Martin, F. Viral internal ribosomal entry sites: four classes for one goal: Viral internal ribosomal entry sites. *WIREs RNA* **9**, e1458 (2018).

Appendix A

The Israeli acute paralysis virus IRES captures host ribosomes by mimicking a ribosomal state with hybrid tRNAs

Francisco Acosta-Reyes[1,†], Ritam Neupane[1,2,†], Joachim Frank[1,2,*] & Israel S Fernández[1,**]

Abstract

Colony collapse disorder (CCD) is a multi-faceted syndrome decimating bee populations worldwide, and a group of viruses of the widely distributed Dicistroviridae family have been identified as a causing agent of CCD. This family of viruses employs non-coding RNA sequences, called internal ribosomal entry sites (IRESs), to precisely exploit the host machinery for viral protein production. Using single-particle cryo-electron microscopy (cryo-EM), we have characterized how the IRES of Israeli acute paralysis virus (IAPV) intergenic region captures and redirects translating ribosomes toward viral RNA messages. We reconstituted two *in vitro* reactions targeting a pre-translocation and a post-translocation state of the IAPV-IRES in the ribosome, allowing us to identify six structures using image processing classification methods. From these, we reconstructed the trajectory of IAPV-IRES from the early small subunit recruitment to the final post-translocated state in the ribosome. An early commitment of IRES/ribosome complexes for global pre-translocation mimicry explains the high efficiency observed for this IRES. Efforts directed toward fighting CCD by targeting the IAPV-IRES using RNA-interference technology are underway, and the structural framework presented here may assist in further refining these approaches.

Keywords internal ribosomal entry sites; Israeli acute paralysis virus; ribosome; translation

Subject Categories Translation & Protein Quality; Structural Biology

DOI 10.15252/embj.2019102226 | Received 10 April 2019 | Revised 2 September 2019 | Accepted 19 September 2019 | Published online 14 October 2019

The EMBO Journal (2019) 38: e102226

Introduction

Apis mellifera, the common western honey bee, is affected worldwide by an enigmatic syndrome characterized by a drastic disappearance of the workforce, causing the accelerated collapse of the hive (Ratnieks & Carreck, 2010). Given the essential role bees play in pollination of economically important crops, the impact of this syndrome, termed colony collapse disorder (CCD), has been estimated to cost the US economy $15 billion in direct loss of crops and $75 billion in indirect losses (Chopra *et al*, 2015).

Though the exact etiology of CCD is unknown (Anderson & East, 2008), a group of viruses belonging to the *Dicistroviridae* family were found in metagenomic studies of CCD-affected hives (Cox-Foster *et al*, 2007; Chen *et al*, 2014). Among this group of viruses, the Israeli acute paralysis virus (IAPV) showed a strong correlation with CCD, revealing a prominent role in the development of the syndrome (Hou *et al*, 2014; Doublet *et al*, 2017).

The *Dicistroviridae* family of viruses exhibits a wide environmental distribution, targeting invertebrates, mainly insects and other arthropods (Shi *et al*, 2016). The genetic architecture of these viruses is composed of a single positive-stranded RNA molecule which contains two open reading frames (ORF1 and ORF2; Wilson *et al*, 2000b; Pisarev *et al*, 2005). ORF1 encodes non-structural proteins: an RNA helicase, a cysteine protease, and an RNA-dependent RNA polymerase (RdRP). ORF2 encodes a single poly-protein that, upon proteolytic digestion, generates the structural proteins that will eventually compose the viral capsid (Kerr & Jan, 2016; Mullapudi *et al*, 2017).

Both ORFs are preceded by non-coding RNA sequences responsible for the regulation of the expression of their downstream genes (Wilson *et al*, 2000a,b; Gross *et al*, 2017). A fine balance between the expression of ORF1 and ORF2 is required for the replication and expansion of the virus (Carrillo-Tripp *et al*, 2016; Khong *et al*, 2016). This is achieved through a precise exploitation of host resources, specially the machinery for protein synthesis (Kerr & Jan, 2016). The non-coding RNA regions preceding both ORFs harbor two different internal ribosomal entry sites (IRESs; Wilson *et al*, 2000b). IRESs are structured RNA sequences able to interfere with canonical translation, capturing host ribosomes in order to redirect them toward the production of viral proteins (Yamamoto *et al*, 2017). Eukaryotic ribosomes are operated by a complex collection of cellular factors that regulate the production of proteins in the cell (Jackson *et al*, 2010). Specially regulated in eukaryotes is the first step of translation, initiation (Aylett & Ban, 2017). During this

---

1 Department of Biochemistry and Molecular Biophysics, Columbia University, New York, NY, USA
2 Department of Biological Sciences, Columbia University, New York, NY, USA
 *Corresponding author. Tel: +1 212 305 9512; E-mail: jf2192@cumc.columbia.edu
 **Corresponding author. Tel: +1 2 2 342 2385; E-mail: isf2106@cumc.columbia.edu
 †These authors contributed equally to this work

initiation phase, the small ribosomal subunit (40S), in partnership with many initiation factors, is able to capture an mRNA, localize its AUG initiation codon, deliver the first aminoacyl-tRNA, and finally recruit the large subunit (60S) in order to assemble an elongation competent ribosome (80S) primed with an aminoacyl-tRNA in the P site and a vacant A site (Hinnebusch & Lorsch, 2012).

The majority of IRES families leverage the complexity of initiation to hijack cellular ribosomes (Yamamoto *et al*, 2017; Jaafar & Kieft, 2019). The IAPV-IRES found in the intergenic region of the IAPV virus belongs to the well-characterized type IV family of viral IRESs. IRES sequences from this group dispense with all canonical initiation factors and are able to assemble by themselves an elongation competent ribosome, successfully redirecting the cellular machinery for viral protein production by an RNA-only mechanism (Hertz & Thompson, 2011). This is accomplished by an elaborate use of intrinsically dynamic elements of the ribosome, naturally involved in translocation (Noller *et al*, 2017a). These IRESs are able to induce an artificial state on the ribosome, mimicking a pre-translocation state with tRNAs. Elongation factors eEF2 and eEF1A can then be recruited to effectively by-pass the highly regulated initiation (Abeyrathne *et al*, 2016; Murray *et al*, 2016; Pisareva *et al*, 2018), jumpstarting directly in the elongation phase (Johnson *et al*, 2017).

The type IV IRES family exhibits a remarkable structural diversity, which remains poorly characterized (Hertz & Thompson, 2011). Two genera, based on phylogenetic analysis of ORF2 as well as the intergenic region, have been defined: Aparaviruses and Cripaviruses. The cricket paralysis virus IRES (CrPV-IRES), the prototypical Cripavirus, has been extensively studied due to its early discovery and use as model mRNA of early studies in translation (Jan *et al*, 2001; Pestova *et al*, 2004). Recently, a divergent IRES sequence of a shrimp-infecting virus, the Taura virus syndrome IRES, has been visualized by cryo-EM in complex with yeast ribosomes (Koh *et al*, 2014; Abeyrathne *et al*, 2016).

The IAPV-IRES presents the prototypical features of an Aparavirus, with an additional stem loop (SL-III) nested within the pseudoknot I (PKI) and an extended L1.1 region (Au *et al*, 2015). Importantly, this IRES can drive translation in two different ORFs, able to produce two different polypeptides from the same mRNA. A frameshift event at the first coding codon is responsible for this multi-coding capacity (Wang & Jan, 2014).

Research efforts directed toward finding the cause of CCD and developing strategies to prevent it are underway (Hunter *et al*, 2010; Chen *et al*, 2014). RNA interference has proved effective in protecting against CCD. Directing double-stranded RNAs complementary to the IRES of the intergenic region of the IAPV virus decreases the probability of hive collapse, preventing effectively the massive death of the workforce, guaranteeing the protection of the queen and thus the survival of the colony (Maori *et al*, 2009).

Using single-particle cryo-electron microscopy (cryo-EM), we have characterized how the IAPV-IRES redirects the host machinery for viral protein synthesis exploiting novel ribosomal sites. An early commitment of IRES/ribosome complexes toward global pre-translocation mimicry explains the high efficiency in ribosome hijacking observed for this IRES. These results may inspire structure-based rational designs for the fight against CCD by RNA-interference technology (Chen *et al*, 2014).

Results

Biochemical set-up and cryo-EM strategy

Previous biochemical and genetic studies of IAPV-IRES established the secondary structure scheme displayed in Fig 1A (Au *et al*, 2015). The prototypical architecture of the type IV IRES family consisting of three nested pseudoknots is extended by a 5′ terminal stem loop (SL-VI) proposed to play functional roles in the early positioning of the IAPV-IRES in the ribosome (Schuler *et al*, 2006; Au *et al*, 2018). Additionally, the genus Aparavirus is characterized by an extended PKI which contains an insertion of a large stem loop (SL-III, Fig 1A, bottom). In the IAPV-IRES, SL-III consists of eight Watson–Crick canonical base pairs and a terminal loop of six nucleotides. Notably, two unpaired adenine residues are placed in a strategic position at the core of the three-way helical junction connecting SL-III, the anti-codon stem loop (ASL)-like element of the PKI (residues 6,546–6,574), and the double-helical segment connecting PKI and PKIII. A variable loop region (VLR) bridges the mRNA-like element of PKI (residues 6,613–6,617) with the helical region connecting PKI and PKIII. This single-stranded RNA loop is poorly conserved in sequence; however, even though its role in IRES functioning remains enigmatic, biochemical experiments have proved its integrity is mandatory for productive IRES-driven translation (Ruehle *et al*, 2015).

In order to understand in structural terms how these constituent units of the IAPV-IRES are involved in ribosome hijacking, we produced a full, wild-type IAPV-IRES, including SL-VI and the first two coding codons. Binary complexes with mammalian ribosomes and IAPV-IRES were generated by incubating IRES with ribosomal subunits. We designed a reaction featuring an excess of 40S over 60S in an overall background of IRES excess, to test the ability of the IAPV-IRES to engage both 40S and full 80S ribosomes in a productive and stable binary interaction. The stability of these interactions was tested through a sucrose gradient run overnight (Fig 1B) where the different complexes could be resolved according to their size differences. Each peak was subjected to RNA extraction and UREA-PAGE analysis where the presence of IAPV-IRES bound in the 80S peak as well as in the 40S peak could be confirmed (Fig 1B).

Leveraging latest cryo-EM maximum-likelihood classification methods implemented in RELION 3.0 (Scheres, 2012; von Loeffelholz *et al*, 2017; Zivanov *et al*, 2018), we decided to directly image the above reaction without the sucrose gradient step. A large dataset ensuing from this experiment was subjected to an optimized classification scheme combining different *in silico* classification approaches (Appendix Fig S1). This allowed us to identify and refine to high-resolution five distinctive classes from a single dataset (Fig 1C–E). The nominal resolution of the maps was calculated to be around 3 Å (Appendix Figs S2, S3, and S6). In the best areas, such as the 60S subunit and the body of the 40S subunit, the maps exhibit characteristics in accordance with this resolution, with very well-resolved side chains in proteins and clear base separation in the ribosomal RNA components (Appendix Fig S4A and B). However, the resolution of the IRES density, due to the intrinsic flexibility of this component, deviates from the nominal resolution. Areas of the IRES stabilized by ribosomal components are well-resolved, with local resolution better than 4 Å (Fig 3D and E), while

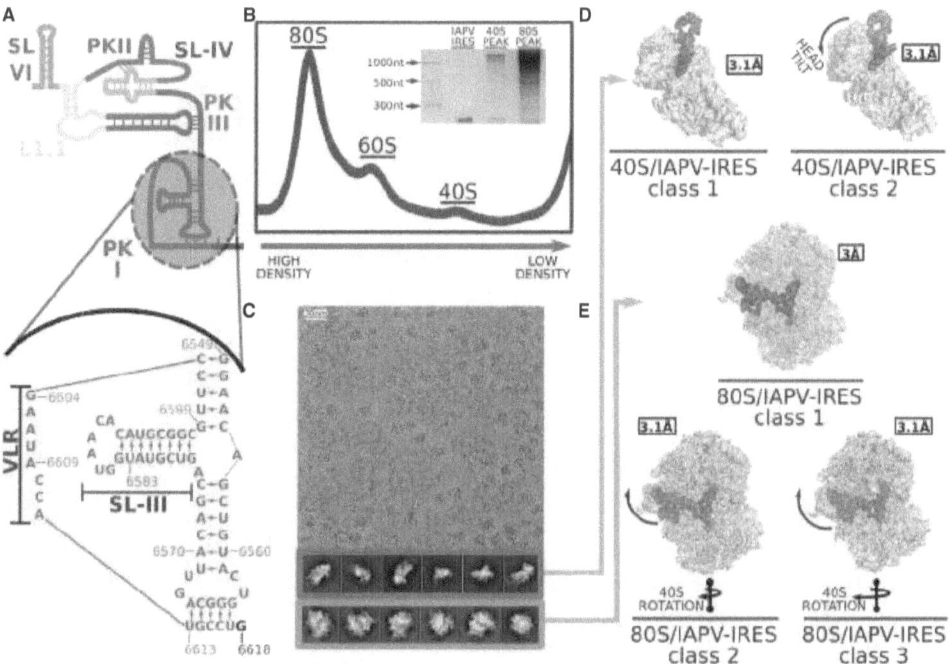

Figure 1. IAPV-IRES secondary structure, experimental set-up, and cryo-EM image processing workflow.

A IAPV-IRES diagram colored according to secondary structure motifs. Bottom, a closer view of the IAPV-IRES PKI highlighting its sequence, with base pairs indicated as well as the variable loop region (VLR) and stem loop III (SL-III).
B Sucrose gradient UV profile of 60S/40S/IAPV-IRES reaction mixture after an overnight run. The peaks corresponding to 80S and 40S were used for RNA extraction and UREA-PAGE shown in the inset.
C Representative cryo-EM image where roughly half of the particles correspond to 40S (blue) and the other half to 80S (orange).
D Two classes with robust density for the IAPV-IRES were found in the 40S group.
E After classification, three classes with clear IAPV-IRES density and small differences in the conformation of the 40S were found in the 80S group.

areas not stabilized by the ribosome or in contact with intrinsically dynamic elements of the ribosome like the L1 stalk exhibit lower local resolution (Appendix Figs S2, S3, and S6). In order to properly visualize the continuity of the maps for the full IRES, we show the unsharpened maps in the figures, especially where large areas of the maps are depicted. In those regions of the maps exhibiting resolution better than 4 Å for the IRES, maps sharpened with B factors reported in Appendix Table S1 are shown.

The IAPV-IRES restricts the conformational freedom of the 40S blocking functional sites

The 40S subunit can be roughly divided into two parts: the body, which forms the bulk of the subunit accounting for two-thirds of its mass, and a more mobile part roughly comprising the remaining third, designated as the head (Fig 2A). The interface between these two components forms the tRNA binding sites of the small subunit.

The head of the 40S subunit is a dynamic component, modifying its relative orientation with respect to the body. This dynamics is of critical importance in two aspects of translation: the positioning of the initiator aminoacyl-tRNA and in the concerted movement of mRNA and tRNAs during elongation (Ramrath *et al*, 2013; Llacer *et al*, 2015). We identified two classes of particles showing robust density for IAPV-IRES in the context of a binary interaction with the 40S (Fig 2B and C). All elements of the IAPV-IRES included in the produced construct were identified in the maps except SL-VI, which proved disordered—no density could be assigned to it even in low-pass filtered maps. The L1.1 region in the context of a binary interaction with the 40S shows a high degree of mobility and can only be modeled in maps filtered to 4 Å.

The IAPV-IRES inserts two elements of its structure between the head and the body of the 40S subunit, effectively restricting the dynamics of the 40S head to specific ranges of conformations. The ASL/mRNA mimicking part of the PKI is inserted in the decoding

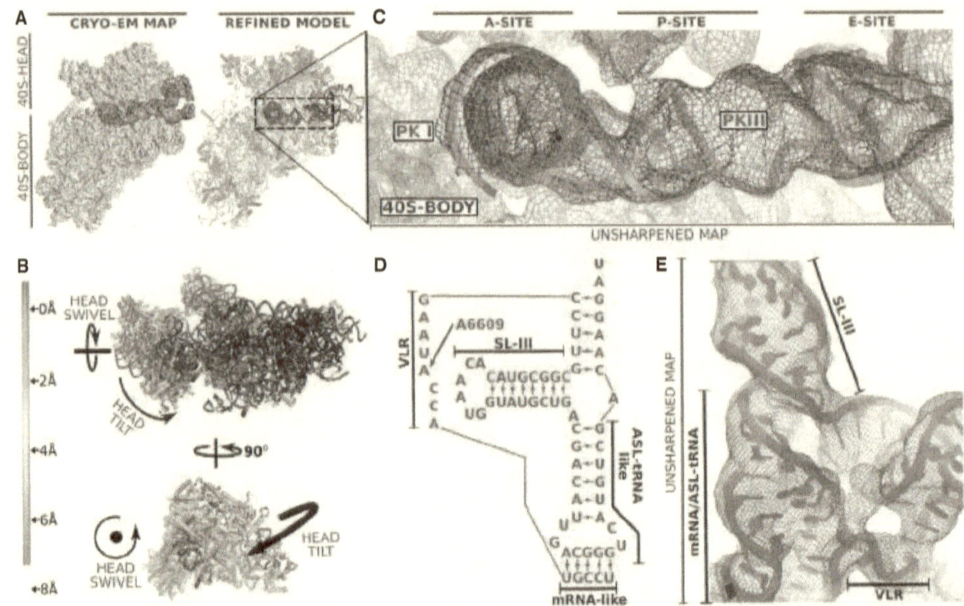

Figure 2. Structure of the IAPV-IRES in complex with the 40S ribosomal subunit.

A Overview of the mammalian 40S in complex with IAPV-IRES. Left, cryo-EM final post-processed map of class 1 with 40S colored yellow and IAPV-IRES maroon. Right, corresponding final refined model with IAPV-IRES colored according to Fig 1A.

B Ribbon diagram of the 40S colored by pairwise root-mean-square deviation displacements observed between the two IAPV-IRES/40S classes. The different position of the 40S head between both classes is a composition of swiveling and tilt movements (indicated by arrows in orthogonal views).

C Close-up view of the ribosomal sites of the 40S for IAPV-IRES/40S class 1 showing cryo-EM unsharpened cryo-EM density.

D Sequence of the PKI three-way helical junction.

E Unsharpened cryo-EM density for the PKI region of the IAPV-IRES in class 1 with the SL-III and the tRNA/mRNA mimicking domain indicated.

site (A site) of the small subunit stabilized by the decoding bases of the 18S rRNA A1824-A1825 and G626 [A1492, A1494, and G530 in *Escherichia coli* (Ogle & Ramakrishnan, 2005)], inducing a decoding event. SL-IV is deeply inserted in the interface of head and body, in the surroundings of the E site, clamped by stacking interactions established with residue A6498 of the IRES and tyrosine 72 from uS7 and arginine 135 from uS11 (Fig 3E). In this conformation, the IAPV-IRES fully blocks all three tRNA binding sites of the 40S subunit, interfering with early steps of canonical initiation (Fig 2C; Aylett & Ban, 2017).

Density for the full PKI, including SL-III, was clearly visible in the maps (Fig 2D and E, and Appendix S3) which allowed accurate modeling of the three-way helical junction characteristic of Aparavirus IRESs. The VLR was also visible in the maps, partially occupying the P site, stabilized by a stacking interaction between A6609 of the IRES and A1085 of the 18S rRNA (Fig 2E).

The two classes present a conformation of the IAPV-IRES nearly identical (r.m.s.d. = 1.12 Å between the two IRES conformations) but in displaced position with respect to the 40S body (Fig 3A and B). The IAPV-IRES seems to follow the movement of the 40S head, pivoting around the anchored PKI and SL-IV which are exceptionally stabilized by ribosomal elements from both head and body, effectively "clamping" the IRES to the 40S subunit (Fig 3D and E).

The three-way helical junction modeled in the PKI of the IAPV-IRES resembles a "hammer" shape, with SL-III coaxially stacked on top of the ASL-like stem (Fig 2D and E). Perpendicular to both and situated in between them, a helical segment connects PKI and PKIII. The coaxially aligned SL-III and ASL-like domain forms a straight unit of shape and dimensions similar to a tRNA, excluding the acceptor stem (Fig 3C). Alignments of structures containing tRNAs in several configurations [canonical tRNA PDBID:4V5D (Voorhees et al, 2009), with A/T-tRNA PDBID:5LZS (Shao et al, 2016) and hybrid tRNA PDBID:3J7R (Voorhees et al, 2014)] with the structure of the IAPV-IRES in complex with the 40S, reveal an interesting positioning of the coaxial unit formed by SL-III and the ASL-like part of PKI (Fig 3C). Notably, the SL-III/ASL-like unit of IAPV-IRES populates a space more similar to a hybrid A/P-tRNA than a canonical, A/A-tRNA or A/T-tRNA [following nomenclature of hybrid

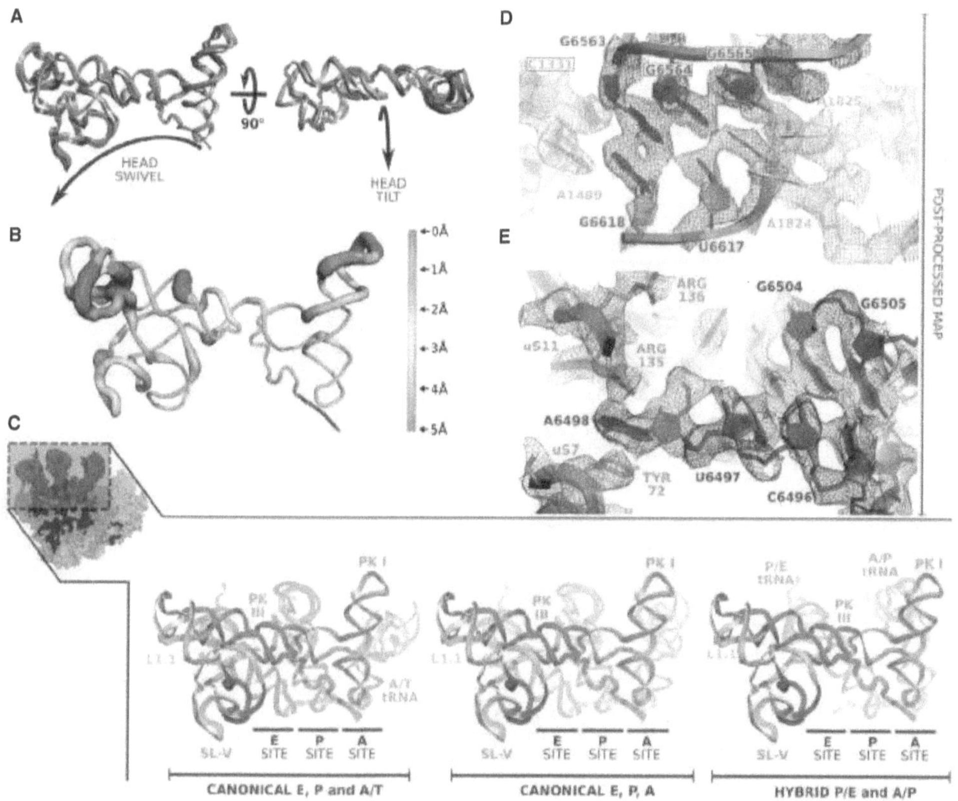

Figure 3. IAPV-IRES conformation in the context of a 40S interaction.

A   Superposition of IAPV-IRES models corresponding to IAPV-IRES/40S class 1 and class 2 after alignments excluding the IRES and the 40S head. A similar conformation can be observed with distinctive relative orientation with respect to the 40S body. The movement characteristic of the 40S head is indicated by arrows in orthogonal views.

B   Ribbon diagram of the IAPV-IRES colored by pairwise root-mean-square deviation displacements observed between the two IAPV-IRES/40S classes. The ASL/mRNA-like regions of the PKI and well as the SL-IV show the lowest degree of displacement (blue), whereas the apical part of SL-III and the L1.1 the highest (red).

C   Superposition of the IAPV-IRES with tRNAs in different configurations indicated at the bottom. IAPV-IRES is depicted as ribbons colored according to the secondary structure elements, and tRNAs are represented as gray ribbons. Alignments of the models were computed with the 40S body, excluding from the computation the ligands (IRES/tRNAs) and the 40S head.

D   Detailed view of the refined model for IAPV-IRES/40S class 1 inserted in the post-processed cryo-EM density focused on the decoding center of the 40S. PKI of IAPV-IRES is depicted green and 18S rRNA yellow.

E   Close-up view of the refined model for IAPV-IRES/40S class 1 inserted in the post-processed cryo-EM density focused on the SL-IV of the IAPV-IRES (depicted blue).

states previously proposed (Ratje *et al*, 2010)]. Similarly, PKIII overlaps with the position occupied by a hybrid P/E-tRNA, mimicking its helical components of the elbow region of a tRNA in this intermediate configuration. Overall, the IAPV-IRES is able to manipulate the 40S subunit in isolation, blocking the functional sites where canonical initiation factors eIF1, eIF1A, and eIF5B bind and, at the same time, steering the intrinsic dynamics of the 40S head toward a configuration reminiscent of an early elongation, pre-translocated state.

SL-III interacts with ribosomal protein uL16 stabilizing the 80S in a pre-translocation configuration mimicking hybrid tRNAs

The binary IAPV-IRES/80S complex populates three major conformations, with limited differences between them (Fig 1E). The majority of particles populated a class where the 40S subunit exhibits a small degree of intersubunit rotation (approx. 1°) compared with the unrotated, canonical configuration (Fig 5A). No major swiveling or tilt of the 40S head is visible in this conformation. The IAPV-IRES maintains a similar global conformation as in the binary complex with 40S, but in the 80S map, both the L1.1 region and the tip of the SL-III show good density as their dynamics are restricted by specific contacts with elements of the 60S: The L1 stalk stabilizes the L1.1 region and the A site finger and the ribosomal protein uL16 the SL-III. The A site finger (28S rRNA helix 38) is a flexible component of the 28S rRNA which plays an important role in translocation of tRNAs from the A to the P site (Nguyen *et al*, 2017). In many structures, it is not visible due to its intrinsic flexibility, required to perform its role escorting in-transit tRNAs (Brown *et al*, 2016; Nguyen *et al*, 2017). The SL-III of IAPV-IRES contacts the A site finger, stabilizing it in a fixed conformation, which allows the apical loop of SL-III to reach deep into the 60S, establishing a novel interaction with the ribosomal protein uL16 (Fig 4A and B). The IAPV-IRES positions the apical loop of SL-III (nucleotides 6,585–6,590) in direct contact with basic residues of uL16, which are in electrostatic interacting distance with negatively charged phosphates of the RNA backbone of the IRES (Fig 4C). The additional anchoring points to the ribosome contributed by SL-III, allows the IAPV-IRES, in the context of an 80S interaction, to be stabilized in a conformation that overlaps with the space occupied by a hybrid A/P-tRNA. The coaxial unit SL-III/ASL-like domain of the IAPV-IRES functionally mimics an A/P-tRNA priming the 80S for eEF2 recruitment, effectively bypassing the initiation stage (Fig 4D). Additionally, the anchoring points provided by the IAPV-IRES along the intersubunit space

probably contribute to an effective recruitment of the 60S in the absence of the dedicated factor responsible for such function in canonical translation, eIF5B (Pestova *et al*, 2000).

Aparavirus IRESs restrict the small subunit rotation dynamics in the pre-translocation state

No populations with wide rotations of the small subunit were identified in our large 80S/IAPV-IRES dataset. This suggests that, in contrast with Cripavirus IRESs (Fernandez *et al*, 2014; Koh *et al*, 2014), the IAPV-IRES is able to restrict the dynamics of the small subunit, channeling it toward a canonical, non-rotated configuration (Fig 5A). This is accomplished by a solid anchoring of the PKI in the A site, which not only mimics the ASL of a tRNA interacting with a cognate codon in the A site, but also, by placing the SL-III in a similar position as a hybrid A/P-tRNA (Moazed & Noller, 1989; Frank, 2012), mimics the T and D arms of a tRNA (Figs 4D and 5B, and Appendix Fig S7). In such position, the PKI of the IAPV-IRES establishes a network of interactions with both the large and the small subunits, effectively restricting the rotation of the 40S. Apart from the interactions established by the apical loop of the SL-III with ribosomal protein uL16 (Fig 4C), the decoding event elicited by the placement of the PKI in the decoding center allows the establishment of an interaction with the 28S rRNA base A3760 (A1913 in *E. coli*), normally involved in decoding (Fig 5C; Demeshkina *et al*, 2010). This interaction is maintained along the

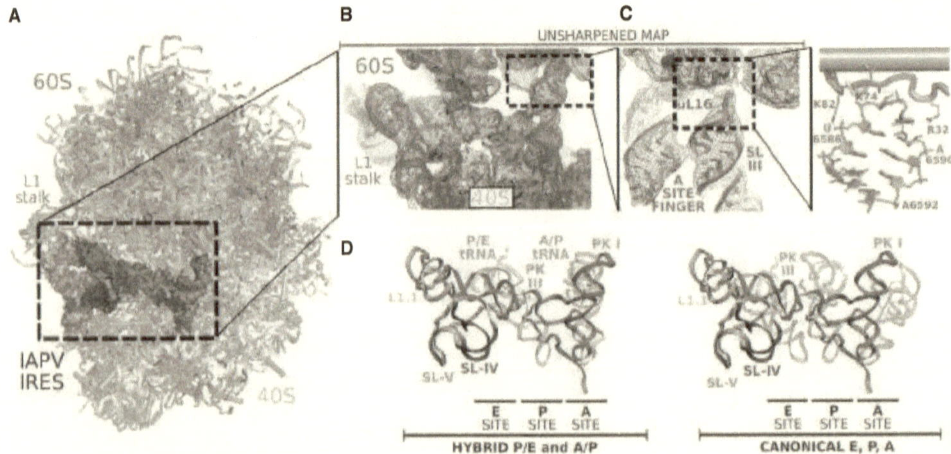

Figure 4. SL-III of IAPV-IRES engages novel sites of the 60S ribosomal subunit.

A   Overall view of the IAPV-IRES/80S complex class 1 with 60S represented as cyan ribbons, the 40S as yellow ribbons, and the IAPV-IRES represented as solid Van der Waals surface colored by secondary structure motifs.

B   Close-up view of the intersubunit space with the IAPV-IRES depicted as cartoons colored as in (A) inserted in the unsharpened cryo-EM density.

C   Zoomed view of the A site finger in interacting distance with the SL-III (green). The apical loop of SL-III reaches deep into the 60S contacting the ribosomal protein uL16.

D   Superposition of the IAPV-IRES in complex with 80S (class 1) with tRNAs in different configurations indicated at the bottom. Alignments of the models were computed with the 40S body, excluding from the computation the ligands (IRES/tRNAs) and the 40S head. IAPV-IRES PKI component SL-III/ASL-like domain populates a space of the intersubunit space similar to a A/P-tRNA. IAPV-IRES PKIII (red) mimics the elbow region of a hybrid P/E-tRNA.

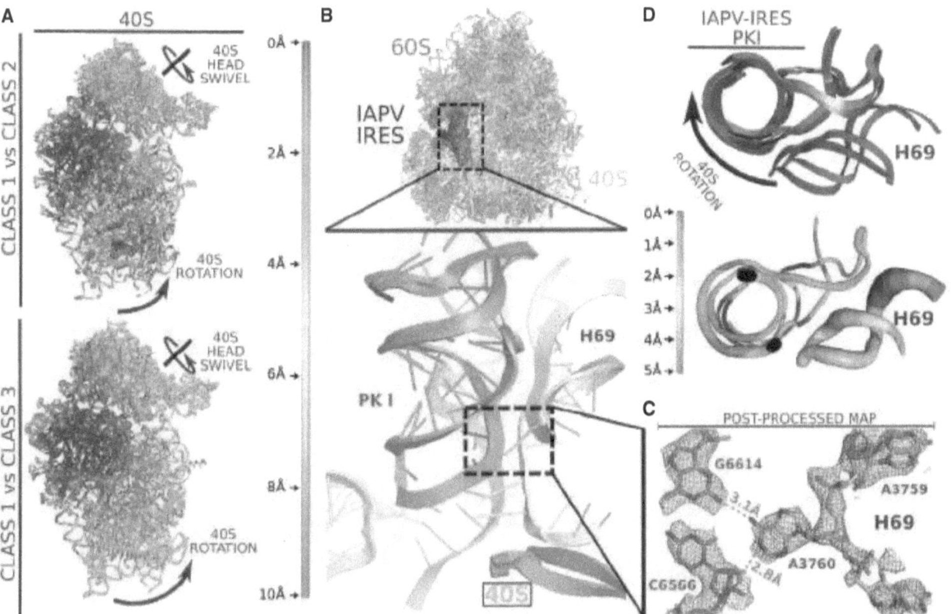

Figure 5. The IAPV-IRES restricts the small subunit rotational dynamics in a pre-translation complex with 80S.

A   Ribbon diagram of IAPV-IRES/80S complex viewed from the 40S colored by pairwise root-mean-square deviation displacements observed between the classes indicated on the left. Class 1 is unrotated, while classes 2 and 3 exhibit a small rotational movement of the 40S.

B   Top, general overview of the non-rotated IAPV-IRES/80S class 1 structure. IAPV-IRES is depicted as solid Van Der Waals surface colored according to the secondary structure motifs. The PKI (green) is solidly anchored to the A site. Bottom, close-up view of the IAPV-IRES PKI inserted in unsharpened map.

C   Zoomed view of A3760, a nucleotide belonging to the helix 69 (H69) of the 28S rRNA interacting with PKI. Final refined model inserted in the post-processed map is shown.

D   This interaction is not disrupted along the small fluctuations of the 40S. An apical view along the axis of the PKI of a superposition of class 1 versus class 2 shows the IRES displacements are minimal and are followed by the H69 which constantly interacts with the IRES.

small fluctuations of the 40S, which, locally, are restricted to displacements of a few Angströms (Fig 5D). The IAPV-IRES seems to bind very tightly in a binary, pre-translation complex with the 80S. The recruitment of elongation factors to commit the ribosome to the production of viral proteins seems to be achieved not through a dynamic manipulation of the 40S rotation, but by directly adopting a configuration reminiscent of a ribosome with tRNAs in hybrid configurations.

Remodeling of specific components of the IAPV-IRES allows its translocation through the ribosome

Due to their intrinsic flexibility, IRESs are able to populate multiple conformational states, while maintaining a basic structural framework dictated by their base-pairing scheme. A combination of rigid elements connected via flexible linkers allows these RNAs to tune their interactions with different ribosomal sites as they transit from an early pre-translocated state to a post-translocated one. Along this vectorial movement, these IRESs take advantage of intrinsic dynamic elements of the ribosome, normally involved in translocation of tRNAs and mRNAs (Voorhees & Ramakrishnan, 2013; Noller et al, 2017a).

In order to visualize the IAPV-IRES in a post-translocated state, we engineered a stop codon in the first coding codon following the IRES sequence (Muhs et al, 2015; Pisareva et al, 2018). By simultaneously incubating a pre-translocated 80S/IAPV-IRES complex with eEF2 and a catalytically inactive version of the eukaryotic release factor 1 (eRF1*) (Alkalaeva et al, 2006) in the presence of GTP, we were able to stabilize the IRES after a single translocation on the ribosome, allowing the visualization of the overall conformation of the IRES in a post-translocated state as well as the specific determinants of the IRES in binding the ribosomal P site (Fig 6A).

Globally, the post-translocated IAPV-IRES exhibits an extended conformation with the ASL/SL-III unit deeply inserted in the P site and the L1.1 region maintaining its original connection with the L1 stalk. The SL-IV and SL-V of the IRES are no longer in contact with

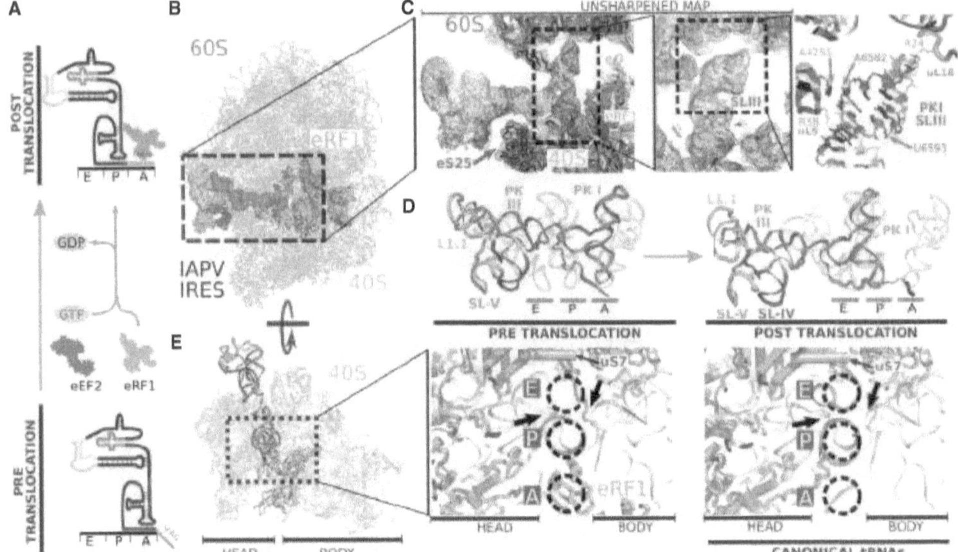

Figure 6. Visualization of IAPV-IRES in a post-translocated state in the ribosome.

A Biochemical strategy employed to trap a post-translocated state of IAPV-IRES in the ribosome.

B General view of the IAPV-IRES in a post-translocated state on the ribosome: 60S depicted as blue ribbons, 40S as yellow ribbons, IAPV-IRES represented as solid Van Der Waals surface colored according to secondary structure elements described in Fig 1A, and eRF1* depicted orange.

C Close-up view of the SL-III inserted in the experimental unsharpened cryo-EM density. On the right, refined model with residues from the 60S (blue) in interacting distance with the SL-III (green) indicated.

D Comparison of the final refined model for IAPV-IRES colored according to the secondary structure described in Fig 1A with canonical tRNAs (PDBID: 4V5D) in the pre-translocated state (left) and after translocation (right).

E Left, overall top view of the intersubunit space of the 40S for the post-translocated state. Inset, close-up view of the 40S tRNA binding sites where it can be appreciated the insertion of PKI of the IAPV-IRES in the P site, projecting the VLR toward the E site. The elements of the 18S rRNA forming the "P site gate" are indicated by solid arrows. On the right, equivalent view for canonical tRNAs.

the 40S (Fig 6B and C left). The SL-III seems to play an important role in orienting the IRES unit formed by the ASL/SL-III to a position that perfectly matches that of a canonical P site tRNA (Fig 6D). Nucleotides belonging to the SL-III establish contacts with several residues of ribosomal proteins uL5 and uL16 as well as with the 28S rRNA nucleotide A4255, all components of the large ribosomal subunit (60S) (Fig 6C, right).

In canonical translocation, the movement of tRNAs and mRNA has to be coordinated in order to vacate the ribosomal A site for the next incoming aminoacyl-tRNA. After peptidyl transfer, the ribosome adopts a rotated configuration of the small ribosomal subunit with tRNAs in hybrid configurations. The movement of the peptidyl-tRNA in the A site to the P site has to be coordinated with the movement of the P site tRNA to the E site, and the tRNA occupying the E site has to be ejected from the ribosome with the assistance of the L1 stalk. During this process, it is of capital importance that the correct reading frame on the mRNA be maintained (Voorhees & Ramakrishnan, 2013; Noller *et al*, 2017a,b). This is accomplished by the participation of specific components of

the ribosome (mainly RNA bases) which interact with tRNAs and mRNA defining specific checkpoints so as to prevent in-transit tRNAs from slipping or loosing contact with the mRNA on the correct frame (Zhou *et al*, 2014). One of the most important checkpoints is defined by the so-called "P site gate" (Zhou *et al*, 2014): a constriction formed by elements of the 18S rRNA (bases 1,054–1,064 and 1,638–1,645, 1,335–1,344, and 785–795 in *E. coli*) that physically block the progression of a translocated peptidyl-tRNA in the P site from slipping into the E site. The "closing" of the P site gate marks the end of a correct translocation cycle, allowing the ribosome to reset to a canonical, non-rotated configuration (Noller *et al*, 2017b).

By perfectly mimicking a canonical P site tRNA, the IAPV-IRES in a post-translocated state is able to position the VLR, a flexible element of its structure, in contacting distance with the E site of the 40S subunit. This is accomplished even with a fully closed P site gate (Fig 6E): Sliding through the P site gate, the single-stranded VLR can contact the ribosomal protein uS7, normally involved in stabilizing E site tRNAs (Fig 6E; Brown *et al*, 2018). The VLR thus

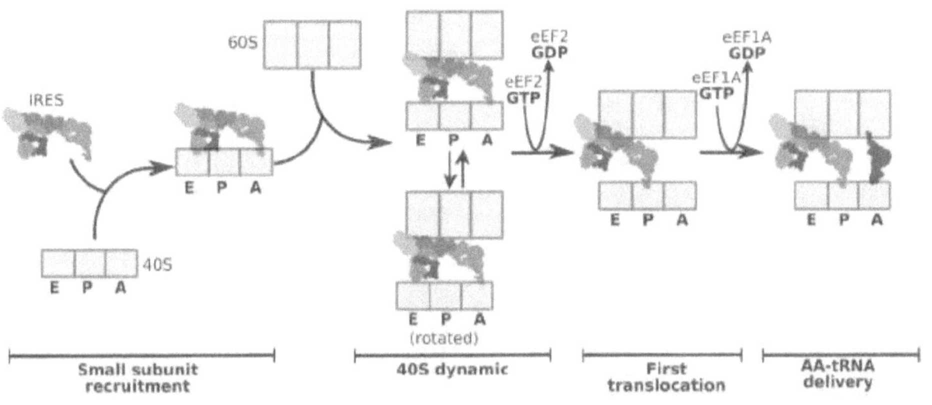

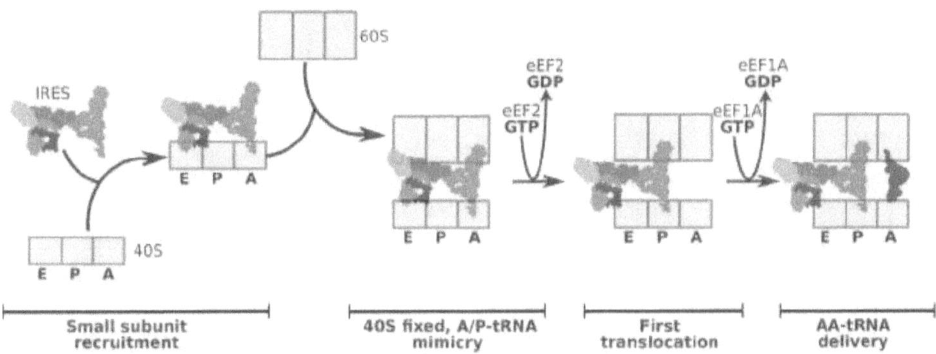

Figure 7. Type IV IRES families exploit different pre-translocation features of canonical translation for ribosome hijacking.

(Top), The Cripavirus family of type IV IRESs is able to capture free 40S subunits and engage them on a pre-translocation complex by recruiting 60S. This complex is extremely dynamic, with the 40S alternating between non-rotated and rotated configurations with respect to the 60S subunit. IRESs belonging to this family, exemplified by the CrPV-IRES, recruit elongation factors by mimicking a rotated stated of the ribosome with tRNAS. (Bottom), The Aparavirus family of IRESs follows a similar pathway in order to assemble a pre-translocation complex; however, specific structural components of this family allow for additional contacts with the 60S, limiting the rotational freedom of the 40S. Elongation factor engagement and thus effective ribosome hijacking are accomplished by mimicking a ribosome state with tRNAs in hybrid configurations.

undergoes a marked remodeling as the IAPV-IRES transitions from the pre-translocation to the post-translocation state. While in the pre-translocation conformation, the VLR establishes interactions with ribosomal components of the P site, after translocation, it populates a more extended conformation that is able to reach the E site even with the P site gate locked. The remodeling of this IRES component seems to be crucial for proper translocation of the IRES, as mutations that alter both its length and its base composition impact negatively on the ability of the IRES to initiate translation (Ruehle *et al*, 2015).

Discussion

Metagenomic studies of environmental samples have recently underscored the pervasive role RNA viruses exert on the biosphere (Shi *et al*, 2016). With estimates of dozens of RNA viruses infecting a single species, the diversity and impact these molecular entities have in biology and evolution is highly under-appreciated (Koonin & Dolja, 2013). It has been discovered that insects and arthropods host the highest diversity of RNA viruses (Shi *et al*, 2016). Within that realm, viruses from the *Dicistrovirideae* family have generated special interest given their wide range and distribution of hosts (Cox-Foster *et al*, 2007; Shi *et al*, 2016).

Using cryo-EM, we comprehensively characterized how the intergenic IRES of IAPV, a virus of the *Dicistrovirideae* family and a causing agent of the colony collapse disorder (CCD), binds and manipulates ribosomes to redirect them toward the production of viral proteins. The IAPV-IRES is able to establish a stable binary interaction with the small ribosomal subunit. An early capturing of free 40S subunits, committing them toward viral protein production, may represent a limiting step in a cellular environment where cellular and viral messages contend for ribosomal access. The IAPV-IRES is able to insert its PKI domain in the A site (decoding site) of the 40S effectively blocking the binding of eIF1/eIF1A, two initiation factors required for canonical initiation (Fig 2C). While blocking the 40S functional sites, the IAPV-IRES is able to steer the intrinsic dynamics of the 40S subunit toward a specific configuration, facilitating the recruitment of the large ribosomal subunit (60S) in the absence of eIF5B, the cellular factor catalyzing this event.

Bypassing the highly regulated initiation stage of translation is a customary requirement for the type IV family of IRES (Jackson *et al*, 2010). This is accomplished by inducing an altered ribosome state which is able to recruit elongation factors (eEF2 and eEF1A) directly and without tRNAs. Cripavirus IRESs such as the CrPV-IRES accomplish this by inducing a wide rotation on the 40S, mimicking a pre-translocation state of the ribosome with tRNAs (Fig 7, top). In marked contrast, Aparavirus IRESs such as the IAPV-IRES capture elongation factors not by inducing a rotated state of the 40S, but by directly mimicking a ribosomal state with hybrid tRNAs (Fig 7, bottom). The limited dynamics of 40S subunit rotation observed in our large cryo-EM dataset reflects a solid anchoring of both ribosomal subunits, mediated mainly by the additional contacts contributed by the SL-III characteristic of Aparavirus IRESs. However, a compromise between rigidity and flexibility is required for these IRESs to operate: In order to place the first coding codon in the decoding center of the 40S, the IAPV-IRES has to move (translocate) from the A site to the P site. We were able to gain structural information of this transition visualizing the post-translocated state of the IAPV-IRES. In the post-translocated state, the PKI mimics a canonical P site tRNA and the SL-III establishes new contacts with proteins and rRNA components of the P site of the 60S (Fig 6C). Additionally, the extremely dynamics of a single-stranded region of the IAPV-IRES termed the VLR (Fig 2D) is able to modify its configuration in a context-specific manner. In the context of a binary interaction with the 40S subunit and in the pre-translocation state with the 80S ribosome, the VLR exhibits a compact configuration, contacting bases of the 40S subunit's P site. After translocation, and once PKI is displaced to the P site, the VLR is extended, contacting ribosomal protein uS7, a component of the 40S subunit's E site (Fig 6E). Biochemical evidence supports a key role of the VLR in IRES function, as mutations and/or shortening impacts the ability of the IRES to efficiently translocate (Ruehle *et al*, 2015).

Using high-resolution single-particle cryo-EM analysis, we show how the prototypical Aparavirus IRES of the intergenic region of the IAPV manipulates the eukaryotic ribosome to position a viral non-AUG codon in the ribosomal A site to effectively hijack the host machinery for protein production (Movie EV1). We have uncovered a new strategy of early pre-translocation mimicry used by this IRES sub-family as well as visualize the conformational dynamics of a strategic single-stranded segment of the IRES, the VLR. The structures presented here will allow structure-based design of new and better RNA-interfering strategies directed toward the vital intergenic IRES of the IAPV. Ongoing efforts in that direction have already proved successful in fighting CCD, protecting bee hives from collapse (Maori *et al*, 2009; Piot *et al*, 2015).

Materials and Methods

Plasmids

Expression vector for His-tagged eRF1*(AGQ mutant) has been previously described (Ratnieks & Carreck, 2010). A pUC19-based transcription vector for IAPV-IRES-WT was constructed by inserting a T7 promoter sequence upstream of IAPV IGR IRES sequence followed by the first two coding triplets (nucleotides 6,372–6,623 from NC009025). An EcoRI site was included after the second codon. Site-directed mutagenesis was employed to change the first coding codon (GGC) to a stop codon (TAG) to create the IAPV-IRES-STOP construct. Both IRES constructs were transcribed using T7 RNA polymerase. Briefly, 0.1 mg/ml of EcoRI-linearized vector was transcribed using 0.1 mg/ml of homemade T7 RNA polymerase in 10 ml transcription buffer (100 mM HEPES-KOH pH 7.4, 10 mM of each NTP, 22 mM MgCl$_2$, 50 mM DTT, 2 mM Spermidine, 1 μl/ml IPP) for 4 h at 37°C. The RNA was then washed, concentrated, and separated on a 6% UREA-PAGE gel. The IAPV-IRES band was cut from the gel, electro-eluted, buffer exchanged into Buffer A (20 mM Tris–HCl, pH 7.5, 100 mM KCl, 8 mM MgCl$_2$, 2 mM DTT), and snap-frozen in liquid nitrogen.

Purification of translation components

Native 40S and 60S ribosomal subunits and eukaryotic elongation factor 2 (eEF2) were prepared from rabbit reticulocytes as previously described (Pisarev *et al*, 2010). Recombinant eRF1* was purified according to a previously described protocol (Alkalaeva *et al*, 2006).

Assembly of ribosomal complexes

To reconstitute ribosomal complexes in the pre-translocation state with IAPV-IRES, we incubated 6 pmol of 40S ribosomal subunits with 60 pmol of IAPV-IRES-WT RNA in a 15 μl reaction mixture in Buffer A for 5 min at 37°C. Then, the reaction mixture was supplemented with 4.5 pmol of 60S ribosomal subunits and incubated for 5 min at 37°C. We maintained an excess of 40S ribosomal subunits to trap both the 40S/IAPV-IRES and the 80S/IAPV-IRES complexes in the same reaction. Assembly of the complexes was verified by running the reaction through an overnight sucrose gradient, phenol extracting the RNA, and running a UREA-PAGE gel using standard methods. The assembly and integrity of the complex was also verified on a F20 screening microscope via negative stain EM and cryo-EM.

To reconstitute ribosomal complexes in a post-translocated state with IAPV-IRES-STOP, we incubated 8.8 pmol of 40S ribosomal subunits with 90 pmol of IAPV-IRES-STOP RNA in a 15 µl reaction mixture containing Buffer A with 1.7 mM GTP for 5 min at 37°C. Then, the reaction mixture was supplemented with 8.7 pmol of 60S ribosomal subunits and incubated for 5 min at 37°C. Next, we added 27 pmol of eEF2 and 90 pmol eRF* and incubated the reaction for an additional 30 min at 37°C.

Cryo-EM sample preparation and data acquisition

For ribosomal complexes in a pre-translocated state: 3 µl aliquots of assembled ribosomal complexes at 240–390 nM concentration were incubated for 15 s either on plasma-treated holey carbon grids (QUANTIFOIL R2/2 with homemade continuous carbon film estimated to be 50 Å thick) or on plasma-treated holey gold grids [UltrAuFoil R1.2/1.3 (Russo & Passmore, 2016)]. Grids were blotted for 2.5–3.0 s and flash-cooled in liquid ethane using an FEI Vitrobot. Grids were then transferred to a Polara-G2 microscope operated at 300 kV and equipped with a Gatan K2 Summit direct detector. 11,234 movies of 40 frames were collected in counting mode at 8e⁻/pix/s at a magnification of 31,000 corresponding to a calibrated pixel size of 1.25 Å. Defocus values specified in Leginon (Carragher *et al*, 2000) ranged from 0.8 to 2.5 µm.

For ribosomal complexes in a post-translocated state: 3 µl aliquots of assembled ribosomal complexes around 600 nM were incubated for 15 s on plasma-treated holey gold grids [UltrAuFoil R1.2/1.3 (Russo & Passmore, 2016)]. Grids were blotted for 2.5 s and flash-cooled in liquid ethane using an FEI Vitrobot. Grids were then transferred to a Titan Krios microscope operated at 300 kV and equipped with an energy filter (slits aperture 20 eV) and a Gatan K2 Summit detector. 6,904 movies of 40 frames were collected in counting mode at 8e⁻/pix/s at a magnification of 130,000 corresponding to a calibrated pixel size of 1.0605 Å. Defocus values specified in Leginon (Carragher *et al*, 2000) ranged from 0.5 to 3.0 µm. About half the movies (3,350) were collected at a 35-degree tilt to compensate for preferred orientation that was first identified during screening sessions.

On both microscopes, movies were recorded in automatic mode using the Leginon (Carragher *et al*, 2000) software and frames were aligned using Motioncor2 (Zheng *et al*, 2017). Data collection was monitored and checked on the fly using APPION (Lander *et al*, 2009).

Image processing and structure determination

For ribosomal complexes in a pre-translocated state, contrast transfer function parameters were estimated using GCTF (Zhang, 2016). For particle picking, a set of templates were generated using density data obtained from a screening session on a F20 microscope that had employed Gaussian picking. Using these templates, particle picking was performed using GAUTOMACH and a particle diameter value of 320 Å. The picked particles were manually screened on the micrographs to remove problematic regions. All 2D and 3D classifications and refinements were performed using RELION (Scheres, 2012). The picked particles were binned four times and subjected to a 2D classification to separate the 40S and 80S particles. We then employed 3D Refine to generate initial

consensus models from both the 40S and 80S particle sets (Appendix Fig S1). We then used our previous CrPV-IRES model (PDB ID 5IT9) to create appropriate masks for these initial models. The mask for the 40S initial model enclosed the putative IRES binding site. The mask for the 80S initial model enclosed the inter-subunit space, the A site finger, the L1 stalk, and ideal helices for SL-III and SL-VI regions of the IAPV-IRES. Using these masks, we performed a round of 3D classification with signal subtraction to remove 3D classes without the IRES. On the classes with the IRES, we performed focused classification without alignment using a new set of masks (Appendix Fig S1). This focused classification step resulted in the final set of classes that were eventually used for modeling.

For ribosomal complexes in a post-translocated state, we employed a similar set of protocols but with a different set of masks that also included regions for eEF2, eRF1*, and the L1 stalk in an extended conformation (Appendix Fig S5).

Final refinements with unbinned data for the selected classed yielded high-resolution maps with density features in agreement with the reported resolution. Local resolution was computed with RESMAP (Kucukelbir *et al*, 2014). New features implemented in Relion 3.0 (Zivanov *et al*, 2018) such as contrast transfer value refinement allowed extending the resolution to close to 3 Å.

Model building and refinement

Models for the mammalian ribosome and eRF1* were docked into the maps using CHIMERA (Pettersen *et al*, 2004), and COOT (Emsley & Cowtan, 2004) was used to manually adjust the L1 stalk and build the IAPV-IRES using our CrPV-IRES model as initial step. An initial round of refinement was performed in Phenix using real-space refinement with secondary structure restraints (Adams *et al*, 2011). A final step of reciprocal-space refinement using REFMAC was performed (Murshudov *et al*, 1997) for all complexes. The fit of the model to the map density was quantified using FSCaverage and Cref.

Data availability

Cryo-EM maps for each reconstruction have been deposited in the Electron Microscopy Databank (EMDB) with accession codes EMD-20248; http://www.ebi.ac.uk/pdbe/entry/EMD-20248 (40S-IAPV class 1), EMD-20249; http://www.ebi.ac.uk/pdbe/entry/EMD-20249 (40S-IAPV class 2), EMD-20255; http://www.ebi.ac.uk/pdbe/entry/EMD-20255 (80S-IAPV class 1), EMD-20256; http://www.ebi.ac.uk/pdbe/entry/EMD-20256 (80S-IAPV class 2), EMD-20257; http://www.ebi.ac.uk/pdbe/entry/EMD-20257 (80S IAPV class 3), and EMD-20258; http://www.ebi.ac.uk/pdbe/entry/EMD-20258 (80S-IAPV-eRF1, post-translocated). Atomic coordinates associated with these maps have been deposited in the Protein Data Bank (PDB) with accession codes 6P4G; http://www.rcsb.org/pdb/explore/explore.do?structureId = 6P4G (40S-IAPV class 1), 6P4H; http://www.rcsb.org/pdb/explore/explore.do?structureId = 6P4H (40S-IAPV class 2), 6P5I; http://www.rcsb.org/pdb/explore/explore.do?structureId = 6P5I (80S-IAPV class 1), 6P5J; http://www.rcsb.org/pdb/explore/explore.do?structureId = 6P5J (80S-IAPV class 2), 6P5K; http://www.rcsb.org/pdb/explore/explore.do?structureId = 6P5K

(80S-IAPV class 3), and 6P5N; http://www.rcsb.org/pdb/explore/explore.do?structureId=6P5N (80S-IAPV-eRF1, post-translocated).

Expanded View for this article is available online.

Acknowledgements

We are thankful to Vera Pisareva and Andrey Pisarev for a generous donation of ribosomal subunits and eRF1*. We acknowledge Bob Grassucci for technical assistance in data acquisition. Part of this work was performed at the Simons Electron Microscopy Center and National Resource for Automated Molecular Microscopy located at the New York Structural Biology Center, supported by grants from the Simons Foundation (SF349247), NYSTAR, and the NIH National Institute of General Medical Sciences (GM103310). We are especially grateful to Ed Eng, Bill Rice, and Laura Kim for support in data acquisition. Density maps have been deposited at the EMDB with accession codes 20248, 20249, 20255, 20256, 20257, and 20258. Atomic coordinates have been deposited in the PDB with accession codes 6P4G, 6P4H, 6P5I, 6P5J, 6P5K, and 6P5N. J.F. is funded by a National Institutes of Health grant (GM029169).

Author contributions

ISF conceived the project, designed the experiments, and collected the data with assistance of FA-R and RN. RN produced all RNA constructs with assistance of FA-R. Data processing was carried out by FA-R with RN assistance and ISF supervision. Modeling and model refinement were done by ISF. Data analysis and manuscript drafting were done by ISF with JF assistance. All authors contributed to article writing.

Conflict of interest

The authors declare that they have no conflict of interest.

References

Abeyrathne PD, Koh CS, Grant T, Grigorieff N, Korostelev AA (2016) Ensemble cryo-EM uncovers inchworm-like translocation of a viral IRES through the ribosome. *Elife* 5: e14874

Adams PD, Afonine PV, Bunkoczi G, Chen VB, Echols N, Headd JJ, Hung LW, Jain S, Kapral GJ, Grosse Kunstleve RW *et al* (2011) The Phenix software for automated determination of macromolecular structures. *Methods* 55: 94–106

Alkalaeva EZ, Pisarev AV, Frolova LY, Kisselev LL, Pestova TV (2006) *In vitro* reconstitution of eukaryotic translation reveals cooperativity between release factors eRF1 and eRF3. *Cell* 125: 1125–1136

Anderson D, East IJ (2008) The latest buzz about colony collapse disorder. *Science* 319: 724–725; author reply 724–5

Au HH, Cornilescu G, Mouzakis KD, Ren Q, Burke JE, Lee S, Butcher SE, Jan E (2015) Global shape mimicry of tRNA within a viral internal ribosome entry site mediates translational reading frame selection. *Proc Natl Acad Sci USA* 112: E6446–E6455

Au HHT, Elspass VM, Jan E (2018) Functional insights into the adjacent stem-loop in honey bee dicistroviruses that promotes internal ribosome entry site-mediated translation and viral infection. *J Virol* 92

Aylett CH, Ban N (2017) Eukaryotic aspects of translation initiation brought into focus. *Philos Trans R Soc Lond B Biol Sci* 372

Brown A, Fernandez IS, Gordiyenko Y, Ramakrishnan V (2016) Ribosome-dependent activation of stringent control. *Nature* 534: 277–280

Brown A, Baird MR, Yip MC, Murray J, Shao S (2018) Structures of translationally inactive mammalian ribosomes. *Elife* 7: e40486

Carragher B, Kisseberth N, Kriegman D, Milligan RA, Potter CS, Pulokas J, Reilein A (2000) Leginon: an automated system for acquisition of images from vitreous ice specimens. *J Struct Biol* 132: 33–45

Carrillo-Tripp J, Dolezal AG, Goblirsch MJ, Miller WA, Toth AL, Bonning BC (2016) *In vivo* and *in vitro* infection dynamics of honey bee viruses. *Sci Rep* 6: 22265

Chen YP, Pettis JS, Corona M, Chen WP, Li CJ, Spivak M, Visscher PK, DeGrandi-Hoffman G, Boncristiani H, Zhao Y *et al* (2014) Israeli acute paralysis virus: epidemiology, pathogenesis and implications for honey bee health. *PLoS Pathog* 10: e1004261

Chopra SS, Bakshi BR, Khanna V (2015) Economic dependence of U.S. Industrial Sectors on Animal-Mediated Pollination Service. *Environ Sci Technol* 49: 14441–14451

Cox-Foster DL, Conlan S, Holmes EC, Palacios G, Evans JD, Moran NA, Quan PL, Briese T, Hornig M, Geiser DM *et al* (2007) A metagenomic survey of microbes in honey bee colony collapse disorder. *Science* 318: 283–287

Demeshkina N, Jenner L, Yusupova G, Yusupov M (2010) Interactions of the ribosome with mRNA and tRNA. *Curr Opin Struct Biol* 20: 325–332

Doublet V, Poeschl Y, Gogol-Doring A, Alaux C, Annoscia D, Aurori C, Barribeau SM, Bedoya-Reina OC, Brown MJ, Bull JC *et al* (2017) Unity in defence: honeybee workers exhibit conserved molecular responses to diverse pathogens. *BMC Genom* 18: 207

Emsley P, Cowtan K (2004) Coot: model-building tools for molecular graphics. *Acta Crystallogr D Biol Crystallogr* 60: 2126–2132

Fernandez IS, Bai XC, Murshudov G, Scheres SH, Ramakrishnan V (2014) Initiation of translation by cricket paralysis virus IRES requires its translocation in the ribosome. *Cell* 157: 823–831

Frank J (2012) Intermediate states during mRNA-tRNA translocation. *Curr Opin Struct Biol* 22: 778–785

Gross L, Vicens Q, Einhorn E, Noireterre A, Schaeffer L, Kuhn L, Imler JL, Eriani G, Meignin C, Martin F (2017) The IRES5′UTR of the dicistrovirus cricket paralysis virus is a type III IRES containing an essential pseudoknot structure. *Nucleic Acids Res* 45: 8993–9004

Hertz MI, Thompson SR (2011) Mechanism of translation initiation by Dicistroviridae IGR IRESs. *Virology* 411: 355–361

Hinnebusch AG, Lorsch JR (2012) The mechanism of eukaryotic translation initiation: new insights and challenges. *Cold Spring Harb Perspect Biol* 4: a011544

Hou C, Rivkin H, Slabezki Y, Chejanovsky N (2014) Dynamics of the presence of Israeli acute paralysis virus in honey bee colonies with colony collapse disorder. *Viruses* 6: 2012–2027

Hunter W, Ellis J, Vanengelsdorp D, Hayes J, Westervelt D, Glick E, Williams M, Sela I, Maori E, Pettis J *et al* (2010) Large-scale field application of RNAi technology reducing Israeli acute paralysis virus disease in honey bees (*Apis mellifera*, Hymenoptera: Apidae). *PLoS Pathog* 6: e1001160

Jaafar ZA, Kieft JS (2019) Viral RNA structure-based strategies to manipulate translation. *Nat Rev Microbiol* 17: 110–123

Jackson RJ, Hellen CU, Pestova TV (2010) The mechanism of eukaryotic translation initiation and principles of its regulation. *Nat Rev Mol Cell Biol* 11: 113–127

Jan E, Thompson SR, Wilson JE, Pestova TV, Hellen CU, Sarnow P (2001) Initiator Met-tRNA-independent translation mediated by an internal ribosome entry site element in cricket paralysis virus-like insect viruses. *Cold Spring Harb Symp Quant Biol* 66: 285–292

Johnson AG, Grosely R, Petrov AN, Puglisi JD (2017) Dynamics of IRES-mediated translation. *Philos Trans R Soc Lond B Biol Sci* 372: 20160177

Kerr CH, Jan E (2016) Commandeering the ribosome: lessons learned from dicistroviruses about translation. *J Virol* 90: 5538–5540

Khong A, Bonderoff JM, Spriggs RV, Tammpere E, Kerr CH, Jackson TJ, Willis AE, Jan E (2016) Temporal regulation of distinct internal ribosome entry sites of the Dicistroviridae cricket paralysis virus. *Viruses* 8: E25

Koh CS, Brilot AF, Grigorieff N, Korostelev AA (2014) Taura syndrome virus IRES initiates translation by binding its tRNA-mRNA-like structural element in the ribosomal decoding center. *Proc Natl Acad Sci USA* 111: 9139–9144

Koonin EV, Dolja VV (2013) A virocentric perspective on the evolution of life. *Curr Opin Virol* 3: 546–557

Kucukelbir A, Sigworth FJ, Tagare HD (2014) Quantifying the local resolution of cryo-EM density maps. *Nat Methods* 11: 63–65

Lander GC, Stagg SM, Voss NR, Cheng A, Fellmann D, Pulokas J, Yoshioka C, Irving C, Mulder A, Lau PW *et al* (2009) Appion: an integrated, database-driven pipeline to facilitate EM image processing. *J Struct Biol* 166: 95–102

Llacer JL, Hussain T, Marler L, Aitken CE, Thakur A, Lorsch JR, Hinnebusch AG, Ramakrishnan V (2015) Conformational differences between open and closed states of the eukaryotic translation initiation complex. *Mol Cell* 59: 399–412

von Loeffelholz O, Natchiar SK, Djabeur N, Myasnikov AG, Kratzat H, Menetret JF, Hazemann I, Klaholz BP (2017) Focused classification and refinement in high-resolution cryo-EM structural analysis of ribosome complexes. *Curr Opin Struct Biol* 46: 140–148

Maori E, Paldi N, Shafir S, Kalev H, Tsur E, Glick E, Sela I (2009) IAPV, a bee-affecting virus associated with colony collapse disorder can be silenced by dsRNA ingestion. *Insect Mol Biol* 18: 55–60

Moazed D, Noller HF (1989) Intermediate states in the movement of transfer RNA in the ribosome. *Nature* 342: 142–148

Muhs M, Hilal T, Mielke T, Skabkin MA, Sanbonmatsu KY, Pestova TV, Spahn CM (2015) Cryo-EM of ribosomal 80S complexes with termination factors reveals the translocated cricket paralysis virus IRES. *Mol Cell* 57: 422–432

Mullapudi E, Fuzik T, Pridal A, Plevka P (2017) Cryo-electron microscopy study of the genome release of the dicistrovirus israeli acute bee paralysis virus. *J Virol* 91: e02060-16

Murray J, Savva CG, Shin BS, Dever TE, Ramakrishnan V, Fernandez IS (2016) Structural characterization of ribosome recruitment and translocation by type IV IRES. *Elife* 5: e13567

Murshudov GN, Vagin AA, Dodson EJ (1997) Refinement of macromolecular structures by the maximum-likelihood method. *Acta Crystallogr D Biol Crystallogr* 53: 240–255

Nguyen K, Yang H, Whitford PC (2017) How the ribosomal A-site finger can lead to tRNA species-dependent dynamics. *J Phys Chem B* 121: 2767–2775

Noller HF, Lancaster L, Mohan S, Zhou J (2017a) Ribosome structural dynamics in translocation: yet another functional role for ribosomal RNA. *Q Rev Biophys* 50: e12

Noller HF, Lancaster L, Zhou J, Mohan S (2017b) The ribosome moves: RNA mechanics and translocation. *Nat Struct Mol Biol* 24: 1021–1027

Ogle JM, Ramakrishnan V (2005) Structural insights into translational fidelity. *Annu Rev Biochem* 74: 129–177

Pestova TV, Lomakin IB, Lee JH, Choi SK, Dever TE, Hellen CU (2000) The joining of ribosomal subunits in eukaryotes requires eIF5B. *Nature* 403: 332–335

Pestova TV, Lomakin IB, Hellen CU (2004) Position of the CrPV IRES on the 40S subunit and factor dependence of IRES/80S ribosome assembly. *EMBO Rep* 5: 906–913

Pettersen EF, Goddard TD, Huang CC, Couch GS, Greenblatt DM, Meng EC, Ferrin TE (2004) UCSF Chimera—a visualization system for exploratory research and analysis. *J Comput Chem* 25: 1605–1612

Piot N, Snoeck S, Vanlede M, Smagghe G, Meeus I (2015) The effect of oral administration of dsRNA on viral replication and mortality in *Bombus terrestris*. *Viruses* 7: 3172–3185

Pisarev AV, Shirokikh NE, Hellen CU (2005) Translation initiation by factor-independent binding of eukaryotic ribosomes to internal ribosomal entry sites. *C R Biol* 328: 589–605

Pisarev AV, Skabkin MA, Pisareva VP, Skabkina OV, Rakotondrafara AM, Hentze MW, Hellen CU, Pestova TV (2010) The role of ABCE1 in eukaryotic postterminaion ribosomal recycling. *Mol Cell* 37: 196–210

Pisareva VP, Pisarev AV, Fernandez IS (2018) Dual tRNA mimicry in the cricket paralysis virus IRES uncovers an unexpected similarity with the hepatitis C virus IRES. *Elife* 7: e34062

Ramrath DJ, Lancaster L, Sprink T, Mielke T, Loerke J, Noller HF, Spahn CM (2013) Visualization of two transfer RNAs trapped in transit during elongation factor G-mediated translocation. *Proc Natl Acad Sci USA* 110: 20964–20969

Ratje AH, Loerke J, Mikolajka A, Brunner M, Hildebrand PW, Starosta AL, Donhofer A, Connell SR, Fucini P, Mielke T *et al* (2010) Head swivel on the ribosome facilitates translocation by means of intra-subunit tRNA hybrid sites. *Nature* 468: 713–716

Ratnieks FL, Carreck NL (2010) Ecology. Clarity on honey bee collapse? *Science* 327: 152–153

Ruehle MD, Zhang H, Sheridan RM, Mitra S, Chen Y, Gonzalez RL, Cooperman BS, Kieft JS (2015) A dynamic RNA loop in an IRES affects multiple steps of elongation factor-mediated translation initiation. *Elife* 4: e08146

Russo CJ, Passmore LA (2016) Ultrastable gold substrates: properties of a support for high-resolution electron cryomicroscopy of biological specimens. *J Struct Biol* 193: 33–44

Scheres SH (2012) RELION: implementation of a Bayesian approach to cryo-EM structure determination. *J Struct Biol* 180: 519–530

Schuler M, Connell SR, Lescoute A, Giesebrecht J, Dabrowski M, Schroeer B, Mielke T, Penczek PA, Westhof E, Spahn CM (2006) Structure of the ribosome-bound cricket paralysis virus IRES RNA. *Nat Struct Mol Biol* 13: 1092–1096

Shao S, Murray J, Brown A, Taunton J, Ramakrishnan V, Hegde RS (2016) Decoding mammalian ribosome-mRNA states by translational GTPase complexes. *Cell* 167: 1229–1240 e15

Shi M, Lin XD, Tian JH, Chen LJ, Chen X, Li CX, Qin XC, Li J, Cao JP, Eden JS *et al* (2016) Redefining the invertebrate RNA virosphere. *Nature* 540: 539–543

Voorhees RM, Weixlbaumer A, Loakes D, Kelley AC, Ramakrishnan V (2009) Insights into substrate stabilization from snapshots of the peptidyl transferase center of the intact 70S ribosome. *Nat Struct Mol Biol* 16: 528–533

Voorhees RM, Ramakrishnan V (2013) Structural basis of the translational elongation cycle. *Annu Rev Biochem* 82: 203–236

Voorhees RM, Fernandez IS, Scheres SH, Hegde RS (2014) Structure of the mammalian ribosome-Sec61 complex to 3.4 A resolution. *Cell* 157: 1632–1643

Wang QS, Jan E (2014) Switch from cap- to factorless IRES-dependent 0 and +1 frame translation during cellular stress and dicistrovirus infection. *PLoS ONE* 9: e103601

Wilson JE, Pestova TV, Hellen CU, Sarnow P (2000a) Initiation of protein synthesis from the A site of the ribosome. *Cell* 102: 511–520

Wilson JE, Powell MJ, Hoover SE, Sarnow P (2000b) Naturally occurring dicistronic cricket paralysis virus RNA is regulated by two internal ribosome entry sites. *Mol Cell Biol* 20: 4990–4999

Yamamoto H, Unbehaun A, Spahn CMT (2017) Ribosomal chamber music: toward an understanding of IRES mechanisms. *Trends Biochem Sci* 42: 655–668

Zhang K (2016) Gctf: real-time CTF determination and correction. *J Struct Biol* 193: 1–12

Zheng SQ, Palovcak E, Armache JP, Verba KA, Cheng Y, Agard DA (2017) MotionCor2: anisotropic correction of beam-induced motion for improved cryo-electron microscopy. *Nat Methods* 14: 331–332

Zhou J, Lancaster L, Donohue JP, Noller HF (2014) How the ribosome hands the A-site tRNA to the P site during EF-G-catalyzed translocation. *Science* 345: 1188–1191

Zivanov J, Nakane T, Forsberg BO, Kimanius D, Hagen WJ, Lindahl E, Scheres SH (2018) New tools for automated high-resolution cryo-EM structure determination in RELION-3. *Elife* 7: e42166

Appendix for

The Israeli Acute Paralysis Virus IRES captures host ribosomes by mimicking a ribosomal state with hybrid tRNAs.

Francisco Acosta-Reyes, Ritam Neupane, Joachim Frank[*] and Israel S. Fernández[*]

[*] Corresponding author email:

jf2192@cumc.columbia.edu (JF) and isf2106@cumc.columbia.edu (ISF)

This file contains:

Appendix Figures S1-S7

Appendix Table S1

Appendix Figures S1-S7

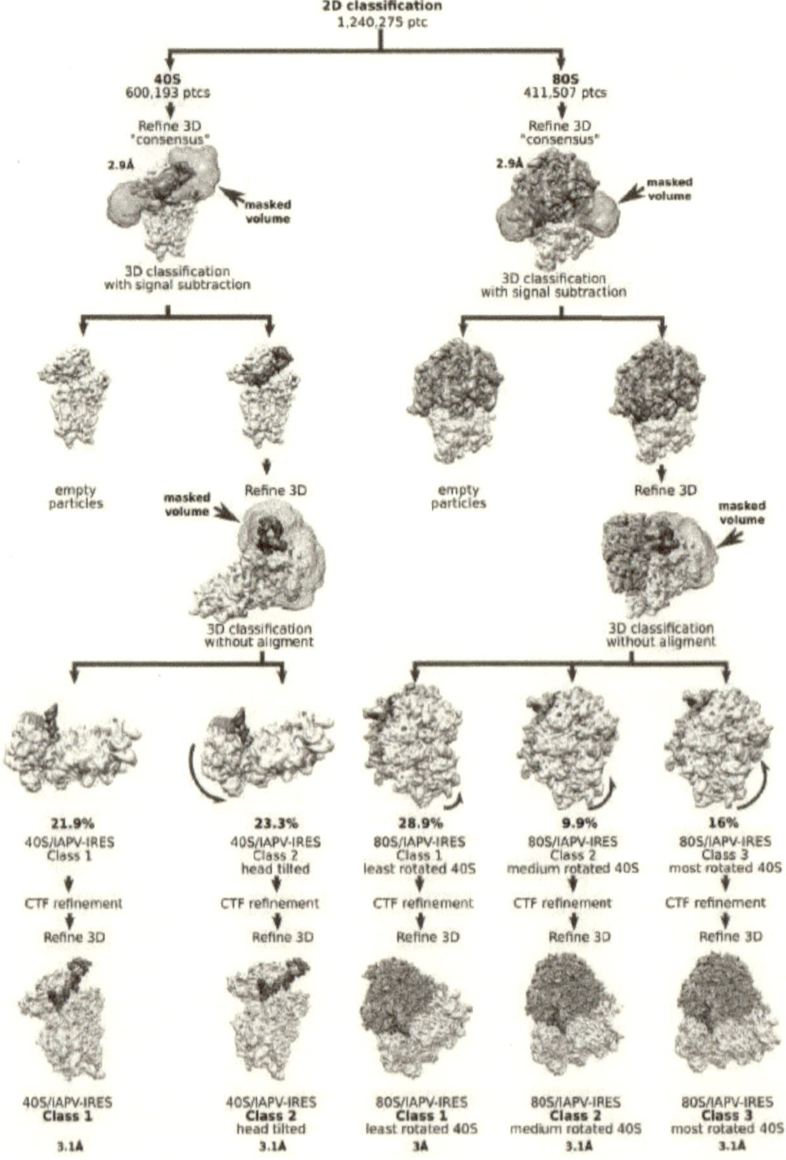

Figure S1. Classification scheme followed for the pre-translocation dataset.

(Top), 40S and 80S particles were separated by reference-free 2D classification. Homogeneous subgroups of 40S and 80S particles were subjected to two steps of masked classification intercalated with refinements to, on a first instance, identified those particles with IAPV-IRES and, in a second instance, distinguish among the groups with IAPV-IRES, different conformations. (Bottom), New features implemented in Relion3.0 (Zivanov, Nakane et al., 2018) such as contrast transfer values refinement allowed extending the resolution to close to 3Å for the five populations.

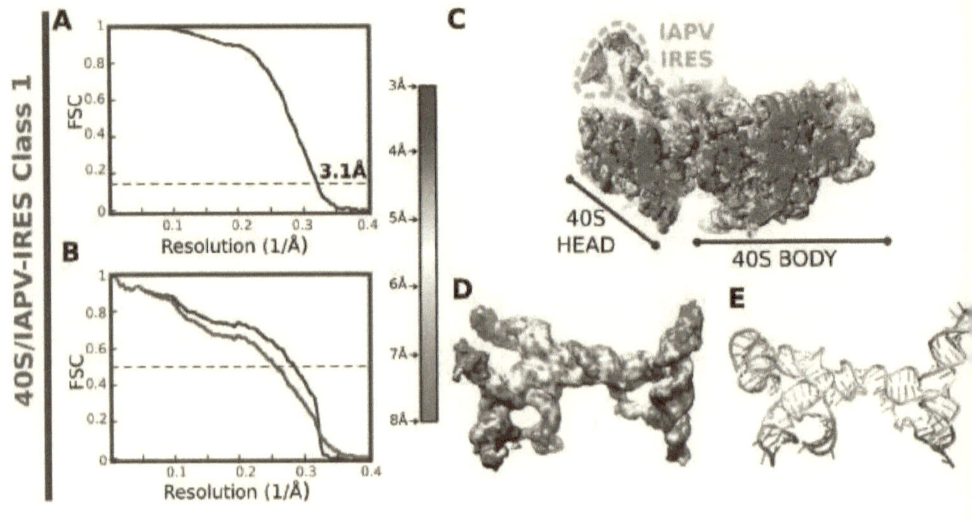

40S/IAPV-IRES Class 1
A
FSC
Resolution (1/Å)
0.1 0.2 0.3 0.4
3.1Å
B
FSC
Resolution (1/Å)
0.1 0.2 0.3 0.4
3Å
4Å
5Å
6Å
7Å
8Å
C
IAPV
IRES
40S
HEAD
40S BODY
D
E

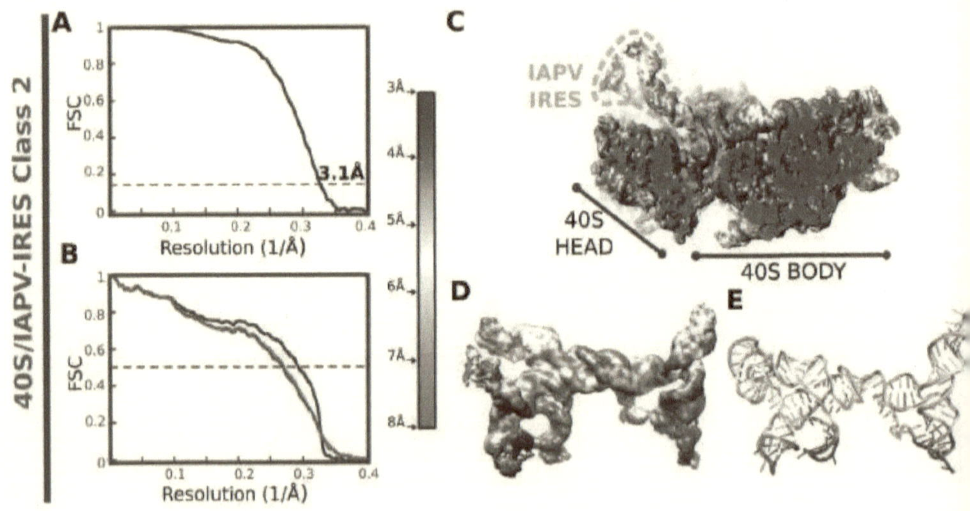

40S/IAPV-IRES Class 2
A
FSC
Resolution (1/Å)
0.1 0.2 0.3 0.4
3.1Å
B
FSC
Resolution (1/Å)
0.1 0.2 0.3 0.4
3Å
4Å
5Å
6Å
7Å
8Å
C
IAPV
IRES
40S
HEAD
40S BODY
D
E

Figure S2: Fourier Shell Correlation curves and local resolution estimation for the 40S/IAPV-IRES complex classes.

For each class:

(A) Fourier Shell Correlation (FSC) computed for the two half maps of the final subset of particles after classification for the first class identified for the 40S/IAPV-IRES complex. The resolution is estimated to be 3.1Å using the 0.143 criterion (Rosenthal & Henderson, 2003). (B) Map-versus-model cross validation FSC. The final model was validated using standard procedures: FSC of the refined model against half map 1 (blue) overlaps with the FSC against half map 2 (red, not included in the refinement). The black curve corresponds to the FSC of the final model against the final map. (C) Slice through the final, unsharpened map colored according to the local resolution as reported by RESMAP (Kucukelbir, Sigworth et al., 2014). (D) Close-up views of the final density colored according to local resolution values as in (C) for the IAPV-IRES. (E) Final refined IAPV-IRES model colored according to the estimated B-factors in Å^2 computed by REFMAC (Murshudov, Vagin et al., 1997). Specific values for estimated resolutions are indicated for each class.

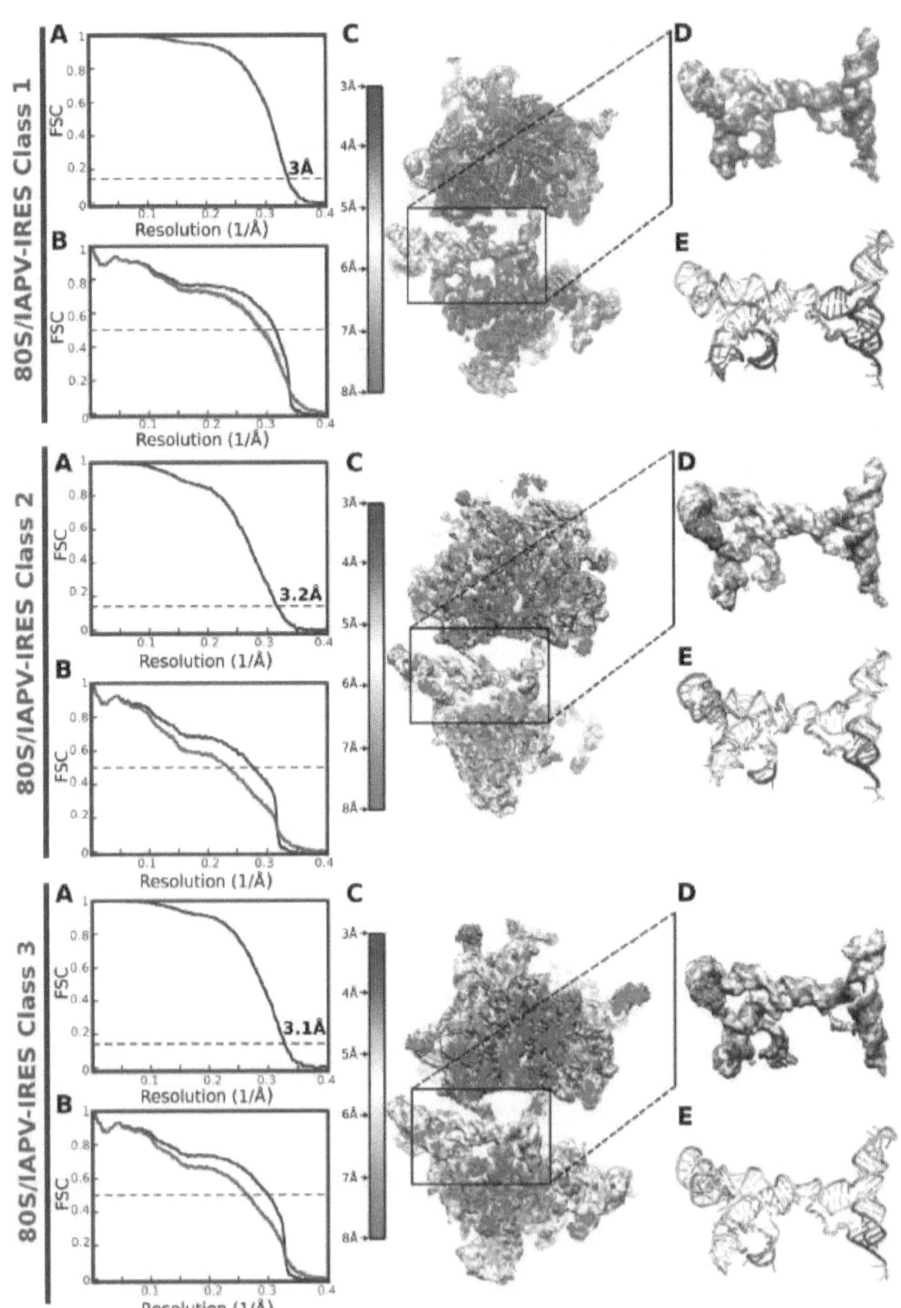

Figure S3: Fourier Shell Correlation curves and local resolution estimation for the 80S/IAPV-IRES complex classes.
For each class:

(A) Fourier Shell Correlation (FSC) computed for the two half maps of the final subset of particles after classification for the first class identified for the 80S/IAPV-IRES complex. The resolution are estimated to be 3.0-3.2Å using the 0.143 criterion (Rosenthal & Henderson, 2003). (B) Map-versus-model cross validation FSC. The final model was validated using standard procedures: FSC of the refined model against half map 1 (blue) overlaps with the FSC against half map 2 (red, not included in the refinement). The black curve corresponds to the FSC of the final model against the final map. (C) Slice through the final, unsharpened map colored according to the local resolution as reported by RESMAP (Kucukelbir et al., 2014). (D) Close-up views of the final density colored according to local resolution values as in (C) for the IAPV-IRES. (E) Final refined IAPV-IRES model colored according to the estimated B-factors in $Å^2$ computed by REFMAC. Specific values for estimated resolutions are indicated for each class.

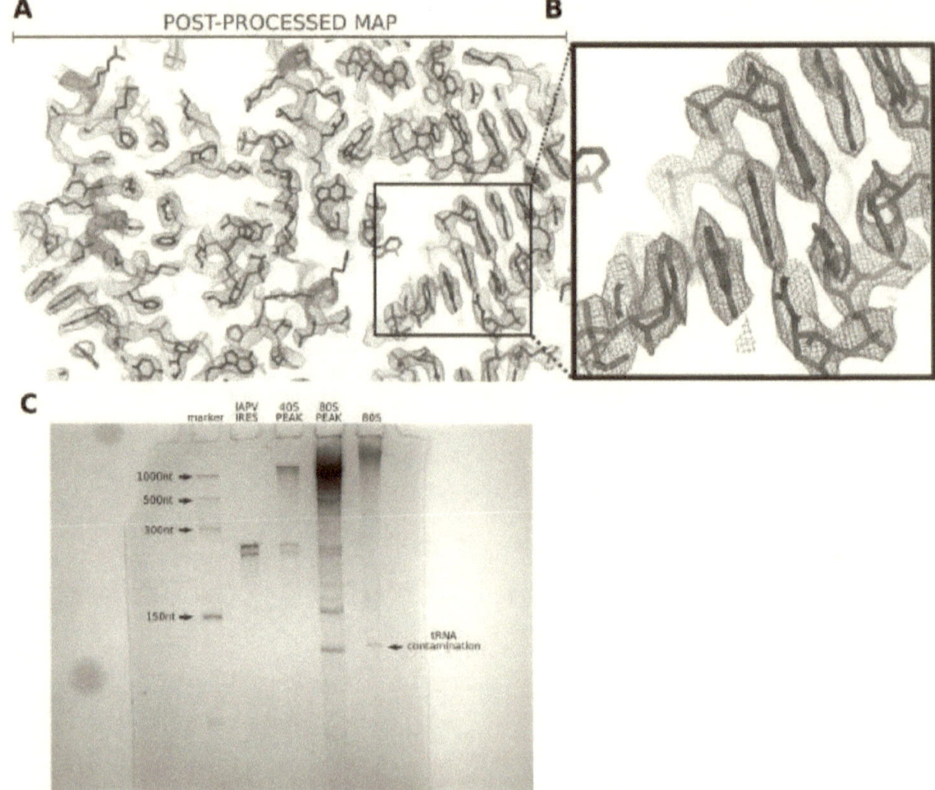

Figure S4: Post-processed map and UREA-PAGE.

(A) and (B) Slice through the 80S/IAPV-IRES class 1 post-processed map centered around the 60S. Map features expected for a 3Å map like protein side chains and base separation for nucleic acids can be appreciated. (C). UREA-PAGE analysis of 40S and 80S peaks resolved in a overnight sucrose gradient run. Bands corresponding to the purified IAPV-IRES RNA can be appreciated in both 40S and 80S peaks.

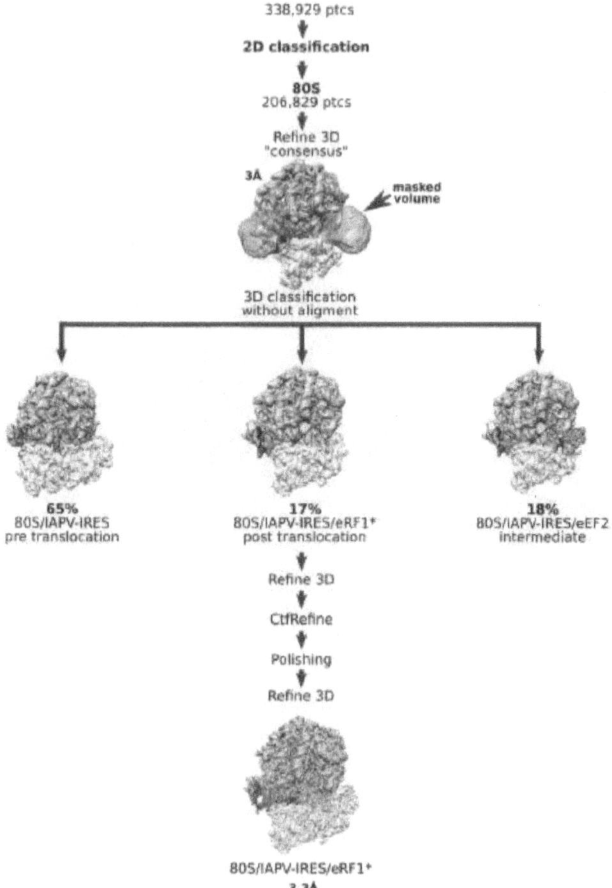

Figure S5: Classification scheme followed for the post-translocation dataset.

(Top), after 2D and 3D masked classifications, three populations of particles corresponding to pre-translocation, post-translocation and an intermediate state with eEF2 could be identified in the dataset. (Bottom), Contrast transfer values refinement and Bayesian particle polishing implemented in Relion3.0 (Zivanov et al., 2018) allowed the extension of the resolution for the post-translocation state to 3.2Å.

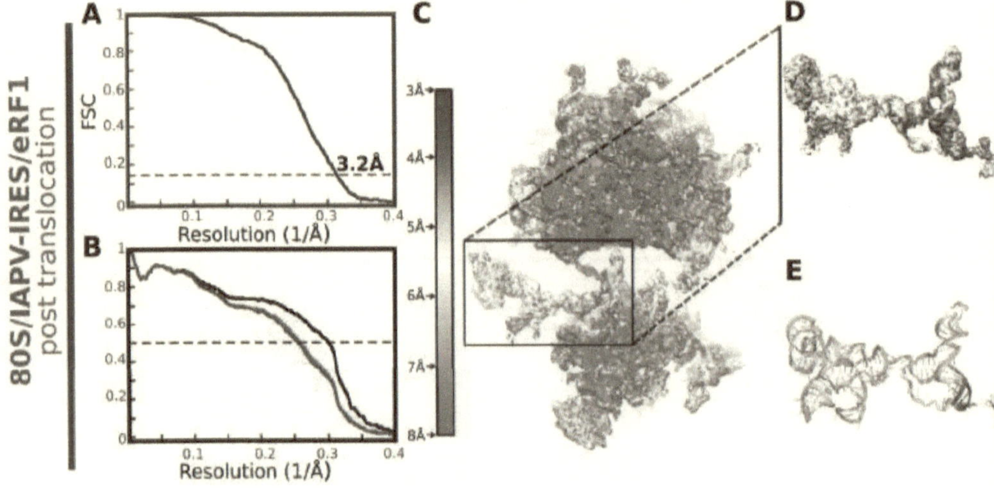

Figure S6: Fourier Shell Correlation curves and local resolution estimation for the post-translocated 80S/IAPV-IRES/eRF1* complex.

(A) Fourier Shell Correlation (FSC) computed for the two half maps of the final subset of particles after classification for the first class identified for the 80S/IAPV-IRES complex. The resolution is estimated to be 3.2Å using the 0.143 criterion (Rosenthal & Henderson, 2003). (B) Map-versus-model cross validation FSC. The final model was validated using standard procedures: FSC of the refined model against half map 1 (blue) overlaps with the FSC against half map 2 (red, not included in the refinement). The black curve corresponds to the FSC of the final model against the final map. (C) Slice through the final, unsharpened map colored according to the local resolution as reported by RESMAP (Kucukelbir et al., 2014). (D) Close-up views of the final density colored according to local resolution values as in (C) for the IAPV-IRES. (E) Final refined IAPV-IRES model colored according to the estimated B-factors in Å^2 computed by REFMAC (Murshudov et al., 1997).

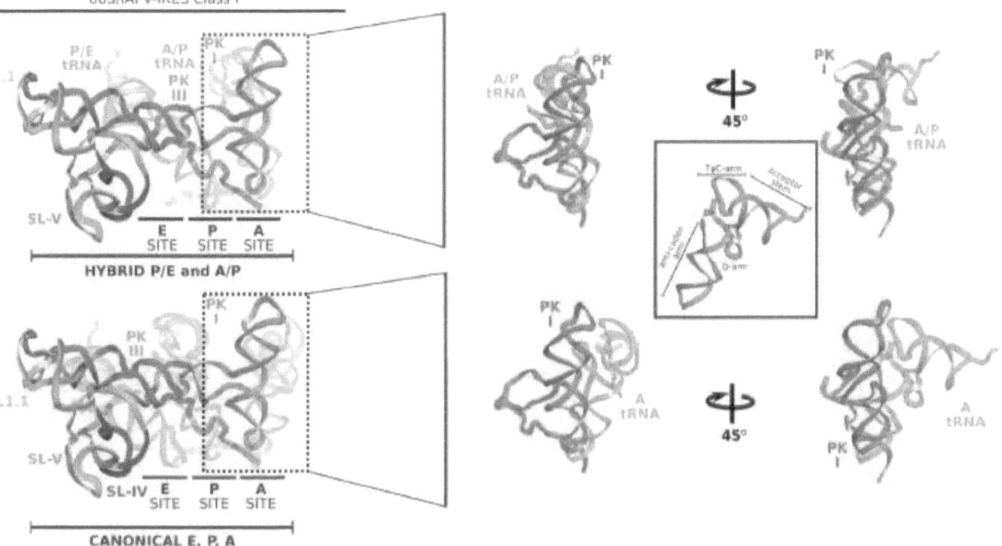

Figure S7: tRNA mimicry in the 80S/IAPV-IRES pre-translocation state.

(A) and (B) Superposition of IAPV-IRES in the pre-translocation state class 1 with tRNAs in hybrid-states (A) and with canonical tRNAs (B). On the right, close-up view of the superposition for the PKI domain of the IAPV- IRES in two orientations. It can be appreciated the co-axial unit formed by SL-III and the tRNA/mRNA-like domain of the IRES occupies a space similar to a hybrid A/P tRNA. The tRNA mimicry is however not complete as the acceptor stem of the tRNA is not accounted by any element of the IRES.

Data collection	Collection Polara-F30					Collection Titan-Krios
Micrographs	9973					4887
Picked Particles	1240275					338929
Class 2D particles	1011700					206829
Voltage (KV)	300					300
Defocus range* (μm)	0.4-3.5					1.2-4.9
Defocus mean (μm)	1.5					2.2
Pixel size (Å/pixel)	1.233					1.0605
Frames / Movie	40					40
Electron dose (e-/Å^2)	42.1					56.9
Electron dose per frame (e-/Å^2)	1.0525					1.4225

Stucture						
Component	40S/IAPV-IRES		80S/IAPV-IRES pre-translocated			80S/IAPV-IRES post-translocated
	Class-1	**Class-2**	**Class-1**	**Class-2**	**Class-3**	**Class-Post T.**
Particles	91056	96826	120176	40701	68697	27658
FSC 0.143 (Å)	3.1	3.1	3	3.1	3.1	3.2
Map sharpening (Å^2)	-70.3	-66.6	-67.5	-54.1	-65.1	-75.4

Refinement						
Program/Protocol	Refmac5 / Phenix	Refmac5 / Phenix	Refmac5 / Phenix	Refmac5 / Phenix	Refmac5 / Phenix	Refmac5 / Phenix
Used in refinement (Å)	3.1	3.1	3.1	3.1	3.1	3.2
Average B-factors (Å^2)	128.29	125.96	77.57	89.34	63.32	123.21
Avg B-fac Prot/RNA (Å^2)	129.77/126.89	127.35/124.62	83.85/72.86	93.28/86.39	65.82/61.44	126.50/120.66
R.m.s deviations:						
Bonds (Å)	0.0037	0.0031	0.0031	0.0027	0.0027	0.0032
Angles (deg)	1.1622	1.1208	1.1038	1.0408	1.0511	1.083

Validation						
Molprobity score	1.86	1.65	1.75	1.68	1.74	1.63
Clashcore, all atoms	2.09	2.29	2.23	2.07	2.25	2.11
Rotamer outliers (%)	2.77	1.77	2.16	1.6	2.04	1.5
Ramachandran plot:						
Outliers (%)	0.86	0.67	0.63	0.52	0.62	0.61
Favored (%)	90.36	92.75	91.79	90.01	91.43	91.32

Composition						
Non hydrogen atoms	79372	79379	215972	215966	215976	219296
Protein residues	4862	4862	11491	11490	11490	11892
RNA bases	1900	1900	5766	5766	5766	5768
Ligands	0	0	0	0	0	0

Accession codes						
EMDB	20248	20249	20255	20256	20257	20258
PDB	6P4G	6P4H	6P5I	6P5J	6P5K	6P5N

Table S1:

Data collection, model refinement and validation statistics. *Defocus range reported by GCT

Supplemental References:

Kucukelbir A, Sigworth FJ, Tagare HD (2014) Quantifying the local resolution of cryo-EM density maps. *Nat Methods* 11: 63-5

Murshudov GN, Vagin AA, Dodson EJ (1997) Refinement of macromolecular structures by the maximum-likelihood method. *Acta Crystallogr D Biol Crystallogr* 53: 240-55

Rosenthal PB, Henderson R (2003) Optimal determination of particle orientation, absolute hand, and contrast loss in single-particle electron cryomicroscopy. *J Mol Biol* 333: 721-45

Zivanov J, Nakane T, Forsberg BO, Kimanius D, Hagen WJ, Lindahl E, Scheres SH (2018) New tools for automated high-resolution cryo-EM structure determination in RELION-3. *Elife* 7

Appendix B

A complex IRES at the 5′-UTR of a viral mRNA assembles a functional 48S complex via an uAUG intermediate

Ritam Neupane[1,2†], Vera P Pisareva[3†], Carlos F Rodriguez[4], Andrey V Pisarev[3*], Israel S Fernández[2*]

[1]Department of Biological Sciences, Columbia University, New York, United States; [2]Department of Biochemistry and Molecular Biophysics, Columbia University, New York, United States; [3]Department of Cell Biology, SUNY Downstate Medical Center, Brooklyn, United States; [4]Structural Biology Programme, Centro Nacional de Investigaciones Oncológicas (CNIO), Madrid, Spain

Abstract Taking control of the cellular apparatus for protein production is a requirement for virus progression. To ensure this control, diverse strategies of cellular mimicry and/or ribosome hijacking have evolved. The initiation stage of translation is especially targeted as it involves multiple steps and the engagement of numerous initiation factors. The use of structured RNA sequences, called Internal Ribosomal Entry Sites (IRES), in viral RNAs is a widespread strategy for the exploitation of eukaryotic initiation. Using a combination of electron cryo-microscopy (cryo-EM) and reconstituted translation initiation assays with native components, we characterized how a novel IRES at the 5′-UTR of a viral RNA assembles a functional initiation complex via an uAUG intermediate. The IRES features a novel extended, multi-domain architecture, that circles the 40S head. The structures and accompanying functional data illustrate the importance of 5′-UTR regions in translation regulation and underline the relevance of the untapped diversity of viral IRESs.

*For correspondence:
andrey.pisarev@downstate.edu
(AVP);
isf2106@cumc.columbia.edu (ISFá)

[†]These authors contributed
equally to this work

Competing interests: The
authors declare that no
competing interests exist.

Funding: See page 14

Received: 19 December 2019
Accepted: 13 April 2020
Published: 14 April 2020

Reviewing editor: Adam Frost,
University of California, San
Francisco, United States

Introduction

Metagenomic studies of environmental samples have uncovered a great diversity of viruses that have a pervasive presence in the biosphere (*Zhang et al., 2019*; *Zhang et al., 2018*; *Greninger, 2018*). This diversity is especially overwhelming in RNA viruses that infect animal hosts (*Shi et al., 2016*; *Dolja and Koonin, 2018*). As strict cellular parasites, viruses rely on capturing cellular ribosomes to gain access to the host machinery for protein production (*Jan et al., 2016*). In eukaryotes, especially in animals, this machinery is complex and sophisticated, involving large, multi-component protein factors that assist in the operation of eukaryotic ribosomes (*Hashem and Frank, 2018*). Although complex, translation in eukaryotes conserves four main phases that are also found in its prokaryotic counterparts, namely: initiation, elongation, termination and recycling (*Schmeing and Ramakrishnan, 2009*). Initiation is significantly expanded in eukaryotes, with two GTP-regulated steps required for the correct positioning of the first aminoacyl-tRNA responsible for setting up the correct reading frame on the messenger RNA (mRNA) (*Jackson et al., 2010*; *Aitken and Lorsch, 2012*; *Myasnikov et al., 2009*; *Hinnebusch, 2014*).

Eukaryotic initiation starts when the 40S subunit together with the initiation factors eIF1, eIF1A, eIF3, eIF5 and the Ternary Complex (TC) (eIF2–Met-tRNA$_i^{Met}$–GTP) form the 43S Pre-Initiation Complex (43S-PIC), which is competent for mRNA recruitment (*Jackson et al., 2010*). Eukaryotic mRNAs are then docked to the 43S-PIC at their 5′ ends, forming the 48S complex (*Hinnebusch, 2017*). Once the AUG codon is detected, a structural transition in the 48S from an open, scanning-competent conformation to a closed, scanning-arrested conformation occurs (*Hussain et al., 2014*). This conformational change is accompanied by the release of eIF1, eIF2 and GDP, leaving the Met-tRNA$_i$^Met at the P-site of the 40S base paired with the AUG codon (*Aitken and Lorsch, 2012*). A second GTP-regulated step, catalyzed by initiation factor eIF5B, is then required for the recruitment of the large (60S) ribosomal subunit (*Pestova et al., 2000*; *Lee et al., 2002*). A full (80S) ribosome primed with mRNA and Met-tRNA$_i$^Met at the P-site then transitions to the elongation phase (*Wang et al., 2019*; *Voorhees and Ramakrishnan, 2013*).

The pathway described above is called the canonical, 5'-end and cap-dependent translation route of initiation (*Hinnebusch, 2014*). The bulk of eukaryotic mRNAs transitions follow this route, but deviations from the canonical route are common, and normally associated with translation under stress conditions (*Starck et al., 2016*; *Shatsky et al., 2010*). Non-canonical initiation is also associated with extended 5' UnTranslated Regions (5'-UTRs) on mRNAs (*Sendoel et al., 2017*; *Young and Wek, 2016*). In complex eukaryotes, 5'-UTRs can be very long and can harbor short Open Reading Frames (ORFs) designated as upstream ORFs (uORFs) (*Young and Wek, 2016*; *Wethmar, 2014*). Well-studied examples of the functional relevance of uORFs at 5'-UTRs can be found in the yeast stress response regulator GCN4 or the mammalian transcription factor ATF4 (*Hinnebusch, 1993*; *Vattem and Wek, 2004*). uAUG codons that are immediately followed by a stop codon (designated as 'start-stop uORFs') are also found in the 5'-UTRs of mammalian mRNAs (*Wethmar, 2014*; *Gunišová et al., 2018*), but little is known about how these 'start-stop uORFs' regulate translation.

Viruses exploit the complexity of eukaryotic initiation to gain access to the host machinery for protein production (*Jaafar and Kieft, 2019*). Strategies such as mimicking the cap structure or transferring caps from cellular mRNAs ('cap-snatching') allow viral mRNAs to hijack host ribosomes, redirecting them towards the production of viral proteins (*Jan et al., 2016*; *Jaafar and Kieft, 2019*). A more prominent viral strategy for ribosome hijacking is the use of structured RNA sequences in viral mRNAs (*Yamamoto et al., 2017*). These sequences are called Internal Ribosomal Entry Sites (IRES), and a tentative classification based on their degree of RNA structure and dependency on canonical initiation factors divided them in four main types (*Filbin and Kieft, 2009*; *Johnson et al., 2017*).

The *Dicistroviridae* family of positive single-stranded RNA ((+)-ssRNA) viruses employs two types of IRESs to express the regulatory versus the structural genes differentially (*Nakashima and Uchiumi, 2009*). The genome architecture of these viruses functionally segregates both kinds of genes in two ORFs (*Figure 1A*; *Hertz and Thompson, 2011*). The first ORF is preceded by an approximately 700-nucleotide 5'-UTR, which harbors an IRES assigned to the type III family (*Gross et al., 2017*). In vitro characterization of the 5'-UTR-IRES of the Cricket Paralysis Virus (CrPV), a prototypical *Dicistrovirus*, narrowed down the region of the 5'-UTR responsible for the IRES activity and established the strict requirement of eIF3 for this IRES to initiate translation. Interestingly, the AUG codon of the CrPV ORF1 is immediately preceded by a 'start-stop uORF' (*Gross et al., 2017*).

We sought to characterize the structure of the 5'-UTR-IRES of the CrPV in its ribosome-bound configuration, to gain insights relating to the ribosome-binding determinants of this peculiar IRES, and to understand how the delivery of Met-tRNA$_i$^Met is accomplished. Two high-resolution cryo-EM reconstructions of 40S–5'-UTR-IRES–eIF3 complexes, combined with biochemical analysis, allowed us to characterize how this IRES uses an extended structure with a modular, multi-domain architecture to bind to and manipulate the 40S.

Results

The 5'-UTR-IRES of the CrPV requires eIF3 for a stable interaction with the 40S

Previous studies of the IRES located at the 5'-UTR of the CrPV (hereafter referred to as 5'-UTR-IRES) precisely defined the region of the 5'-UTR that is responsible for the IRES activity (residues 357 to 709), as well as its dependency on initiation factor eIF3 for efficient translation initiation (*Gross et al., 2017*). In contrast to the well-characterized type IV family of IRESs found in the Inter-Genic Region (IGR-IRES) of these viruses, the 5'-UTRs of *Dicistroviruses* seem to harbor divergent sequences, making structural modelling based on sequence conservation difficult (*Kieft, 2009*). In order to address this gap in knowledge, we produced a truncated version of the 5'-UTR region of the genomic RNA of the CrPV that contains the IRES (residues 357 to 728, *Figure 1A*) in order to obtain structural information about its 40S-bound conformation by electron cryo-microscopy (cryo-

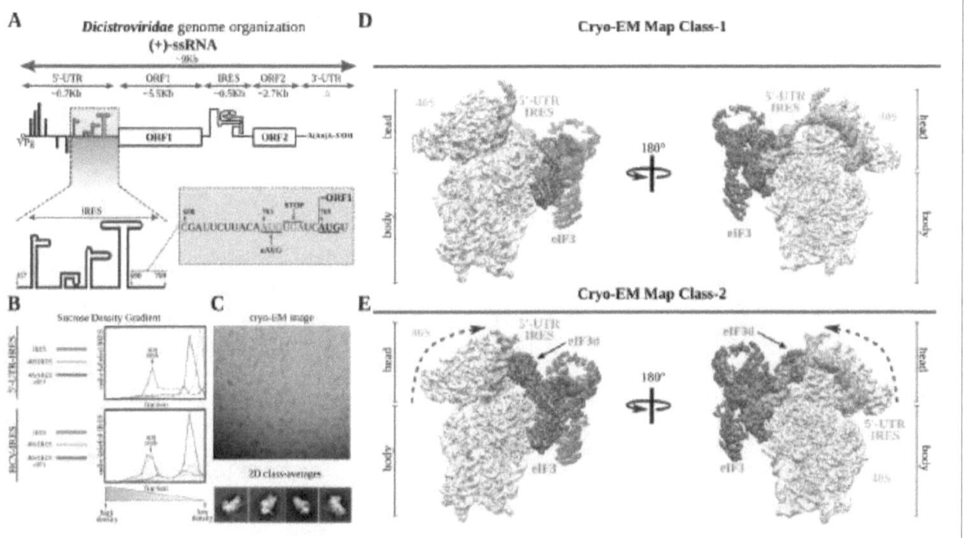

Figure 1. *Dicistroviridae* genome organization, in vitro complex formation and cryo-EM maps. (A) Top, schematic representation of the genome organization of *Dicistroviruses*. The approximate genomic lengths of the different components are indicated by the arrows. Bottom, detailed view of the region described as harboring the IRES activity of the 5'-UTR of the CrPV. On the right, the sequence adjacent to the initiation AUG codon of ORF1, located at nucleotide 709 and preceded by a 'start-stop uORF' indicated in red. (B) Sucrose-gradient analysis showed that the 5'-UTR-IRES is dependent on eIF3 in order to form a stable complex with the 40S. 5'-UTR-IRES co-migrates with the 40S only in the presence of eIF3 (top). By contrast, HCV IRES does not require eIF3 for 40S binding (bottom). (C) Representative cryo-EM micrograph of the 40S–5'-UTR-IRES–eIF3 complex. Bottom, representative reference-free 2D class averages used for further image processing. (D, E) After 3D classifications, two classes showing density for 40S (yellow), eIF3 (red) and 5'-UTR-IRES (blue) could be identified in the dataset. Class-1 (D) presents a non-swiveled configuration of the 40S head and the density for eIF3d is absent. Class-2 (E) shows a swiveled configuration of the 40S head (arrows) with eIF3d (indicated) contacting eIF3's core subunits.

The online version of this article includes the following figure supplement(s) for figure 1:

Figure supplement 1. Cryo-EM representative images and classification workflow.
Figure supplement 2. FSC correlation curves, local resolution and model validation.

EM). We initially tested the in vitro dependency of 5'-UTR-IRES on eIF3 when engaging purified 40S ribosomal subunits in a stable interaction. We assayed the ability of the 5'-UTR-IRES to co-migrate with purified 40S in sucrose density gradients as a test for the presence of a stable complex that is suitable for structural studies (*Figure 1B*). Unexpectedly, the 5'-UTR-IRES does not form a stable complex with the 40S in the absence of eIF3, in contrast to the HCV-IRES, which is able to form stable complexes with the 40S subunit alone and even with full (80S) ribosomes (*Figure 1B*; *Yokoyama et al., 2019*). Sucrose density gradients were manually fractionated from the bottom, where density is heavier. This caused small variations in the position of the 5'-UTR-IRES at the top of the gradients among different experiments. We do not believe such shifts have functional implications.

In the presence of eIF3, however, the 5'-UTR-IRES co-migrates with purified 40S subunits (*Figure 1B*). This complex revealed clear particles in cryo-EM images, rendering detailed two-dimensional class averages in which density for eIF3 could be identified, albeit at lower threshold (*Figure 1C* and *Figure 1—figure supplement 1*). The 40S–5'-UTR-IRES–eIF3 complex exhibited a delicate behavior under cryo-EM conditions, with a strong tendency to disassemble in thin ice. Extensive screening for suitable ice areas was essential to obtain particles of the fully assembled

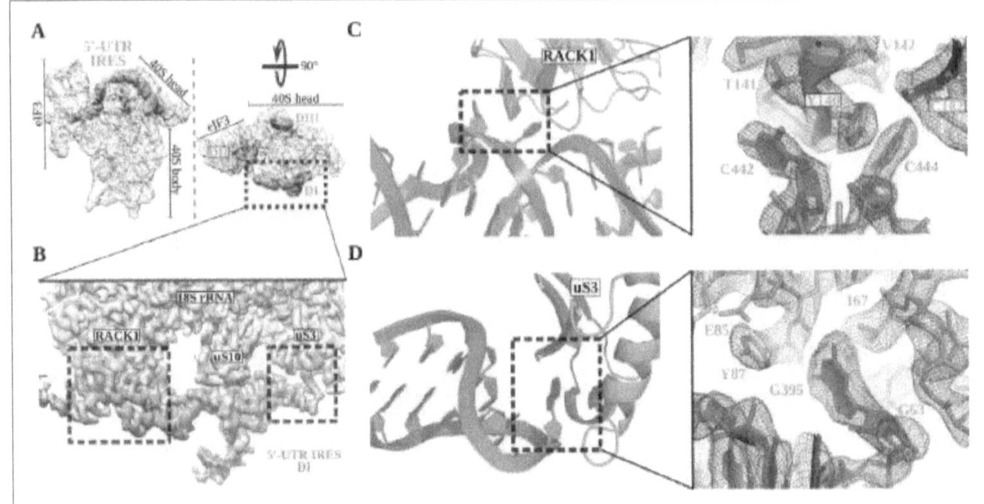

Figure 2. The 5′-UTR-IRES domain I engages ribosomal proteins RACK1 and uS3. **(A)** Overview of the 40S–5′-UTR-IRES–eIF3 map, with 40S and eIF3 depicted gray and 5′-UTR-IRES in blue. **(B)** Detailed view of the cryo-EM density of the 40S–5′-UTR-IRES–eIF3 map centered around 5′-UTR-IRES domain I (DI). Ribosomal proteins are colored brown, 18S rRNA yellow and 5′-UTR-IRES blue. Contacts between 5′-UTR-IRES domain I and ribosomal proteins RACK1 **(C)** and uS3 **(D)** could be defined thanks to well-resolved local cryo-EM densities.

The online version of this article includes the following figure supplement(s) for figure 2:

Figure supplement 1. Structurally derived secondary structure diagram for the 5′-UTR-IRES.

complex (*Figure 1C* and *Figure 1—figure supplement 1*). The sample also exhibited a high degree of heterogeneity, which could be resolved by image processing in Relion (*Scheres, 2012*; *Scheres, 2016*; *Figure 1D,E* and *Figure 1—figure supplement 2*).

Two main classes of particles containing density for 5′-UTR-IRES, 40S and eIF3 were found in the dataset (*Figure 1D and E*). Both classes contain density for the 40S, the IRES and the core subunit of eIF3 (a/c/e/k/l/f/m), and class-2 also presents density for eIF3 subunit d (*Figure 1E*, eIF3d). Class 2 exhibits a 40S head in a swiveled configuration. Owing to this swiveled configuration of the 40S head, eIF3d establishes interactions with eIF3a, a core subunit of eIF3 (see below).

Robust density ascribable to the 5′-UTR-IRES could be found in both classes of complex (*Figure 1D and E*, blue). The ribosome-bound conformation of 5′-UTR-IRES shows an extended configuration, almost circling the 40S head (*Figure 2A* and *Figure 2—figure supplement 1*). Three domains connected by flexible linkers could be defined: an elongated domain I (DI) at the back of the 40S head contacting ribosomal proteins uS3 and RACK1 (*Figure 2*), a second domain (DII) formed by a dual hairpin at the back of the 40S body interacting with eIF3 (*Figure 3*), and a third large helical domain (DIII) placed at the periphery of the 40S E-site, contacting ribosomal protein uS7 and uS11 (*Figure 4*).

Domain I of the 5′-UTR-IRES contacts the ribosomal proteins RACK1 and uS3

The 5′ proximal segment of the 5′-UTR-IRES (residues 357 to 486) forms domain which is characterized by an elongated T-shaped structure anchored to the back of the 40S head (*Figure 2A and B*). A long helical segment in this domain 'wraps' around the apical part of ribosomal protein RACK1. Two bases of this helical segment of domain I, C442 and C444, are extruded from the body of the double helix to establish hydrophobic interactions with tyrosine residue 140 of

RACK1 (*Figure 2C*). These interactions bend the main helical segment of the 5'-UTR-IRES DI, directing the tip of this domain towards ribosomal protein uS3 (*Figure 2D*). Guanine residue 395 is inserted deep into a hydrophobic pocket of ribosomal protein uS3, establishing contacts with mainchain atoms of this ribosomal protein. In this location, 5'-UTR-IRES DI is found adjacent to the mRNA entry channel of the 40S, overlapping with the space previously described as being occupied by the helicase DHX29 involved in canonical initiation (*Figure 2D*; *Hashem et al., 2013a*).

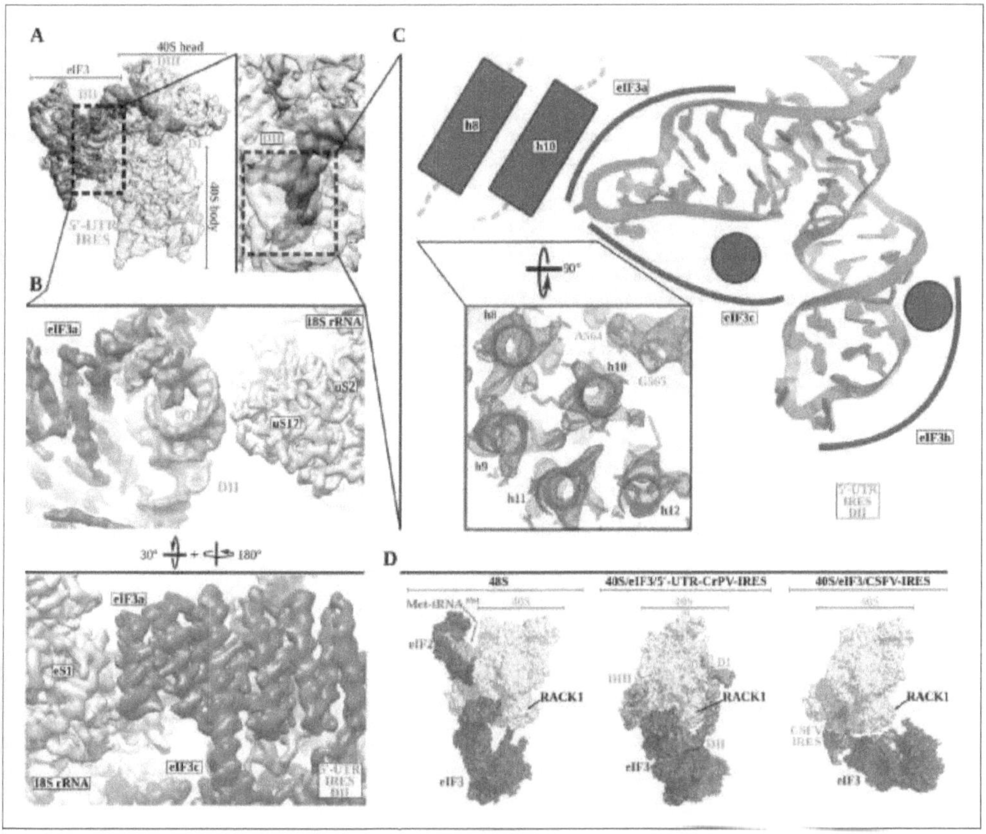

Figure 3. The 5'-UTR-IRES domain II is formed by a dual hairpin that mediates eIF3 recruitment. (A) Overview of the 40S–5'-UTR-IRES–eIF3 cryo-EM map with 40S colored gray, eIF3 red and 5'-UTR-IRES blue. On the right, a zoomed view centers around 5'-UTR-IRES domain II. (B) Detailed view of the cryo-EM map for the region occupied by 5'-UTR-IRES domain II, with 40S components colored gold, eIF3 red and 5'-UTR-IRES blue. Domain II is sandwiched between ribosomal protein uS17 (located at the back of the 40S body) and eIF3 core subunits a and c. (C) 5'-UTR-IRES domain II is formed by a dual hairpin that establishes interactions with α-helices 8 and 10 from eIF3a. These contacts are mediated mainly by basic residues of eIF3 and the phosphate backbone of the IRES. (D) superposition of the 40S–5'-UTR-IRES–eIF3 complex with the canonical 48S complex (left, PDB ID 6FEC) and with the CSFV-IRES–40S complex (right, PDB ID 4c4q). The 5'-UTR-IRES binds to the 40S with a conformation that is compatible with the canonical position described for eIF3 in the 48S complex.

5'-UTR-IRES binding to the 40S is compatible with a canonical configuration of eIF3

The second domain of 5'-UTR-IRES (DII) is connected to domain I by a flexible linker that is poorly defined in our maps as it is exposed to the solvent. This second domain of the 5'-UTR-IRES is formed by a dual hairpin and is wedged between the back of the body of the 40S and eIF3 subunits a and c (*Figure 3A and B*). Ribosomal protein uS17 peripherally contacts this domain, establishing interactions with the phosphate backbone of the IRES (*Figure 3B*). A network of interactions involving eIF3 subunits a, c and h anchors DII to this position (*Figure 3C*). These interactions are also established through contacts between positively charged residues on eIF3 and the phosphate backbone of the IRES. No contacts with specific IRES bases could be observed.

Currently, medium-resolution cryo-EM reconstructions for 40S complexes containing eIF3 and the rest of the components of the canonical 48S complex are available (*Eliseev et al., 2018*) (PDB ID 6FEC), as are such reconstructions for the 40S in complex with eIF3 and the CSFV-IRES (*Hashem et al., 2013b*) (PDB ID 4c4q). Comparisons of these structures with our complex reveal a positioning of eIF3 relative to the 40S that is very similar to the canonical 48S complex and different from the position adopted by eIF3 in the CSFV-IRES–40S complex (*Figure 3D*). In the 48S canonical configuration, eIF3 contacts the 40S through helix 1 of eIF3a and helix 22 of eIF3c, as well as through eIF3d, which is isolated in its 40S interaction, away from the core subunits of eIF3 (*Figure 3D*, left). The CSFV-IRES engages the 40S, displacing eIF3 from its position in the canonical 48S (*Figure 3D*, right). In addition, in the canonical 48S complex, eIF3 interacts with the 40S peripherally, allowing the formation of cavities between eIF3 and the back of the 40S. These cavities are exploited by the 5'-UTR-IRES, which inserts its domain II into one of them, adopting a configuration that is compatible with the binding of eIF3 to the 40S in the canonical 48S complex (*Figure 3D*, middle). No major rearrangement of eIF3 (compared to its position in the canonical 48S complex) is required for the binding of the 5'-UTR-IRES, so there could be an advantage in hijacking preformed cellular 48S complexes that are ready to transit the cap-dependent route of initiation.

Non-canonical base pairing in the 5'-UTR-IRES DIII places the uAUG codon near the P-site

Threading through the 40S channel formed by ribosomal proteins uS7 and uS11, a flexible single stranded linker connects DII with DIII (*Figure 4A*). DIII forms a prominent, helical mass in the surroundings of the E-site of the small subunit at the inter-subunit face of the 40S. The helical segment is very well defined in our maps because it is stabilized by numerous contacts with ribosomal proteins uS7 and uS11 and with 18S ribosomal RNA (rRNA) bases (*Figure 4* and *Figure 2—figure supplement 1*). However, the distal part of this domain forms two short stem loops that, given their flexibility, could only be modelled at low resolution.

Inspection of the cryo-EM density reveled a distortion in the canonical double helix of the main segment of this domain as it approaches the E-site. The quality of the maps in this area allowed de novo modelling of these residues, revealing a set of non-canonical interactions between the RNA bases (*Figure 4A and B*). In-plane triple-base interactions involving sugar and the Hoogsteen edges of the bases, as well as purine–purine Hoogsteen base pairs, could be found in this stretch of residues of the helical segment of DIII (*Figure 4B*; *Leontis and Westhof, 1998*). Overall, these non-canonical base pairs induce a distortion at the base of DIII that helps to position the single-stranded segment of the 5'-UTR-IRES harboring the uAUG codon at position 701 in the mRNA-binding channel of the 40S (*Figure 4C*, middle). The 5'-UTR-IRES accesses the 40 S P-site through the E-site, blocking a concurrent recruitment of the TC (eIF2–Met-tRNA$_i^{Met}$–GTP, *Figure 4C*). Interestingly, a similar strategy is followed by the HCV-IRES. A superposition of the structure of the HCV-IRES in complex with the 40S (*Yamamoto et al., 2015*; *Quade et al., 2015*) (PDB ID 5A2Q) with our structure reveals a very similar positioning of the domain II of HCV-IRES, accessing the P-site through the E-site to position the AUG codon in the surroundings of the P-site (*Figure 4C*, bottom). Even though both IRESs differ markedly in their interaction with the back of the 40S and eIF3, both converge to similar structural solutions for the placement of the AUG initial codon close to the 40 S P-site.

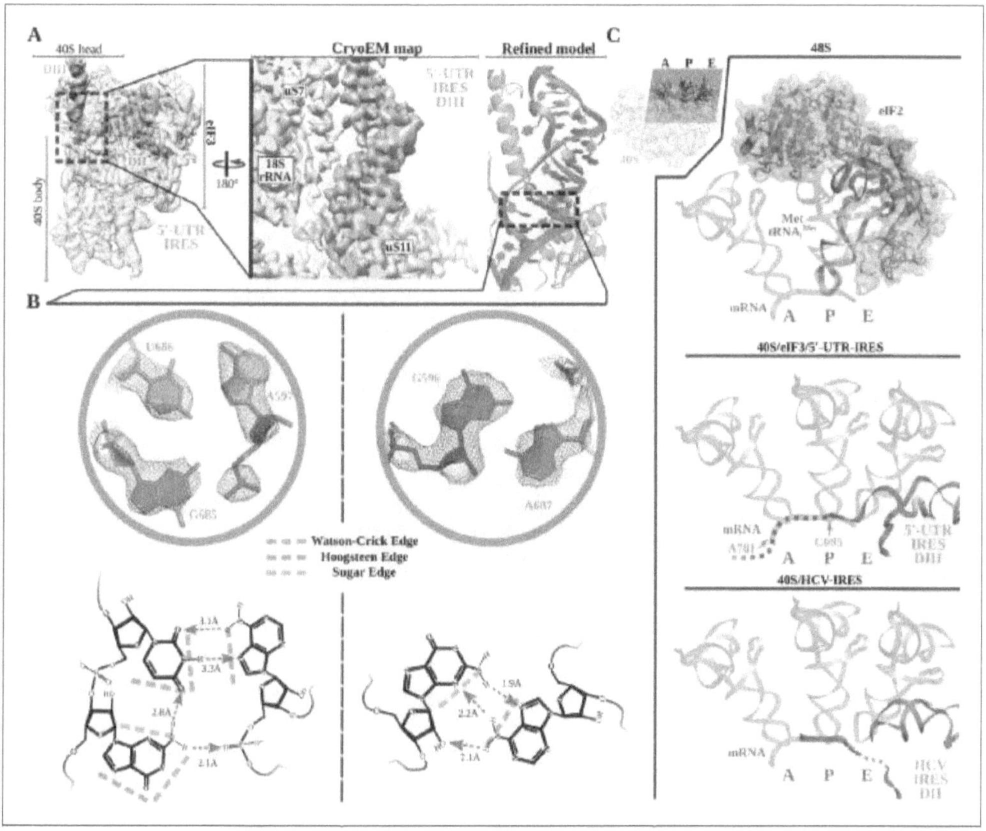

Figure 4. Non-canonical base pairing in 5'-UTR-IRES domain III assists on P-site access. A) Overview of the 40S–5'-UTR-IRES–eIF3 cryo-EM map with 40S and eIF3 colored gray and 5'-UTR-IRES blue. On the right, a detailed view of the E-site, where 5'-UTR-IRES domain III is placed, shows the cryo-EM map with the 5'-UTR-IRES colored blue and 40S components brown. On the far right, the final refined model is colored following the same color scheme. Ribosomal proteins uS7 and uS11 as well as several 18S rRNA bases contact 5'-UTR-IRES domain III. (B) Non-canonical base pairs found in 5'-UTR-IRES domain III induce a distortion of the double helix near the E-site. At the top, two examples are shown with the refined model inserted in the experimental cryo-EM density. The corresponding chemical diagrams below show the base edges and the hydrogen bonds involved in interactions. (C) Superpositions of the canonical 48S complex (PDB ID 6FEC, top) with the HCV-IRES/40S complex (PDB ID 5A2Q, bottom) and the 40S/5'-UTR-IRES/eIF3 (middle) models, focused on the tRNA A, P and E binding sites. 5'-UTR-IRES domain III and HCV-IRES occupies a space on the E-site that overlaps with the position described for eIF2 in the canonical 48S complex. In the middle, the last residue of 5'-UTR-IRES is indicated (C695), as is the putative path along the mRNA binding channel followed by the IRES (dashed line).

Swiveling of the 40S head locks the 5'-UTR-IRES, inducing a compact conformation of eIF3

Initial processing of the cryo-EM data revealed flexibility of the 40S head. Masked classification and refinement in Relion3 (*Zivanov et al., 2018*) revealed two major populations of particles, which differ in the degrees of 40S head swiveling (*Figure 1*). The 40S head is attached to the body by a single RNA helix, making this component of the ribosome extremely flexible (*Johnson et al., 2017*). Intrinsic and independent movements of the 40S head are instrumental in tRNA translocation

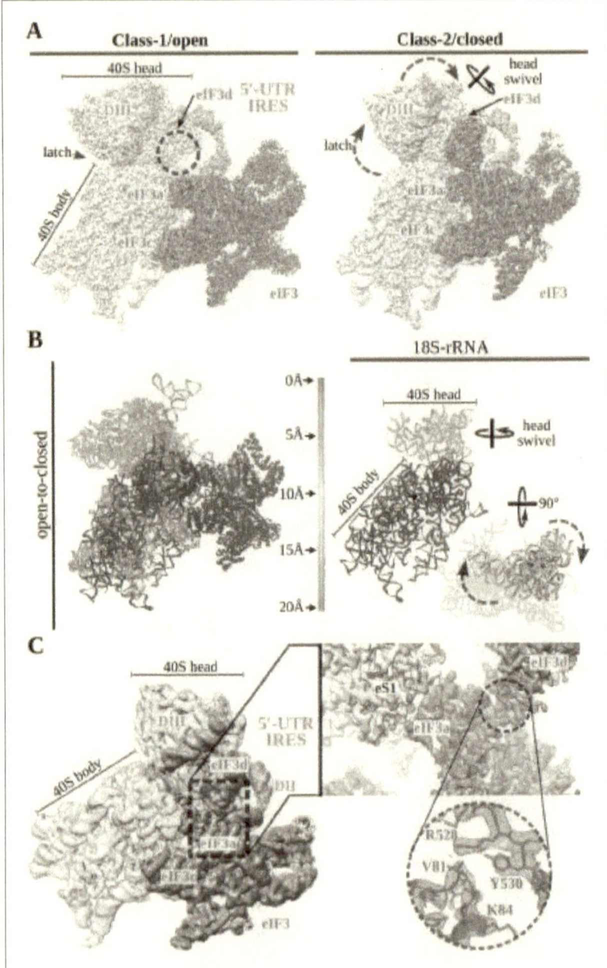

Figure 5. A 40S head swiveling movement 'locks' 5'-UTR-IRES on the 40S, inducing a compact eIF3 configuration. (A) Cryo-EM maps obtained for the two classes present in the 40S–5'-UTR-IRES–eIF3 dataset with 40S colored gold, eIF3 red and 5'-UTR-IRES blue. The positions of the latch, eIF3d and the swiveling rotation axis are indicated. A 40S head swiveling movement in class-2 brings eIF3d closer to the core subunit of eIF3 (i. e. eIF3a), establishing interactions that stabilize its conformation. (B) Left, ribbon diagram colored according to pairwise root mean square deviation (r.m.s.d.) displacements for the open-to-closed transition, with displacements scale at the center. On the right, a simplified diagram shows only the 18S rRNA colored according to the used on the left. Two orthogonal views are shown, in which it can be appreciated that the main displacement is localized at the 40S head. (C) Overview of the closed class with 40S colored gray, eIF3 red and 5'-UTR-IRES blue. Inset, detail of the experimental density obtained for the eIF3a–eIF3d interface for this class. Clear information on side chains was present in the maps, allowing proper model building and refinement.

and also in canonical initiation (*Ratje et al., 2010*; *Flis et al., 2018*). The 5′-UTR-IRES seems to exploit this intrinsic dynamic to bind to the 40S and then to 'lock' the IRES in a specific conformation that commits the complex towards viral translation (*Figure 5*). In class-1 (open conformation), the head of the 40S shows an almost canonical configuration with very little swiveling and no tilt. In this conformation, the latch of the 40S (an early defined contact between the head and the body of the 40S [*Frank et al., 1995*]) is closed. At the other side of the 40S head, access to the channel formed by ribosomal proteins uS7 and uS11 is exposed and eIF3d density is not well defined, probably because of a high degree of flexibility or low occupancy (*Figure 5A*, left). In class-2 (closed conformation), the head of the 40S exhibits a medium-range degree of swiveling when compared to the widest displacement reported (*Ratje et al., 2010*).

In the open and closed classes, the positions of the 5′-UTR-IRES relative to the 40S head are very similar (*Figure 5A*, right and B). In the swiveled conformation, the latch is open, and the channel formed by ribosomal proteins uS7 and uS11 is plugged by eIF3d, which in this class presents robust density (*Figure 5C*). The main subunits of eIF3 (a/c/e/k/l/f/m) show a similar conformation in both classes, having a similar orientation with respect to the 40S body (*Figure 5B*). In the swiveled configuration (class-2), the 40S head brings eIF3d close to eIF3a, one of the core subunits of eIF3 (*Figure 5C*). Well-defined density in this area could be observed for the eIF3a–eIF3d interface (*Figure 5C*, right). This compact state of eIF3 represent a hitherto unknown conformation (*Lee et al., 2016*).

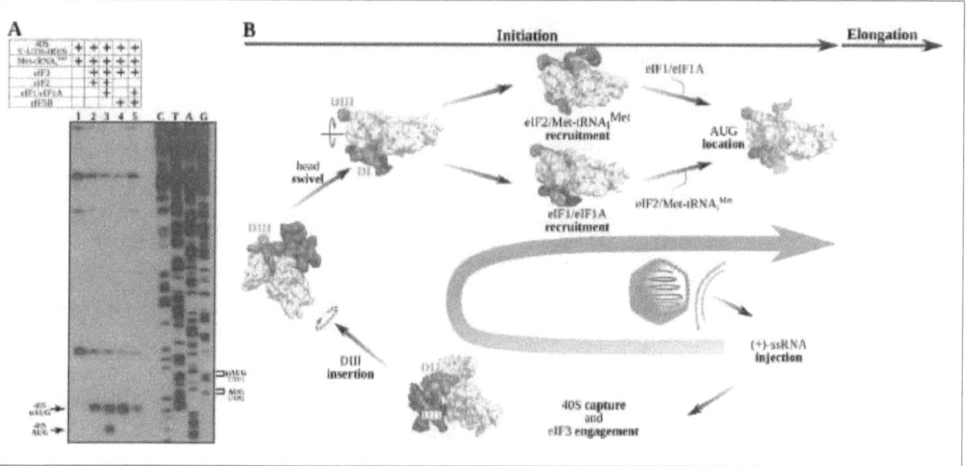

Figure 6. The 5′-UTR-IRES requires TC, eIF1 or eIF1A to assemble a functional initiation complex via an uAUG intermediate. (A) Toe-print analysis of 48S initiation complexes assembled on 5′-UTR-IRES in an in vitro reconstituted system. eIF2 delivers Met-tRNA$_i^{Met}$ to the uAUG (lane 2) and requires the presence of eIF1 or eIF1A to transition to the *bona fide* AUG codon of ORF1 (lane 3). Under some conditions, eIF5B can substitute for eIF2 in Met-tRNA$_i^{Met}$ delivery (*Terenin et al., 2008*). In the absence of eIF1 or eIF1A, a robust toe-print signal is detected in the presence of eIF5B (lane 4), however, Met-tRNA$_i^{Met}$ is delivered to the uAUG, and thus eIF5B is unable to find the annotated AUG even in the presence of eIF1 or eIF1A (lane 5). (B) A model for 5′-UTR-IRES-mediated translation initiation. From the bottom right: injection of the genomic (+)-ssRNA of the CrPV into the cytosol allows the 5′-UTR-IRES to capture free 40S and eIF3, which is recruited to the complex with an initial canonical conformation of the 40S head without tilt or swivel. Insertion of 5′-UTR-IRES DIII in the vicinity of the E-site and a swiveling movement of the 40S head induces a 'locked' conformation of the complex, with the uAUG at 701 in the vicinity of the P-site. Delivery of Met-tRNA$_i^{Met}$ and location of ORF1 AUG is achieved by the concerted action of eIF2, eIF1 and eIF1A. Large subunit recruitment mediated by eIF5B will allow transitioning into elongation.

The online version of this article includes the following figure supplement(s) for figure 6:

Figure supplement 1. Control toe-print experiments.

TC delivers Met-tRNA$_i^{Met}$ to uAUG at position 701, and initiation factors eIF1 and eIF1A assist in AUG location

Our structures of the 40S–5′-UTR-IRES–eIF3 complex revealed a positioning of the DIII of the IRES that overlaps with the position that the TC populates at the E-site in canonical initiation (*Figure 4C*; *Hussain et al., 2014*; *Eliseev et al., 2018*). In addition, in our maps, we could only confidently identify density for the single-stranded segment of RNA of the IRES placed close to the P-site until residue 695, whereas the canonical AUG of ORF1 is found at nucleotide 709. These facts prompted us to wonder how the delivery of Met-tRNA$_i^{Met}$ to the AUG is accomplished. Making use of an in vitro reconstituted mammalian initiation assay with native components and toe-printing analysis (*Kolupaeva et al., 2007*), we analyzed the different steps followed by the 5′-UTR-IRES in order to place Met-tRNA$_i^{Met}$ based paired with the AUG codon at the P-site (*Figure 6A*). Translation initiation factors in mammals and insects are highly homologous. In particular, eIF2-alpha shares 57% identity and 74% similarity between human and *Drosophila*, eIF2-beta 74% identity and 83% similarity, eIF2-gamma 82% identity and 88% similarity, eIF5B 71% identity and 85% similarity, eIF3a 46% identity and 63% similarity, and eIF3c 51% identity and 66% similarity. This high level of homology justifies the utilization of mammalian initiation factors for CrPV analysis, as has been done before for the CrPV IGR-IRES.

Toe-print assay permits identification of the location of functional ribosomal complexes assembled on mRNAs by reverse transcription of a primer annealed to the mRNA. The length of the resulting extended DNA fragment provides information about the position of the ribosome on the mRNA. Due to its large size, the paused ribosome protects a segment of the mRNA, precluding further primer extension and generating toe-print signals approximately 15–17 nucleotides downstream of the P-site of 40S. Cognate aminoacyl or peptidyl-tRNAs in the P-site or post-termination complexes with eRF1 in the A-site yield robust toe-print signals (*Skabkin et al., 2013*). The 40S–5′-UTR-IRES–eIF3 complex produces signal that is ascribable to the secondary structure of the 5′-UTR-IRES (*Figure 6—figure supplement 1*, lanes 1–3), indicating no measurable pausing of the ribosome on the mRNA around any of the AUG codons. In isolation, the TC (eIF2–Met-tRNA$_i^{Met}$–GTP) is able to load Met-tRNA$_i^{Met}$ onto the P-site of the 40S–5′-UTR-IRES–eIF3 complex, producing a robust toe-print (*Figure 6A*, lane 2, label 48S–uAUG) 15–17 nucleotides away from the uAUG located at 701. A similar uAUG delivery of Met-tRNA$_i^{Met}$ can be accomplished by eIF5B which, under stress conditions, has been described as substituting for eIF2 in Met-tRNA$_i^{Met}$ delivery (*Terenin et al., 2008*; *Pestova et al., 2008*; *Yamamoto et al., 2014*; *Kenner et al., 2019*), with eukaryotic initiation then following a 'bacterial-like' mode (*Figure 6A*, lane 4). Transitioning to the correct AUG could only be detected in the presence of eIF1and eIF1A, and only when the TC was present, and not for eIF5B (*Figure 6A*, lanes 3 and 5). Notably, the presence of eIF1and eIF1A seems to be detrimental to uAUG Met-tRNA$_i^{Met}$ loading by eIF5B, as their presence significantly reduces the toe-print signal that can be observed for eIF5B in isolation. However, no concomitant increase in toe-print signal for the canonical AUG could be observed for the eIF1/eIF1A/eIF5B reaction.

Only eIF2 as part of the TC and assisted by eIF1 and eIF1A can properly locate the *bona fide* AUG of ORF1. The role of the uAUG located at nucleotide 701 is not clear, but the fact that eIF5B can deliver Met-tRNA$_i^{Met}$ only to this codon points towards an important role for this uAUG in initiation when eIF2 is unavailable.

Discussion

Ribosome-profiling datasets have revealed the presence of translating ribosomes paused on 5′ UTRs, implying a decisive role of these sequences in regulating translation, especially under stress conditions (*Sendoel et al., 2017*; *Archer et al., 2016*; *Andreev et al., 2015*; *Ingolia et al., 2009*; *Brar and Weissman, 2015*; *Resch et al., 2009*).

The 5′-UTR of the (+)-ssRNA of the *Dicistovirus* CrPV harbors an IRES that is able to direct initiation towards ORF1 in the early phase of infection (*Hertz and Thompson, 2011*; *Garrey et al., 2010*). Expression of ORF1 is instrumental for virus replication because the RNA-dependent RNA polymerase (RdRp), and the protease responsible for the proteolytic digestion of the polyprotein containing the structural proteins, are encoded in ORF1 (*Jan et al., 2003*).

The 5′-UTR-IRES features a novel multi-domain, extended architecture that encircles three quarters of the 40S head, exploiting binding sites not previously described for any IRESs (*Figures 2*

3 and 4). Ribosomal proteins uS3 and RACK1 are used by the IRES to anchor its DI to the back of the 40S head (*Figure 2*). The structure thus rationalizes previous data showing a preeminent role of RACK1 in CrPV and related viruses that infect *Drosophila* (*Majzoub et al., 2014*). The interaction of DI with RACK1 is also instrumental in positioning DII at the back of the 40S body, sandwiched in between ribosomal protein uS17 and eIF3 (*Figure 3*). Interestingly and in contrast with the HCV-IRES, the conformation observed for eIF3 in the complex with 5′-UTR-IRES is very similar to that observed for eIF3 in the 48S complex, with the IRES 'filling up' cavities that are present between the 40S and eIF3 in this canonical complex (*Eliseev et al., 2018*). The HCV-IRES and related IRESs, such as the CSFV-IRES, displace eIF3 from its canonical location using a very different mechanism for IRES docking to the 40S (*Hashem et al., 2013b*).

In order to place the AUG of ORF1 in the surroundings of the P-site, the 5′-UTR-IRES accesses the P-site through the E-site, in a manner similar to that of the HCV-IRES (*Figure 4C*; *Yamamoto et al., 2015*). In this aspect, the 5′-UTR-IRES recapitulates binding strategies that are known for other IRESs, such as the IGR-CrPV-IRES that also makes use of ribosomal protein uS7 for its binding to the ribosome or the HCV-IRES that places its domains II and IV in the surroundings of the P-site, sliding the elongated DII from the back of the 40S to the P-site through the E-site (*Pisareva et al., 2018*).

The placement of the AUG of ORF1 in the surroundings of the P-site seems to be exerted by a mechanism involving the intrinsic dynamics of the 40S head (*Johnson et al., 2017*; *Figure 5*). The 5′-UTR-IRES exploits the characteristic swiveling movement of the 40S head to bind and progress towards a conformation that 'locks' the IRES onto the 40S, and at the same time, induces a compact conformation of eIF3 that has subunit eIF3d in close contact with the core subunits of eIF3 (*Lee et al., 2016*). These dynamics are probably instrumental for the ability of the 5′-UTR-IRES–40S complex to localize the annotated AUG, in a genomic context where a uAUG-stop is physically close. The capacity of the 40S to scan an mRNA bidirectionally upon termination on a stop codon has been previously reported (*Skabkin et al., 2013*). It is thus plausible that the peculiar genetic configuration of the CrPV around the annotated AUG of ORF1 (*Figure 1A*) evolved to leverage these re-initiation mechanisms already present in the translation of cellular messengers. However, these considerations are highly speculative, as the particular role that the uAUG exerts in Met-tRNA$_i^{Met}$ recruitment, or more generally its involvement in initiation of viral messengers, remains enigmatic. A comprehensive understanding of the role of uAUG and the start-stop configuration will demand further studies, ideally in vivo.

We propose the following model for how the 5′-UTR-IRES of the CrPV operates: immediately after the (+)-ssRNA genomic molecule of the CrPV is injected into the cytoplasm of the host cell, the IRES harbored at the 5′-UTR captures 40S subunits (*Figure 6B*, bottom). Recruitment of eIF3 is mediated by DII, allowing the sliding of the flexible linker connecting DII and DIII between the head and the platform of the 40S to place DIII in the surroundings of the E-site (*Figure 6B*, bottom and left). A swivel movement of the 40S head closes the channel between the head and the platform of the 40S, effectively 'locking' the 5′-UTR-IRES into the 40S and inducing a compact conformation of eIF3 with the eIF3d subunit in interacting distance with eIF3's core subunit a (*Figure 6B*, left top). With this configuration, eIF2 as part of the TC can deliver Met-tRNA$_i^{Met}$ to the uAUG located at nucleotide 701, and further assistance by initiation factors eIF1 and eIF1A allows for a downstream location of the AUG codon of ORF1 at nucleotide 709. Large subunit recruitment grants transitioning towards elongation, committing the ribosome to the production of viral proteins (*Figure 6B*, right top).

In summary, we have structurally characterized the 5′-UTR-IRES of the CrPV in its ribosome-bound state and have characterized the delivery of Met-tRNA$_i^{Met}$ by eIF2 and eIF5B. Given the rich diversity of viral sequences in the animal virome, new IRESs exploiting different aspects of animal translation will probably be discovered.

Materials and methods

5′-UTR-IRES and HCV IRES production

For cryo-EM analysis, a transcription vector for 5′-UTR-IRES (nucleotides 357–728) was constructed by inserting a T7 promoter sequence upstream of the 5′-UTR-IRES sequence followed by a BamHI restriction site, using pUC19 as a scaffold vector. For toe-print assays, 5′-UTR-IRES with the extended ORF part for primer annealing was cloned by a similar strategy. The uACG-AGA mutant was obtained by site-directed mutagenesis of 5'-UTR-IRES. T7 RNA polymerase in vitro transcription and purification on Spin-50 mini-column (USA Scientific) were used to obtain highly purified RNAs.

Purification of translation components and ribosomal subunits

Native 40S subunits, eIF2, eIF3, eIF5B and rabbit aminoacyl-tRNA synthetases were prepared as previously described (*Pestova and Hellen, 2005*). Recombinant eIF1 and eIF1A were purified according to a previously described protocol (*Kolupaeva et al., 2007*). In vitro transcribed Met-tRNA$_i^{Met}$ was aminoacylated with methionine in the presence of rabbit aminoacyl-tRNA synthetases as previously described (*Pisarev et al., 2010*).

Assembly of ribosomal complexes

To reconstitute different ribosomal complexes for toe-print assays, we incubated 0.3 pmol 5'-UTR-IRES RNA with 1.8 pmol 40S subunits, 10 pmol eIF1, 10 pmol eIF1A, 10 pmol eIF2, 5 pmol eIF3, 5 pmol eIF5B, and 5 pmol Met-tRNA$_i^{Met}$, as indicated, in a 20 µL reaction mixture containing buffer A (20 mM Tris-HCl [pH 7.5], 100 mM KCl, 2.5 mM MgCl$_2$ and 1 mM DTT) with 0.4 mM GTP and 0.4 mM ATP for 10 min at 37∘C. We analyzed the assembled ribosomal complexes via a toe-print assay, essentially as described by *Pestova and Hellen (2005)*. For the sucrose density gradient experiment, we incubated [^{32}P]-labelled 5'-UTR-IRES or HCV-IRES RNAs co-transcriptionally with 3.7 pmol 40S subunits and 11 pmol eIF3, as indicated, in a 60 µL reaction mixture containing buffer A for 10 min at 37°C, subjected the samples to a 10–30% sucrose density gradient centrifugation, and analyzed the gradient fractions by radioactivity counting.

Cryo-EM sample preparation and data acquisition

Aliquots of 3 µl of assembled ribosome complexes at a concentration range of 250–350 nM were incubated for 30 s on glow-discharged holey gold grids (*Russo and Passmore, 2014*) (UltrAuFoil R1.2/1.3). Grids were blotted for 2.5 s and flash cooled in liquid ethane using a FEI Vitrobot. Grids were transferred to a FEI Titan Krios microscope equipped with an energy filter (slits aperture 20 eV) and a Gatan K2 detector operated at 300 kV. Data were recorded in counting mode at a magnification of 130,000, corresponding to a calibrated pixel size of 1.08 Å. Defocus values ranged from 1 µm to 3.6 µm. Images were recorded in automatic mode using the Leginon (*Carragher et al., 2000*) and APPION (*Lander et al., 2009*) software and frames were aligned using the Relion3 (*Zivanov et al., 2018*) implementation of the Motioncor2 algorithm (*Zheng et al., 2017*).

Image processing and structure determination

Contrast transfer function parameters were estimated using GCTF (*Zhang, 2016*), and particle picking was performed using GAUTOMACH without the use of templates and with a diameter value of 260 pixels. All 2D and 3D classifications and refinements were performed using RELION. An initial 2D classification with a 4 times binned dataset identified all ribosome particles. A consensus reconstruction with all 40S particles was computed using the AutoRefine tool of RELION. Next, 3D classification without alignment (four classes, T parameter 4) identified a class with unambiguous density for eIF3. This class was independently refined, and further masked classification allowed the identification of two subclasses that are distinguishable by a different degree of 40S head swiveling and by the presence or absence of eIF3d density. Final refinements with unbinned data for the selected classes yielded high-resolution maps with density features in agreement with the reported resolution. Local resolution was computed with RESMAP (*Kucukelbir et al., 2014*).

Model building and refinement

Models for the mammalian 40S and eIF3 docked into the maps using CHIMERA (*Pettersen et al., 2004*) and COOT (*Emsley et al., 2010*) were used to adjust these initial models manually. 5'-UTR-IRES was built manually using COOT. An initial round of refinement was performed in Phenix using real-space refinement (*Afonine et al., 2018*) with secondary structure restraints and a final step of reciprocal-space refinement with REFMAC (*Murshudov et al., 1997*). The fit of the model to the map density was quantified using FSCaverage and Cref and model-to-maps over-fitting tests were performed following standard protocols in the field (*Brown et al., 2015*; *Amunts et al., 2014*).

Cryo-EM data collection, refinement and validation statistics

	Class-1 (open) (EMDB-21529) (PDB 6W2S)	Class-2 (closed) (EMDB-21530) (PDB 6W2T)
Data collection and processing		
Magnification	130,000	
Voltage (kV)	300	
Electron exposure (e–/Å^2)	59.55	
Defocus range (μm)	−1 /− 3	
Pixel size (Å)	1.06	
Symmetry imposed	C1	
Initial particle images (no.)	915,647	
Final particle images (no.)	14,257	23,444
Map resolution (Å)	3.3	3.3
FSC threshold	0.143	0.143
Map resolution range (Å)	3–8	3–8
Refinement		
Initial model used (PDB code)	5A2Q	5A2Q
Model resolution (Å)	3.6	3.6
FSC threshold	0.5	0.5
Model resolution range (Å)	3.3–8	3.3–8
Map sharpening B factor (Å^2)	−31.94	−43.41
Model composition	106,817	109,684
Non-hydrogen atoms	-	-
Ligands		
B factors (Å^2)	92.47	96.5
Protein	114.1	117.4
RNA		
R.m.s. deviations	0.014	0.014
Bond lengths (Å)	1.77	1.78
Bond angles (°)		
Validation	2.12	1.99
MolProbity score	6.13	4.92
Clashscore	1.62	1.39
Poor rotamers (%)		
Ramachandran plot	88.92	90.25
Favored (%)	98.37	98.50
Allowed (%)	1.63	1.50
Disallowed (%)	0.18	0.17
RNA validation	2.35	2.25
Angles outliers (%)	0.442	0.428
Sugar puckers outliers (%)		
Average suit		

Acknowledgements

We are grateful to Dr Jean-Luc Imler for a generous donation of a CrPV-5'-UTR-IRES plasmid. We are thankful to Prof. Jennifer Doudna for a generous donation of an HCV-IRES transcription vector. We are thankful to Prof. Kathrin Lang for the identification of an error in *Figure 4* in the pre-print version of this manuscript. We are thankful to Bob Grassucci and Zhening Zhang for assistance in cryo-EM data acquisition. This work was supported by the NIH National Institute of General Medical Sciences (GM097014 to AVP). Part of this work was performed at the Simons Electron Microscopy Center and National Resource for Automated Molecular Microscopy located at the New York Structural Biology Center, supported by grants from the Simons Foundation (SF349247), NYSTAR, and the NIH National Institute of General Medical Sciences (GM103310).

Additional information

Funding

Funder	Grant reference number	Author
Columbia University	Start package	Israel S Fernández
National Institute of General Medical Sciences	GM097014	Andrey V Pisarev

The funders had no role in study design, data collection and interpretation, or the decision to submit the work for publication.

Author contributions

Ritam Neupane, Software, Validation, Investigation, Visualization, Writing - review and editing; Vera P Pisareva, Investigation, Writing - review and editing; Carlos F Rodriguez, Investigation, carried out experiments; Andrey V Pisarev, Conceptualization, Investigation, Writing - original draft, Writing - review and editing; Israel S Fernández, Conceptualization, Resources, Supervision, Funding acquisition, Validation, Investigation, Visualization, Methodology, Project administration

Author ORCIDs

Ritam Neupane https://orcid.org/0000-0003-4787-279X
Carlos F Rodriguez http://orcid.org/0000-0001-9166-0132
Israel S Fernández https://orcid.org/0000-0001-7218-1603

Decision letter and Author response

Decision letter https://doi.org/10.7554/eLife.54575.sa1
Author response https://doi.org/10.7554/eLife.54575.sa2

Additional files

Supplementary files

• Transparent reporting form

Data availability

Atomic coordinates have been deposited in the PDB with accession numbers and 6W2S and 6W2T for the open and closed classes , respectively . CryoEM maps have been deposited at the EMDB with accession numbers EMDB 21529 and 21530 for the open and closed classes respectively.

The following datasets were generated:

Author(s)	Year	Dataset title	Dataset URL	Database and Identifier
Neupane R, Pisareva VP, Rodriguez CF, Pisarev AV, Fernández IS	2020	CryoEM map open class	https://www.ebi.ac.uk/pdbe/entry/emdb/EMD-21529	Electron Microscopy Data Bank, EMD-21529
Neupane R, Pisareva VP, Rodriguez CF, Pisarev AV, Fernández IS	2020	CryoEM map closed class	https://www.ebi.ac.uk/pdbe/entry/emdb/EMD-21530	Electron Microscopy Data Bank, EMD-21530
Neupane R, Pisareva VP, Rodriguez CF, Pisarev AV, Fernández IS	2020	Structure of the Cricket Paralysis Virus 5-UTR IRES (CrPV 5-UTR-IRES) bound to the small ribosomal subunit in the open state (Class 1)	https://www.rcsb.org/structure/6W2S	RCSB Protein Data Bank, 6W2S
Neupane R, Pisareva V, Rodriguez CF, Pisarev A, Fernandez IS	2020	Structure of the Cricket Paralysis Virus 5-UTR IRES (CrPV 5-UTR-IRES) bound to the small ribosomal subunit in the closed state (Class 2)	https://www.rcsb.org/structure/6W2T	RCSB Protein Data Bank, 6W2T

References

Afonine PV, Poon BK, Read RJ, Sobolev OV, Terwilliger TC, Urzhumtsev A, Adams PD. 2018. Real-space refinement in *PHENIX* for cryo-EM and crystallography. *Acta Crystallographica Section D Structural Biology* **74**: 531–544. DOI: https://doi.org/10.1107/S2059798318006551

Aitken CE, Lorsch JR. 2012. A mechanistic overview of translation initiation in eukaryotes. *Nature Structural & Molecular Biology* **19**:568–576. DOI: https://doi.org/10.1038/nsmb.2303, PMID: 22664984

Amunts A, Brown A, Bai XC, Llácer JL, Hussain T, Emsley P, Long F, Murshudov G, Scheres SHW, Ramakrishnan V. 2014. Structure of the yeast mitochondrial large ribosomal subunit. *Science* **343**:1485–1489. DOI: https://doi.org/10.1126/science.1249410, PMID: 24675956

Andreev DE, O'Connor PB, Fahey C, Kenny EM, Terenin IM, Dmitriev SE, Cormican P, Morris DW, Shatsky IN, Baranov PV. 2015. Translation of 5' leaders is pervasive in genes resistant to eIF2 repression. *eLife* **4**:e03971. DOI: https://doi.org/10.7554/eLife.03971, PMID: 25621764

Archer SK, Shirokikh NE, Beilharz TH, Preiss T. 2016. Dynamics of ribosome scanning and recycling revealed by translation complex profiling. *Nature* **535**:570–574. DOI: https://doi.org/10.1038/nature18647, PMID: 27437580

Brar GA, Weissman JS. 2015. Ribosome profiling reveals the what, when, where and how of protein synthesis. *Nature Reviews Molecular Cell Biology* **16**:651–664. DOI: https://doi.org/10.1038/nrm4069, PMID: 26465719

Brown A, Long F, Nicholls RA, Toots J, Emsley P, Murshudov G. 2015. Tools for macromolecular model building and refinement into electron cryo-microscopy reconstructions. *Acta Crystallographica Section D Biological Crystallography* **71**:136–153. DOI: https://doi.org/10.1107/S1399004714021683, PMID: 25615868

Carragher B, Kisseberth N, Kriegman D, Milligan RA, Potter CS, Pulokas J, Reilein A. 2000. Leginon: an automated system for acquisition of images from vitreous ice specimens. *Journal of Structural Biology* **132**:33–45. DOI: https://doi.org/10.1006/jsbi.2000.4314, PMID: 11121305

Dolja VV, Koonin EV. 2018. Metagenomics reshapes the concepts of RNA virus evolution by revealing extensive horizontal virus transfer. *Virus Research* **244**:36–52. DOI: https://doi.org/10.1016/j.virusres.2017.10.020, PMID: 29103997

Eliseev B, Yeramala L, Leitner A, Karuppasamy M, Raimondeau E, Huard K, Alkalaeva E, Aebersold R, Schaffitzel C. 2018. Structure of a human cap-dependent 48S translation pre-initiation complex. *Nucleic Acids Research* **46**:2678–2689. DOI: https://doi.org/10.1093/nar/gky054, PMID: 29401259

Emsley P, Lohkamp B, Scott WG, Cowtan K. 2010. Features and development of *coot*. *Acta Crystallographica. Section D, Biological Crystallography* **66**:486–501. DOI: https://doi.org/10.1107/S0907444910007493, PMID: 20383002

Filbin ME, Kieft JS. 2009. Toward a structural understanding of IRES RNA function. *Current Opinion in Structural Biology* **19**:267–276. DOI: https://doi.org/10.1016/j.sbi.2009.03.005, PMID: 19362464

Flis J, Holm M, Rundlet EJ, Loerke J, Hilal T, Dabrowski M, Bürger J, Mielke T, Blanchard SC, Spahn CMT, Budkevich TV. 2018. tRNA translocation by the eukaryotic 80S ribosome and the impact of GTP hydrolysis. *Cell Reports* **25**:2676–2688. DOI: https://doi.org/10.1016/j.celrep.2018.11.040, PMID: 30517857

Frank J, Zhu J, Penczek P, Li Y, Srivastava S, Verschoor A, Radermacher M, Grassucci R, Lata RK, Agrawal RK. 1995. A model of protein synthesis based on cryo-electron microscopy of the *E. coli* ribosome. *Nature* **376**: 441–444. DOI: https://doi.org/10.1038/376441a0, PMID: 7630422

Garrey JL, Lee YY, Au HH, Bushell M, Jan E. 2010. Host and viral translational mechanisms during cricket paralysis virus infection. *Journal of Virology* **84**:1124–1138. DOI: https://doi.org/10.1128/JVI.02006-09, PMID: 19889774

Greninger AL. 2018. A decade of RNA virus metagenomics is (not) enough. *Virus Research* **244**:218–229. DOI: https://doi.org/10.1016/j.virusres.2017.10.014, PMID: 29055712

Gross L, Vicens Q, Einhorn E, Noireterre A, Schaeffer L, Kuhn L, Imler J-L, Eriani G, Meignin C, Martin F. 2017. The IRES5'UTR of the dicistrovirus cricket paralysis virus is a type III IRES containing an essential pseudoknot structure. *Nucleic Acids Research* **45**:8993–9004. DOI: https://doi.org/10.1093/nar/gkx622

Gunišová S, Hronová V, Mohammad MP, Hinnebusch AG, Valášek LS. 2018. Please do not recycle! translation reinitiation in microbes and higher eukaryotes. *FEMS Microbiology Reviews* **42**:165–192. DOI: https://doi.org/10.1093/femsre/fux059, PMID: 29281028

Hashem Y, des Georges A, Dhote V, Langlois R, Liao HY, Grassucci RA, Hellen CU, Pestova TV, Frank J. 2013a. Structure of the mammalian ribosomal 43S preinitiation complex bound to the scanning factor DHX29. *Cell* **153**:1108–1119. DOI: https://doi.org/10.1016/j.cell.2013.04.036, PMID: 23706745

Hashem Y, des Georges A, Dhote V, Langlois R, Liao HY, Grassucci RA, Pestova TV, Hellen CU, Frank J. 2013b. Hepatitis-C-virus-like internal ribosome entry sites displace eIF3 to gain access to the 40S subunit. *Nature* **503**: 539–543. DOI: https://doi.org/10.1038/nature12658, PMID: 24185006

Hashem Y, Frank J. 2018. The jigsaw puzzle of mRNA translation initiation in eukaryotes: a decade of structures unraveling the mechanics of the process. *Annual Review of Biophysics*:125–151. DOI: https://doi.org/10.1146/annurev-biophys-070816-034034, PMID: 29494255

Hertz MI, Thompson SR. 2011. Mechanism of translation initiation by Dicistroviridae IGR IRESs. *Virology* **411**: 355–361. DOI: https://doi.org/10.1016/j.virol.2011.01.005, PMID: 21284991

Hinnebusch AG. 1993. Gene-specific translational control of the yeast *GCN4* gene by phosphorylation of eukaryotic initiation factor 2. *Molecular Microbiology* **10**:215–223. DOI: https://doi.org/10.1111/j.1365-2958.1993.tb01947.x, PMID: 7934812

Hinnebusch AG. 2014. The scanning mechanism of eukaryotic translation initiation. *Annual Review of Biochemistry* **83**:779–812. DOI: https://doi.org/10.1146/annurev-biochem-060713-035802

Hinnebusch AG. 2017. Structural insights into the mechanism of scanning and start Codon recognition in eukaryotic translation initiation. *Trends in Biochemical Sciences* **42**:589–611. DOI: https://doi.org/10.1016/j.tibs.2017.03.004, PMID: 28442192

Hussain T, Llácer JL, Fernández IS, Munoz A, Martin-Marcos P, Savva CG, Lorsch JR, Hinnebusch AG, Ramakrishnan V. 2014. Structural changes enable start Codon recognition by the eukaryotic translation initiation complex. *Cell* **159**:597–607. DOI: https://doi.org/10.1016/j.cell.2014.10.001, PMID: 25417110

Ingolia NT, Ghaemmaghami S, Newman JR, Weissman JS. 2009. Genome-wide analysis in vivo of translation with nucleotide resolution using ribosome profiling. *Science* **324**:218–223. DOI: https://doi.org/10.1126/science.1168978, PMID: 19213877

Jaafar ZA, Kieft JS. 2019. Viral RNA structure-based strategies to manipulate translation. *Nature Reviews Microbiology* **17**:110–123. DOI: https://doi.org/10.1038/s41579-018-0117-x, PMID: 30514982

Jackson RJ, Hellen CU, Pestova TV. 2010. The mechanism of eukaryotic translation initiation and principles of its regulation. *Nature Reviews Molecular Cell Biology* **11**:113–127. DOI: https://doi.org/10.1038/nrm2838, PMID: 20094052

Jan E, Kinzy TG, Sarnow P. 2003. Divergent tRNA-like element supports initiation, elongation, and termination of protein biosynthesis. *PNAS* **100**:15410–15415. DOI: https://doi.org/10.1073/pnas.2535183100, PMID: 14673072

Jan E, Mohr I, Walsh D. 2016. A Cap-to-Tail guide to mRNA translation strategies in Virus-Infected cells. *Annual Review of Virology* **3**:283–307. DOI: https://doi.org/10.1146/annurev-virology-100114-055014, PMID: 27501262

Johnson AG, Grosely R, Petrov AN, Puglisi JD. 2017. Dynamics of IRES-mediated translation. *Philosophical Transactions of the Royal Society B: Biological Sciences* **372**:20160177. DOI: https://doi.org/10.1098/rstb.2016.0177

Kenner LR, Anand AA, Nguyen HC, Myasnikov AG, Klose CJ, McGeever LA, Tsai JC, Miller-Vedam LE, Walter P, Frost A. 2019. eIF2B-catalyzed nucleotide exchange and phosphoregulation by the integrated stress response. *Science* **364**:491–495. DOI: https://doi.org/10.1126/science.aaw2922, PMID: 31048491

Kieft JS. 2009. Comparing the three-dimensional structures of Dicistroviridae IGR IRES RNAs with other viral RNA structures. *Virus Research* **139**:148–156. DOI: https://doi.org/10.1016/j.virusres.2008.07.007

Kolupaeva VG, de Breyne S, Pestova TV, Hellen CU. 2007. In vitro reconstitution and biochemical characterization of translation initiation by internal ribosomal entry. *Methods in Enzymology* **430**:409–439. DOI: https://doi.org/10.1016/S0076-6879(07)30016-5, PMID: 17913647

Kucukelbir A, Sigworth FJ, Tagare HD. 2014. Quantifying the local resolution of cryo-EM density maps. *Nature Methods* **11**:63–65. DOI: https://doi.org/10.1038/nmeth.2727, PMID: 24213166

Lander GC, Stagg SM, Voss NR, Cheng A, Fellmann D, Pulokas J, Yoshioka C, Irving C, Mulder A, Lau PW, Lyumkis D, Potter CS, Carragher B. 2009. Appion: an integrated, database-driven pipeline to facilitate EM image processing. *Journal of Structural Biology* **166**:95–102. DOI: https://doi.org/10.1016/j.jsb.2009.01.002, PMID: 19263523

Lee JH, Pestova TV, Shin BS, Cao C, Choi SK, Dever TE. 2002. Initiation factor eIF5B catalyzes second GTP-dependent step in eukaryotic translation initiation. *PNAS* **99**:16689–16694. DOI: https://doi.org/10.1073/pnas.262569399, PMID: 12471154

Lee AS, Kranzusch PJ, Doudna JA, Cate JH. 2016. eIF3d is an mRNA cap-binding protein that is required for specialized translation initiation. *Nature* **536**:96–99. DOI: https://doi.org/10.1038/nature18954, PMID: 27462815

Leontis NB, Westhof E. 1998. Conserved geometrical base-pairing patterns in RNA. *Quarterly Reviews of Biophysics* **31**:399–455. DOI: https://doi.org/10.1017/S0033583599003479, PMID: 10709244

Majzoub K, Hafirassou ML, Meignin C, Goto A, Marzi S, Fedorova A, Verdier Y, Vinh J, Hoffmann JA, Martin F, Baumert TF, Schuster C, Imler JL. 2014. RACK1 controls IRES-mediated translation of viruses. *Cell* **159**:1086–1095. DOI: https://doi.org/10.1016/j.cell.2014.10.041, PMID: 25416947

Murshudov GN, Vagin AA, Dodson EJ. 1997. Refinement of macromolecular structures by the maximum-likelihood method. *Acta Crystallographica Section D Biological Crystallography* **53**:240–255. DOI: https://doi.org/10.1107/S0907444996012255, PMID: 15299926

Myasnikov AG, Simonetti A, Marzi S, Klaholz BP. 2009. Structure-function insights into prokaryotic and eukaryotic translation initiation. *Current Opinion in Structural Biology* **19**:300–309. DOI: https://doi.org/10.1016/j.sbi.2009.04.010, PMID: 19493673

Nakashima N, Uchiumi T. 2009. Functional analysis of structural motifs in dicistroviruses. *Virus Research* **139**:137–147. DOI: https://doi.org/10.1016/j.virusres.2008.06.006, PMID: 18621089

Pestova TV, Lomakin IB, Lee JH, Choi SK, Dever TE, Hellen CU. 2000. The joining of ribosomal subunits in eukaryotes requires eIF5B. *Nature* **403**:332–335. DOI: https://doi.org/10.1038/35002118, PMID: 10659855

Pestova TV, de Breyne S, Pisarev AV, Abaeva IS, Hellen CU. 2008. eIF2-dependent and eIF2-independent mode of initiation on the CSFV IRES: a common role of domain II. *The EMBO Journal* **27**:1060–1072. DOI: https://doi.org/10.1038/emboj.2008.49, PMID: 18337746

Pestova TV, Hellen CU. 2005. Reconstitution of eukaryotic translation elongation in vitro following initiation by internal ribosomal entry. *Methods* **36**:261–269. DOI: https://doi.org/10.1016/j.ymeth.2005.04.004, PMID: 16076452

Pettersen EF, Goddard TD, Huang CC, Couch GS, Greenblatt DM, Meng EC, Ferrin TE. 2004. UCSF chimera–a visualization system for exploratory research and analysis. *Journal of Computational Chemistry* **25**:1605–1612. DOI: https://doi.org/10.1002/jcc.20084, PMID: 15264254

Pisarev AV, Skabkin MA, Pisareva VP, Skabkina OV, Rakotondrafara AM, Hentze MW, Hellen CU, Pestova TV. 2010. The role of ABCE1 in eukaryotic posttermination ribosomal recycling. *Molecular Cell* **37**:196–210. DOI: https://doi.org/10.1016/j.molcel.2009.12.034, PMID: 20122402

Pisareva VP, Pisarev AV, Fernández IS. 2018. Dual tRNA mimicry in the cricket paralysis virus IRES uncovers an unexpected similarity with the hepatitis C virus IRES. *eLife* **7**:e34062. DOI: https://doi.org/10.7554/eLife.34062, PMID: 29856316

Quade N, Boehringer D, Leibundgut M, van den Heuvel J, Ban N. 2015. Cryo-EM structure of hepatitis C virus IRES bound to the human ribosome at 3.9-Å resolution. *Nature Communications* **6**:7646. DOI: https://doi.org/10.1038/ncomms8646, PMID: 26155016

Ratje AH, Loerke J, Mikolajka A, Brünner M, Hildebrand PW, Starosta AL, Dönhöfer A, Connell SR, Fucini P, Mielke T, Whitford PC, Onuchic JN, Yu Y, Sanbonmatsu KY, Hartmann RK, Penczek PA, Wilson DN, Spahn CM. 2010. Head swivel on the ribosome facilitates translocation by means of intra-subunit tRNA hybrid sites. *Nature* **468**:713–716. DOI: https://doi.org/10.1038/nature09547, PMID: 21124459

Resch AM, Ogurtsov AY, Rogozin IB, Shabalina SA, Koonin EV. 2009. Evolution of alternative and constitutive regions of mammalian 5'UTRs. *BMC Genomics* **10**:162. DOI: https://doi.org/10.1186/1471-2164-10-162, PMID: 19371439

Russo CJ, Passmore LA. 2014. Electron microscopy: ultrastable gold substrates for electron cryomicroscopy. *Science* **346**:1377–1380. DOI: https://doi.org/10.1126/science.1259530, PMID: 25504723

Scheres SH. 2012. RELION: implementation of a bayesian approach to cryo-EM structure determination. *Journal of Structural Biology* **180**:519–530. DOI: https://doi.org/10.1016/j.jsb.2012.09.006, PMID: 23000701

Scheres SH. 2016. Processing of structurally heterogeneous Cryo-EM data in RELION. *Methods in Enzymology* **579**:125–157. DOI: https://doi.org/10.1016/bs.mie.2016.04.012, PMID: 27572726

Schmeing TM, Ramakrishnan V. 2009. What recent ribosome structures have revealed about the mechanism of translation. *Nature* **461**:1234–1242. DOI: https://doi.org/10.1038/nature08403, PMID: 19838167

Sendoel A, Dunn JG, Rodriguez EH, Naik S, Gomez NC, Hurwitz B, Levorse J, Dill BD, Schramek D, Molina H, Weissman JS, Fuchs E. 2017. Translation from unconventional 5' start sites drives tumour initiation. *Nature* **541**:494–499. DOI: https://doi.org/10.1038/nature21036, PMID: 28077873

Shatsky IN, Dmitriev SE, Terenin IM, Andreev DE. 2010. Cap- and IRES-independent scanning mechanism of translation initiation as an alternative to the concept of cellular IRESs. *Molecules and Cells* **30**:285–293. DOI: https://doi.org/10.1007/s10059-010-0149-1, PMID: 21052925

Shi M, Lin XD, Tian JH, Chen LJ, Chen X, Li CX, Qin XC, Li J, Cao JP, Eden JS, Buchmann J, Wang W, Xu J, Holmes EC, Zhang YZ. 2016. Redefining the invertebrate RNA virosphere. *Nature* **540**:539–543. DOI: https://doi.org/10.1038/nature20167, PMID: 27880757

Skabkin MA, Skabkina OV, Hellen CU, Pestova TV. 2013. Reinitiation and other unconventional posttermination events during eukaryotic translation. *Molecular Cell* **51**:249–264. DOI: https://doi.org/10.1016/j.molcel.2013.05.026, PMID: 23810859

Starck SR, Tsai JC, Chen K, Shodiya M, Wang L, Yahiro K, Martins-Green M, Shastri N, Walter P. 2016. Translation from the 5' untranslated region shapes the integrated stress response. *Science* **351**:aad3867. DOI: https://doi.org/10.1126/science.aad3867, PMID: 26823435

Terenin IM, Dmitriev SE, Andreev DE, Shatsky IN. 2008. Eukaryotic translation initiation machinery can operate in a bacterial-like mode without eIF2. *Nature Structural & Molecular Biology* **15**:836–841. DOI: https://doi.org/10.1038/nsmb.1445

Vattem KM, Wek RC. 2004. Reinitiation involving upstream ORFs regulates ATF4 mRNA translation in mammalian cells. *PNAS* **101**:11269–11274. DOI: https://doi.org/10.1073/pnas.0400541101, PMID: 15277680

Voorhees RM, Ramakrishnan V. 2013. Structural basis of the translational elongation cycle. *Annual Review of Biochemistry* **82**:203–236. DOI: https://doi.org/10.1146/annurev-biochem-113009-092313, PMID: 23746255

Wang J, Johnson AG, Lapointe CP, Choi J, Prabhakar A, Chen DH, Petrov AN, Puglisi JD. 2019. eIF5B gates the transition from translation initiation to elongation. *Nature* **573**:605–608. DOI: https://doi.org/10.1038/s41586-019-1561-0, PMID: 31534220

Wethmar K. 2014. The regulatory potential of upstream open reading frames in eukaryotic gene expression. *Wiley Interdisciplinary Reviews: RNA* **5**:765–768. DOI: https://doi.org/10.1002/wrna.1245, PMID: 24995549

Yamamoto H, Unbehaun A, Loerke J, Behrmann E, Collier M, Bürger J, Mielke T, Spahn CM. 2014. Structure of the mammalian 80S initiation complex with initiation factor 5B on HCV-IRES RNA. *Nature Structural & Molecular Biology* **21**:721–727. DOI: https://doi.org/10.1038/nsmb.2859, PMID: 25064512

Yamamoto H, Collier M, Loerke J, Ismer J, Schmidt A, Hilal T, Sprink T, Yamamoto K, Mielke I, Bürger J, Shaikh TR, Dabrowski M, Hildebrand PW, Scheerer P, Spahn CM. 2015. Molecular architecture of the ribosome-bound hepatitis C virus internal ribosomal entry site RNA. *The EMBO Journal* **34**:3042–3058. DOI: https://doi.org/10.15252/embj.201592469, PMID: 26604301

Yamamoto H, Unbehaun A, Spahn CMT. 2017. Ribosomal chamber music: toward an understanding of IRES mechanisms. *Trends in Biochemical Sciences* **42**:655–668. DOI: https://doi.org/10.1016/j.tibs.2017.06.002, PMID: 28684008

Yokoyama T, Machida K, Iwasaki W, Shigeta T, Nishimoto M, Takahashi M, Sakamoto A, Yonemochi M, Harada Y, Shigematsu H, Shirouzu M, Tadakuma H, Imataka H, Ito T. 2019. HCV IRES captures an actively translating

80S ribosome. *Molecular Cell* **74**:1205–1214. DOI: https://doi.org/10.1016/j.molcel.2019.04.022, PMID: 310 80011

Young SK, Wek RC. 2016. Upstream open reading frames differentially regulate Gene-specific translation in the integrated stress response. *Journal of Biological Chemistry* **291**:16927–16935. DOI: https://doi.org/10.1074/jbc.R116.733899, PMID: 27358398

Zhang K. 2016. Gctf: real-time CTF determination and correction. *Journal of Structural Biology* **193**:1–12. DOI: https://doi.org/10.1016/j.jsb.2015.11.003, PMID: 26592709

Zhang YZ, Wu WC, Shi M, Holmes EC. 2018. The diversity, evolution and origins of vertebrate RNA viruses. *Current Opinion in Virology* **31**:9–16. DOI: https://doi.org/10.1016/j.coviro.2018.07.017, PMID: 30114593

Zhang Y-Z, Chen Y-M, Wang W, Qin X-C, Holmes EC. 2019. Expanding the RNA Virosphere by Unbiased Metagenomics. *Annual Review of Virology* **6**:119–139. DOI: https://doi.org/10.1146/annurev-virology-092818-015851

Zheng SQ, Palovcak E, Armache JP, Verba KA, Cheng Y, Agard DA. 2017. MotionCor2: anisotropic correction of beam-induced motion for improved cryo-electron microscopy. *Nature Methods* **14**:331–332. DOI: https://doi.org/10.1038/nmeth.4193, PMID: 28250466

Zivanov J, Nakane T, Forsberg BO, Kimanius D, Hagen WJ, Lindahl E, Scheres SH. 2018. New tools for automated high-resolution cryo-EM structure determination in RELION-3. *eLife* **7**:e42166. DOI: https://doi.org/10.7554/eLife.42166, PMID: 30412051

Figures and figure supplements

A complex IRES at the 5'-UTR of a viral mRNA assembles a functional 48S complex via an uAUG intermediate

Ritam Neupane *et al*

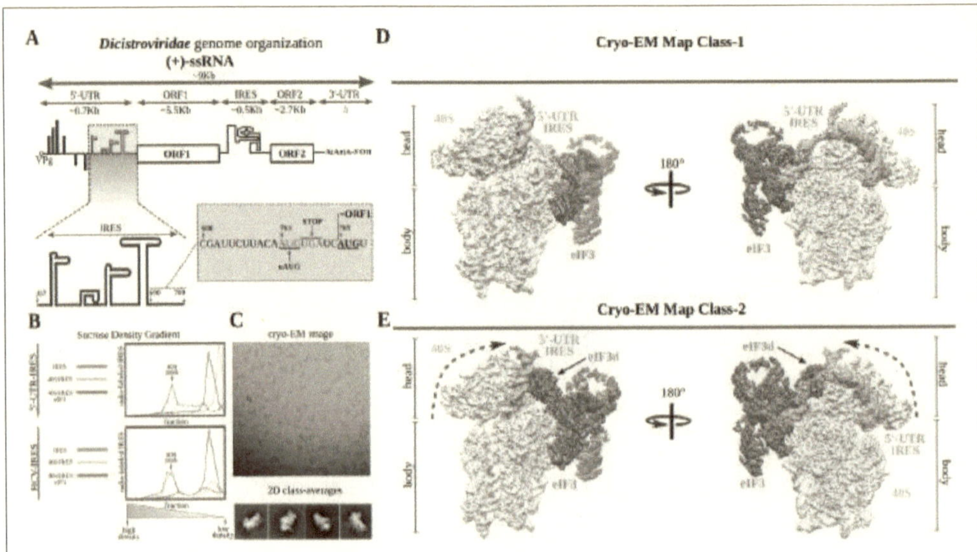

Figure 1. *Dicistroviridae* genome organization, in vitro complex formation and cryo-EM maps. (A) Top, schematic representation of the genome organization of *Dicistroviruses*. The approximate genomic lengths of the different components are indicated by the arrows. Bottom, detailed view of the region described as harboring the IRES activity of the 5'-UTR of the CrPV. On the right, the sequence adjacent to the initiation AUG codon of ORF1, located at nucleotide 709 and preceded by a 'start-stop uORF' indicated in red. (B) Sucrose-gradient analysis showed that the 5'-UTR-IRES is dependent on eIF3 in order to form a stable complex with the 40S. 5'-UTR-IRES co-migrates with the 40S only in the presence of eIF3 (top). By contrast, HCV IRES does not require eIF3 for 40S binding (bottom). (C) Representative cryo-EM micrograph of the 40S–5'-UTR-IRES–eIF3 complex. Bottom, representative reference-free 2D class averages used for further image processing. (D, E) After 3D classifications, two classes showing density for 40S (yellow), eIF3 (red) and 5'-UTR-IRES (blue) could be identified in the dataset. Class-1 (D) presents a non-swiveled configuration of the 40S head and the density for eIF3d is absent. Class-2 (E) shows a swiveled configuration of the 40S head (arrows) with eIF3d (indicated) contacting eIF3's core subunits.

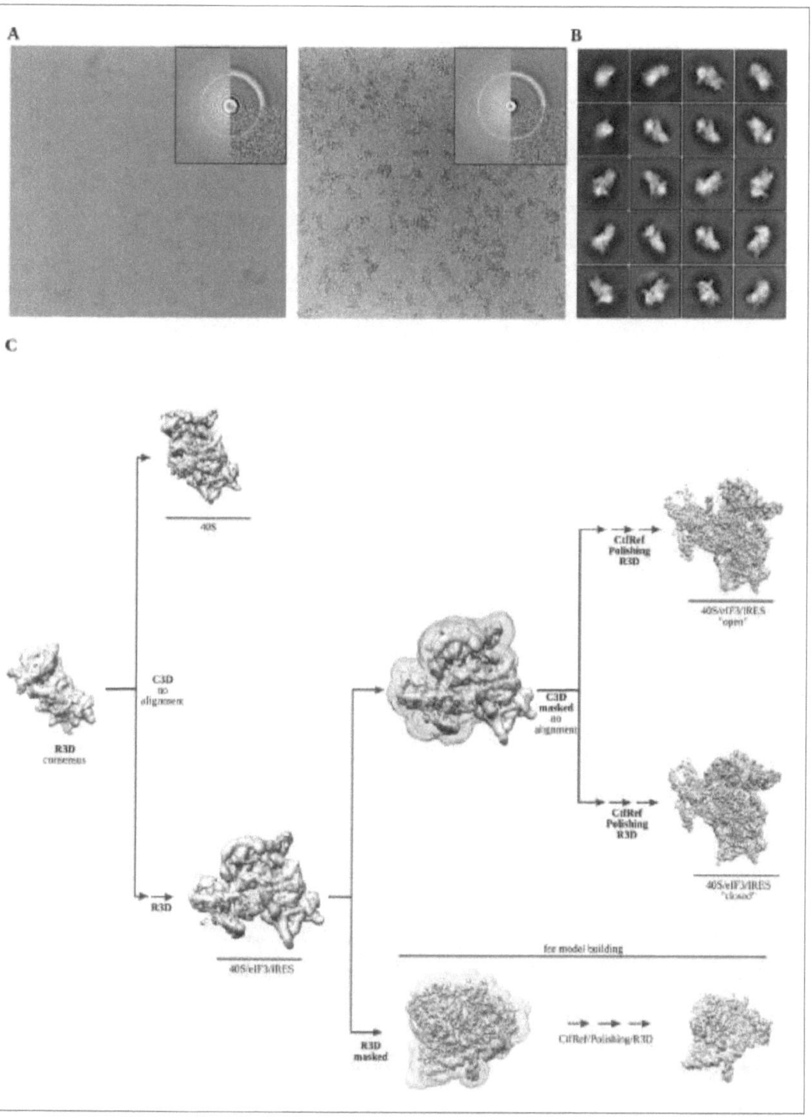

A

B

C

Figure 1—figure supplement 1. Cryo-EM representative images and classification workflow. (A) Two examples of aligned micrographs used for image processing. Data collection in thick ice was instrumental to avoid complex disassembly and preferential orientation. (B) Representative reference-free 2D averages. (C) The classification scheme followed to identify the two described complex classes.

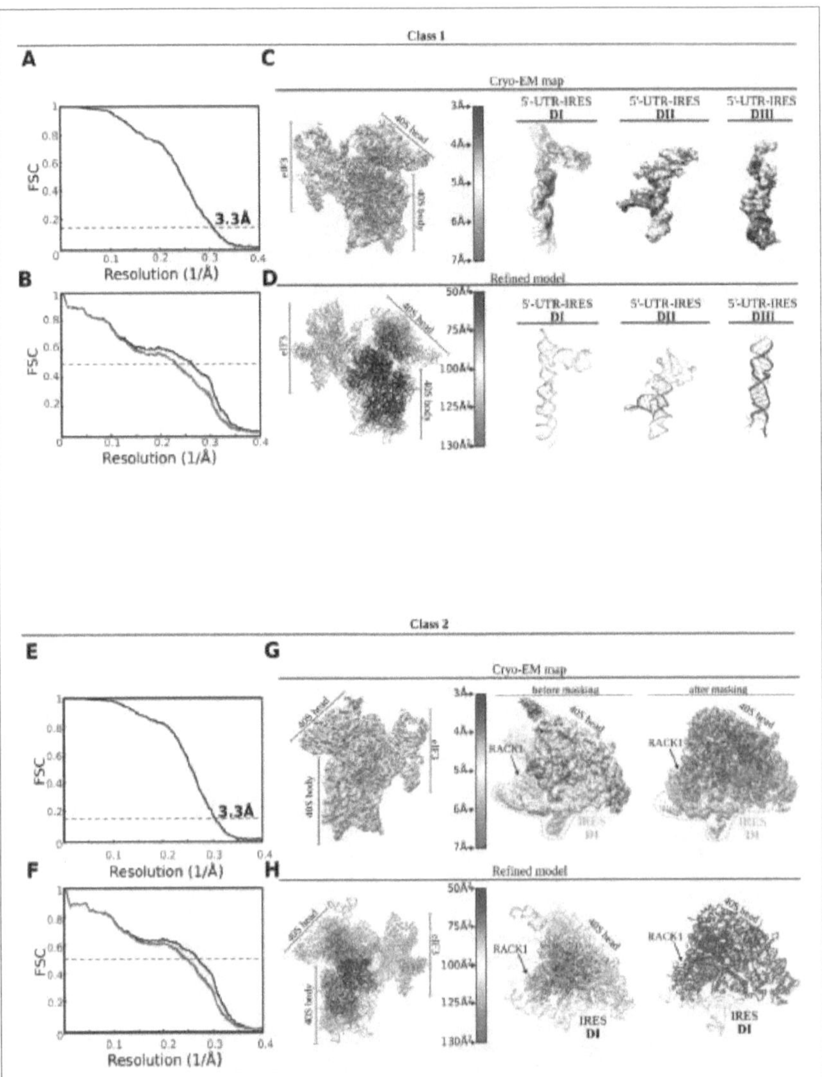

Figure 1—figure supplement 2. FSC correlation curves, local resolution and model validation. Top, class-1. (A) Gold-standard Fourier Shell Correlation (FSC) curves between half-maps independently refined in Relion3. Global resolution by the 0.143 cutoff criterium was estimated to be 3.3 Å. (B) FSC between the final refined model and the final unsharpened map (black curve). The absence of model overfitting is demonstrated by the overlapping of FSC curves between half-map 1 (included in the refinement, blue) and the model and half-map 2 (not included in the refinement, red). (C) Unsharpened map colored according to local resolution as computed by ResMap. On the right, detailed views for the three 5'-UTR-IRES domains. (D) Refined model colored according to estimated B-factors computed by Refmac. Bottom, class-2. (E) Gold-standard FSC curves between half-maps independently

Figure 1—figure supplement 2 continued on next page

Figure 1—figure supplement 2 continued

refined in Relion3. Global resolution by the 0.143 cutoff criterium was estimated to be 3.3 Å. (**F**) FSC between the final refined model and the final unsharpened map (black curve). The absence of model overfitting is demonstrated by the overlapping of FSC curves between half-map 1 (included in the refinement, blue) and the model and half-map 2 (not included in the refinement, red). (**G**) Unsharpened map colored according to local resolution as computed by ResMap. On the right, detailed views for the 40S head before and after masked refinements. (**H**) Refined model colored according to estimated B-factors computed by Refmac.

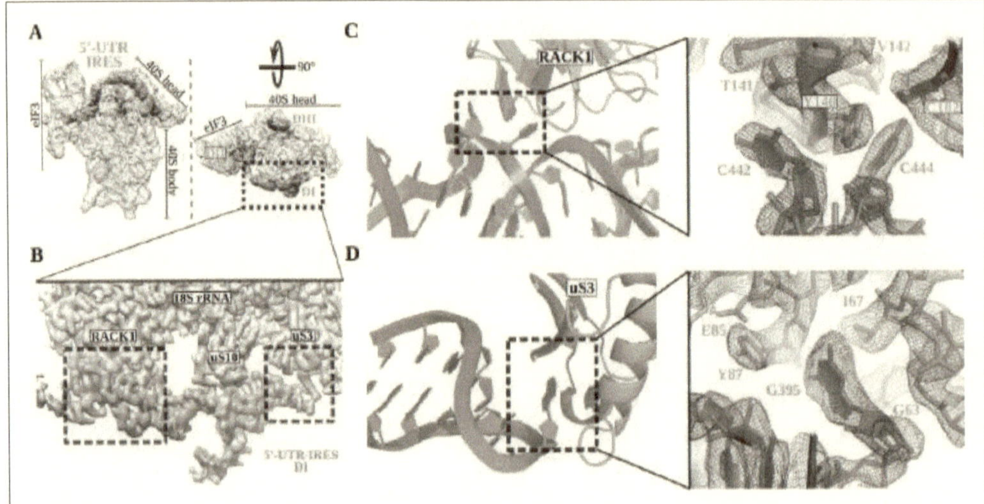

Figure 2. The 5'-UTR-IRES domain I engages ribosomal proteins RACK1 and uS3. (**A**) Overview of the 40S–5'-UTR-IRES–eIF3 map, with 40S and eIF3 depicted gray and 5'-UTR-IRES in blue. (**B**) Detailed view of the cryo-EM density of the 40S–5'-UTR-IRES–eIF3 map centered around 5'-UTR-IRES domain I (DI). Ribosomal proteins are colored brown, 18S rRNA yellow and 5'-UTR-IRES blue. Contacts between 5'-UTR-IRES domain I and ribosomal proteins RACK1 (**C**) and uS3 (**D**) could be defined thanks to well-resolved local cryo-EM densities.

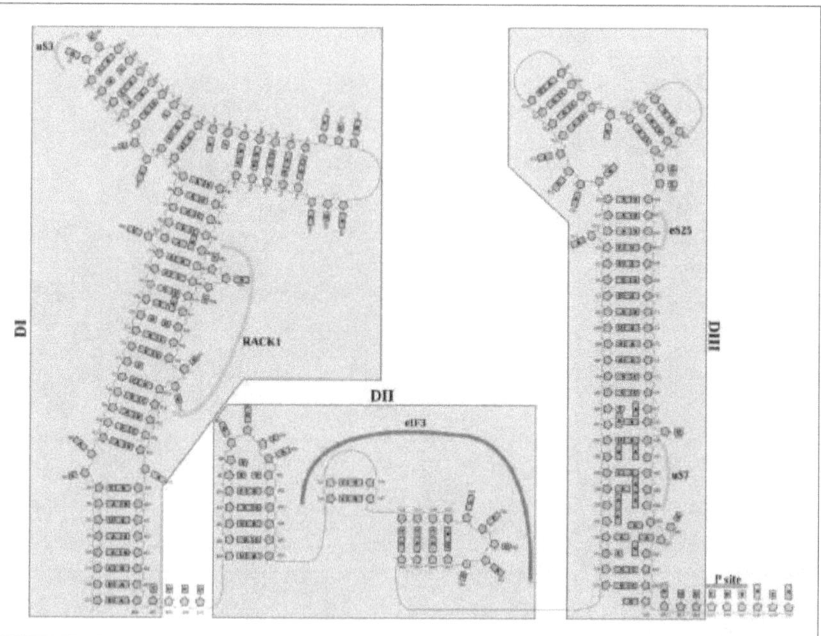

Figure 2—figure supplement 1. Structurally derived secondary structure diagram for the 5′-UTR-IRES. Secondary structure diagram for the 5′-UTR-CrPV IRES derived from the cryo-EM structure.

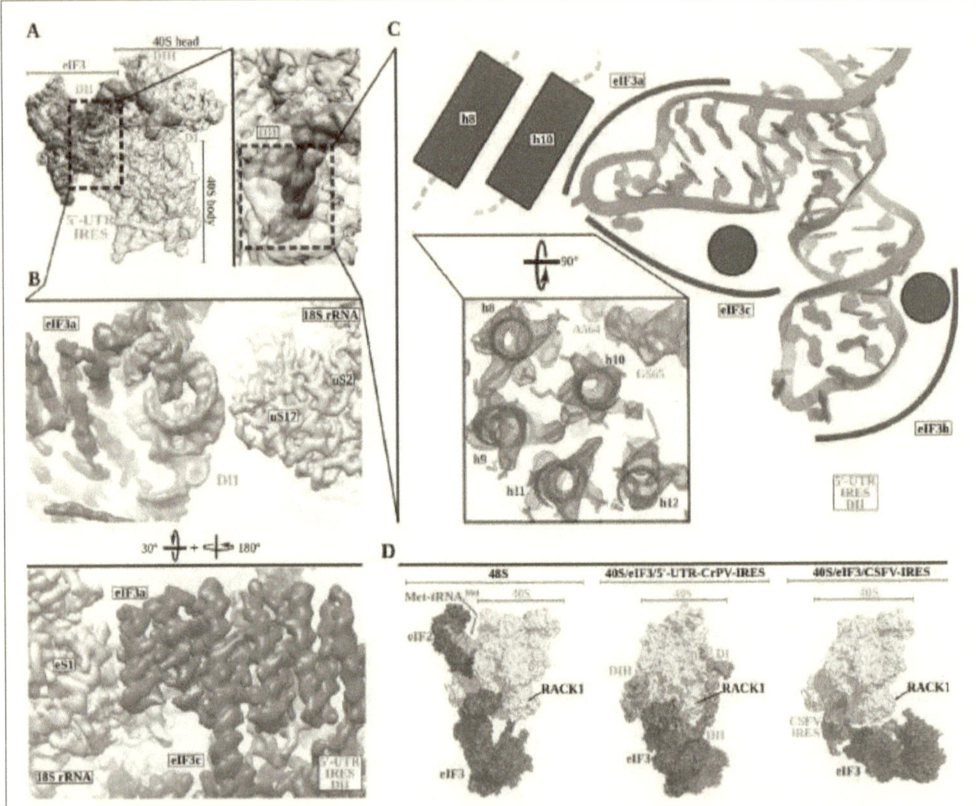

Figure 3. The 5'-UTR-IRES domain II is formed by a dual hairpin that mediates eIF3 recruitment. (**A**) Overview of the 40S–5'-UTR-IRES–eIF3 cryo-EM map with 40S colored gray, eIF3 red and 5'-UTR-IRES blue. On the right, a zoomed view centers around 5'-UTR-IRES domain II. (**B**) Detailed view of the cryo-EM map for the region occupied by 5'-UTR-IRES domain II, with 40S components colored gold, eIF3 red and 5'-UTR-IRES blue. Domain II is sandwiched between ribosomal protein uS17 (located at the back of the 40S body) and eIF3 core subunits a and c. (**C**) 5'-UTR-IRES domain II is formed by a dual hairpin that establishes interactions with α-helices 8 and 10 from eIF3a. These contacts are mediated mainly by basic residues of eIF3 and the phosphate backbone of the IRES. (**D**) superposition of the 40S–5'-UTR-IRES–eIF3 complex with the canonical 48S complex (left, PDB ID 6FEC) and with the CSFV-IRES–40S complex (right, PDB ID 4c4q). The 5'-UTR-IRES binds to the 40S with a conformation that is compatible with the canonical position described for eIF3 in the 48S complex.

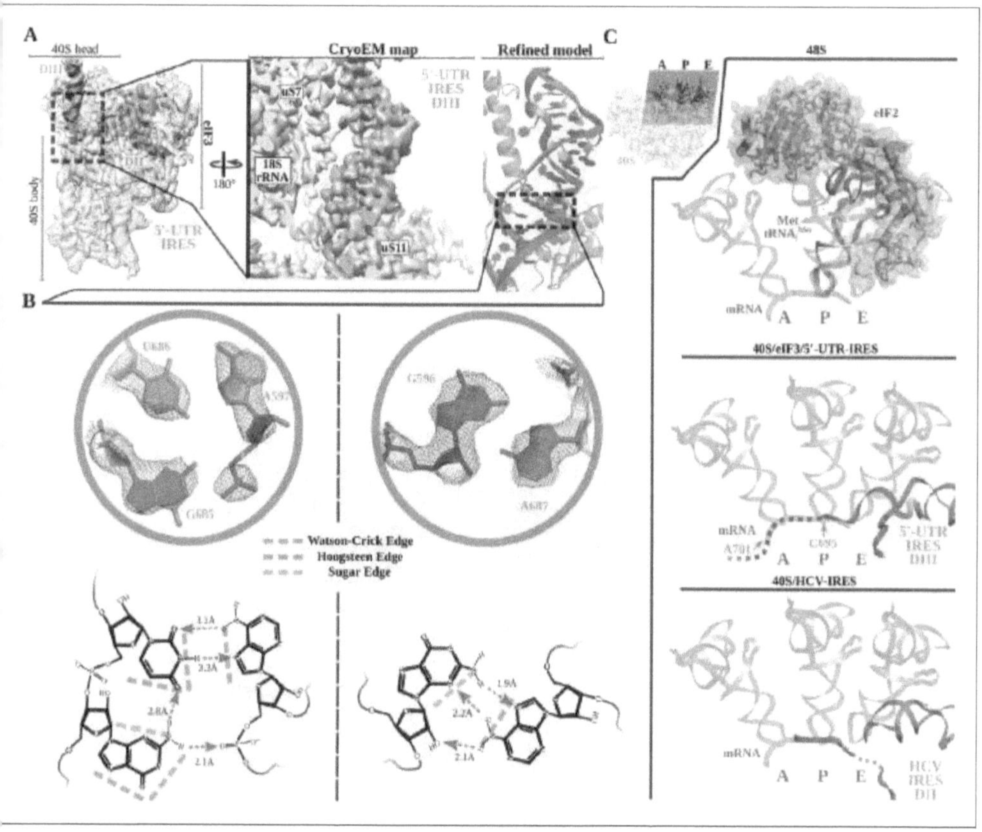

Figure 4. Non-canonical base pairing in 5'-UTR-IRES domain III assists on P-site access. **A)** Overview of the 40S–5'-UTR-IRES–eIF3 cryo-EM map with 40S and eIF3 colored gray and 5'-UTR-IRES blue. On the right, a detailed view of the E-site, where 5'-UTR-IRES domain III is placed, shows the cryo-EM map with the 5'-UTR-IRES colored blue and 40S components brown. On the far right, the final refined model is colored following the same color scheme. Ribosomal proteins uS7 and uS11 as well as several 18S rRNA bases contact 5'-UTR-IRES domain III. **(B)** Non-canonical base pairs found in 5'-UTR-IRES domain III induce a distortion of the double helix near the E-site. At the top, two examples are shown with the refined model inserted in the experimental cryo-EM density. The corresponding chemical diagrams below show the base edges and the hydrogen bonds involved in interactions. **(C)** Superpositions of the canonical 48S complex (PDB ID 6FEC, top) with the HCV-IRES/40S complex (PDB ID 5A2Q, bottom) and the 40S/5'-UTR-IRES/eIF3 (middle) models, focused on the tRNA A, P and E binding sites. 5'-UTR-IRES domain III and HCV-IRES occupies a space on the E-site that overlaps with the position described for eIF2 in the canonical 48S complex. In the middle, the last residue of 5'-UTR-IRES is indicated (C695), as is the putative path along the mRNA binding channel followed by the IRES (dashed line).

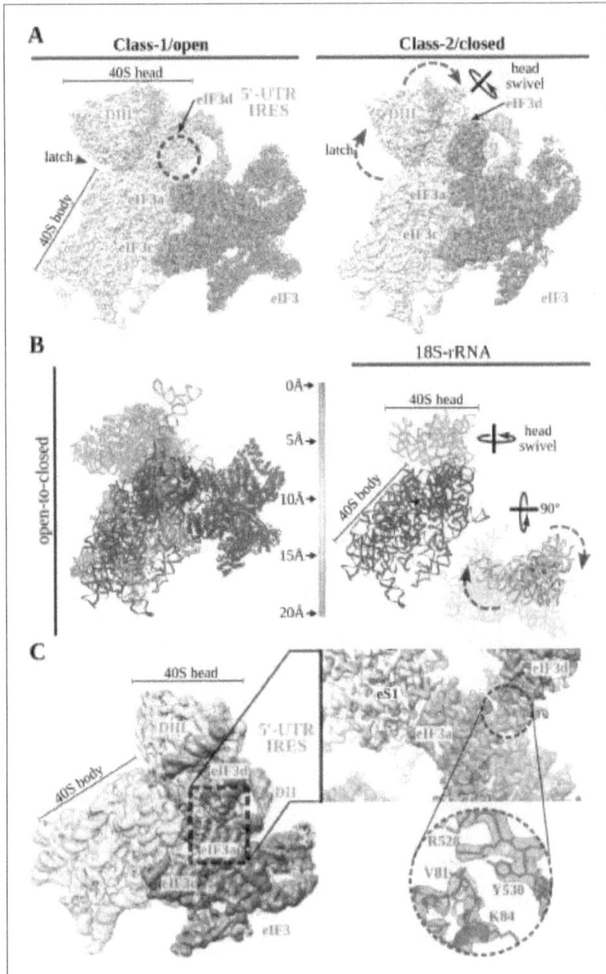

Figure 5. A 40S head swiveling movement 'locks' 5'-UTR-IRES on the 40S, inducing a compact eIF3 configuration.
(A) Cryo-EM maps obtained for the two classes present in the 40S–5'-UTR-IRES–eIF3 dataset with 40S colored
gold, eIF3 red and 5'-UTR-IRES blue. The positions of the latch, eIF3d and the swiveling rotation
axis are indicated. A 40S head swiveling movement in class-2 brings eIF3d closer to the core subunit of eIF3 (i.
e. eIF3a), establishing interactions that stabilize its conformation. (B) Left, ribbon diagram colored according to
pairwise root mean square deviation (r.m.s.d.) displacements for the open-to-closed transition, with displacements
scale at the center. On the right, a simplified diagram shows only the 18S rRNA colored according to the used on
the left. Two orthogonal views are shown, in which it can be appreciated that the main displacement is localized at
the 40S head. (C) Overview of the closed class with 40S colored gray, eIF3 red and 5'-UTR-IRES blue. Inset, detail
of the experimental density obtained for the eIF3a–eIF3d interface for this class. Clear information on side chains
was present in the maps, allowing proper model building and refinement.

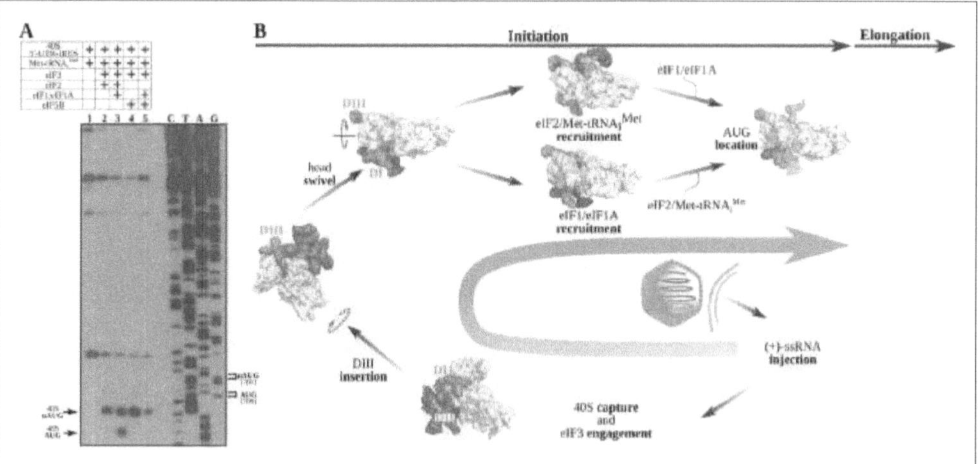

Figure 6. The 5'-UTR-IRES requires TC, eIF1 or eIF1A to assemble a functional initiation complex via an uAUG intermediate. (A) Toe-print analysis of 48S initiation complexes assembled on 5'-UTR-IRES in an in vitro reconstituted system. eIF2 delivers Met-tRNA$_i^{Met}$ to the uAUG (lane 2) and requires the presence of eIF1 or eIF1A to transition to the *bona fide* AUG codon of ORF1 (lane 3). Under some conditions, eIF5B can substitute for eIF2 in Met-tRNA$_i^{Met}$ delivery (*Terenin et al., 2008*). In the absence of eIF1 or eIF1A, a robust toe-print signal is detected in the presence of eIF5B (lane 4), however, Met-tRNA$_i^{Met}$ is delivered to the uAUG, and thus eIF5B is unable to find the annotated AUG even in the presence of eIF1 or eIF1A (lane 5). (B) A model for 5'-UTR-IRES-mediated translation initiation. From the bottom right: injection of the genomic (+)-ssRNA of the CrPV into the cytosol allows the 5'-UTR-IRES to capture free 40S and eIF3, which is recruited to the complex with an initial canonical conformation of the 40S head without tilt or swivel. Insertion of 5'-UTR-IRES DIII in the vicinity of the E-site and a swiveling movement of the 40S head induces a 'locked' conformation of the complex, with the uAUG at 701 in the vicinity of the P-site. Delivery of Met-tRNA$_i^{Met}$ and location of ORF1 AUG is achieved by the concerted action of eIF2, eIF1 and eIF1A. Large subunit recruitment mediated by eIF5B will allow transitioning into elongation.

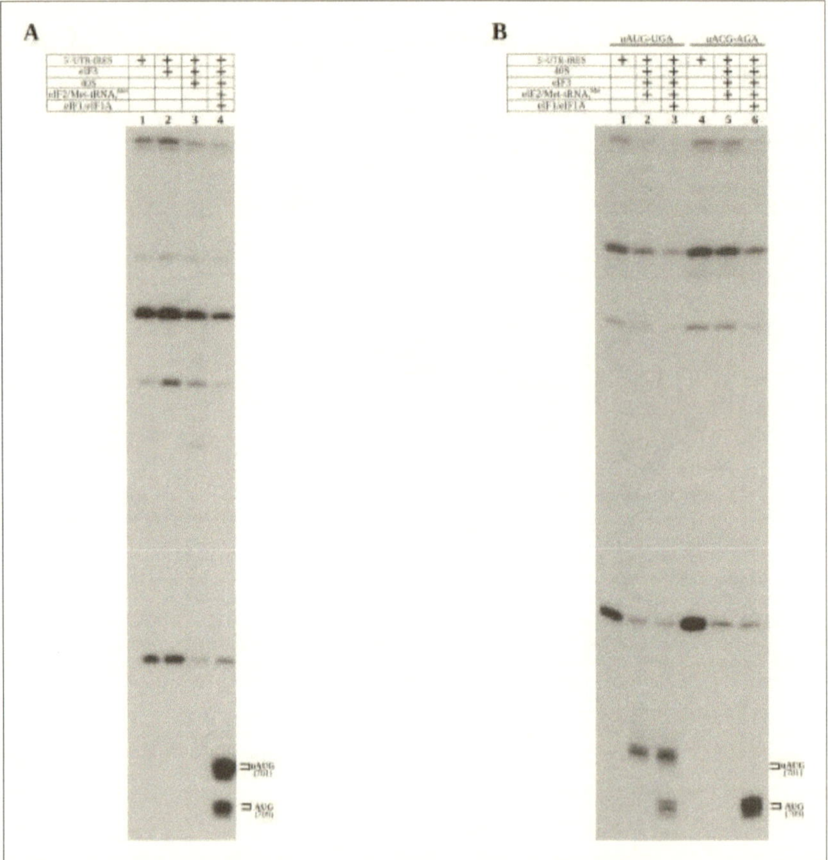

Figure 6—figure supplement 1. Control toe-print experiments. (A) Toe-print analysis for 5'-UTR-IRES, 5'-UTR-IRES–eIF3 and 40S/5'-UTR-IRES–eIF3 complexes (lanes 1, 2 and 3, respectively). No toe-print signal was detected that is ascribable to uAUG or annotated AUG in any of these reactions. Robust toe-print signal could be detected 15–17 nucleotides downstream of uAUG and AUG only for 40S–5'-UTR-IRES –eIF3, and in the presence of TC (eIF2–Met-tRNA$_i^{Met}$–GTP) and eIF1 or eIF1A. (B) The uAUG toe-print signal can be abolished by mutation of the uAUG to ACG, redirecting the Met-tRNA$_i^{Met}$ loading event exclusively to the annotated AUG in the presence of TC and eIF1 or eIF1A.

Appendix C

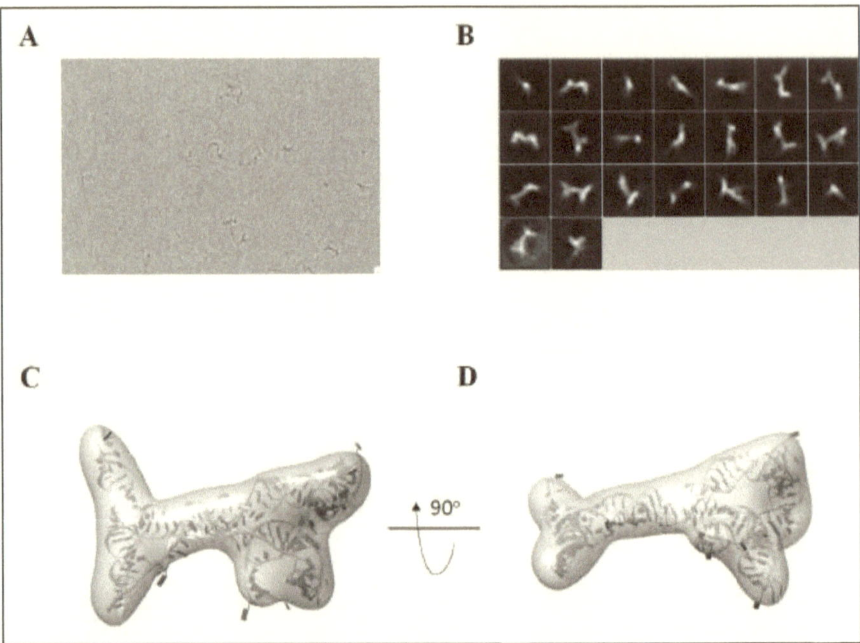

Figure C1 | 3D reconstruction of the IAPV-IRES from an IAPV-IRES RNA only cryo-EM experiment

A: A cropped area from a representative image with low concentration of the IAPV-IRES particles. The IRES used in this experiment had its first four codons mutated to glycine, glycine, leucine, and STOP respectively. The IAPV-IRES RNA only particles are circled in green and were picked using reference-free method from 139 micrographs.
B: A cleaned up 2D class averages of the IAPV-IRES RNA after a few rounds of 2D classification performed in Relion 3.0. These 23 classes were selected for 3D reconstruction.
C, D: 3D reconstruction of the IAPV-IRES RNA was performed in Relion 3.0 and the resulting density had resolution of 19 Å. The density was clearly distinguishable as an IRES shape similar to the one we see in binary complexes with the ribosome. An IAPV-IRES model was fitted into the sawhorse shaped cryo-EM density to demonstrate the fit.

www.ingramcontent.com/pod-product-compliance
Lightning Source LLC
LaVergne TN
LVHW042352190726
843493LV00005B/982